HISTORY OF BRITISH SPACE SCIENCE

HISTORY OF
BRITISH SPACE SCIENCE

SIR HARRIE MASSEY FRS

Late Emeritus Professor of Physics, University College London

M.O. ROBINS CBE

Formerly Director of Astronomy Space and Radio, and of Science,
Science Research Council

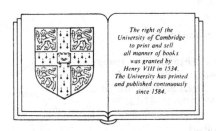

The right of the
University of Cambridge
to print and sell
all manner of books
was granted by
Henry VIII in 1534.
The University has printed
and published continuously
since 1584.

CAMBRIDGE UNIVERSITY PRESS

Cambridge

London   New York   New Rochelle

Melbourne   Sydney

CAMBRIDGE UNIVERSITY PRESS
Cambridge, New York, Melbourne, Madrid, Cape Town, Singapore,
São Paulo, Delhi, Dubai, Tokyo

Cambridge University Press
The Edinburgh Building, Cambridge CB2 8RU, UK

Published in the United States of America by Cambridge University Press, New York

www.cambridge.org
Information on this title: www.cambridge.org/9780521123389

First published 1986
This digitally printed version 2009

A catalogue record for this publication is available from the British Library

Library of Congress Cataloguing in Publication data
Massey, Harrie Stewart Wilson, Sir
  History of British space science.

  Commissioned by the Science and Engineering Research
Council of Great Britain.
  1. Space sciences – Research – Great Britain – History.
I. Robins, M. O. (Malcolm Owen)   II. Science and
Engineering Research Council (Great Britain)   III. Title.
QB500.266.G7M37   1985   500.5'0941   85-13279

ISBN 978-0-521-30783-3 Hardback
ISBN 978-0-521-12338-9 Paperback

CONTENTS

Contents   vii

Glossary of abbreviations in the text and annexes

Where the initials refer to a French title, the English version is given.

ABRC   Advisory Board for the Research Councils
ACSP   Advisory Council on Scientific Policy
AMPTE   Active Magnetospheric Particle Tracer Explorers
APGC   Astronomy Policy and Grants Committee
ARD   Astrophysics Research Division
ASRB   Astronomy Space and Radio Board
AURA   American Association for Research in Astronomy
AWG   Administrative Working Group
AWRE   Atomic Weapons Research Establishment

BAC   British Aircraft Corporation
BNCSR   British National Committee on Space Research
CERN   European Organization for Nuclear Research
COPERS European Preparatory Commission for Space Research
COSPAR Special Committee on Space Research
CSAGI   Special Committee for the International Geophysical Year
CSP   Council for Scientific Policy

DOE   Design of Experiments Sub-committee
DSIR   Department of Scientific and Industrial Research

ELDO   European Launcher Development Organization
ESA   European Space Agency
ESC   European Space Conference
ESDAC   European Space Data Centre
ESF   European Science Foundation
ESLAB   European Space Laboratory
ESRANGE European Space Range
ESTEC   European Space Technology Centre
ESRO   European Space Research Organization

ESTRACK   European Space Tracking Network
EUV   Extreme Ultra-violet
EXOSAT   European X-ray Observatory Satellite

FOC   Faint Object Camera

GEOS   Geostationary Magnetospheric Satellite
GSFC   Goddard Space Flight Center
GTST   Scientific and Technical Working Group

HEAO   High Energy Astronomical Observatory
HEOS   Highly Eccentric Orbit Satellite

IAF   International Astronautical Federation
IAU   International Astronomical Union
ICSU   International Council of Scientific Unions
IGY   International Geophysical Year
INCOSPAR Indian Committee on Space Research
IQSY   International Year of the Quiet Sun
IRAS   Infra-red Astronomy Satellite
ISEE   International Sun Earth Explorer
IUB   International Union of Biochemistry
IUBS   International Union of Biological Sciences
IUE   International Ultra-violet Explorer
IUGG   International Union of Geodesy and Geophysics
IUPAC   International Union of Pure and Applied Chemistry
IUPAP   International Union of Pure and Applied Physics
IUPS   International Union of Physiological Sciences
IUTAM   International Union of Theoretical and Applied Mechanics

JPL   Jet Propulsion Laboratory

LAS   Large Astronomical Satellite
LPAC   Launching Programmes Advisory Committee
LST   Large Space Telescope

NACA   National Advisory Committee for Aeronautics
NASA   National Aeronautics and Space Administration
NATO   North Atlantic Treaty Organization
NEDC   National Economic Development Council
NIRNS   National Institute for Research in Nuclear Science
NRDC   National Research Development Corporation

OAO   Orbiting Astronomical Observatory
OSO   Orbiting Solar Observatory

PSAC   President's Science Advisory Committee

PSSAB   Provisional Space Science Advisory Board
PSSB   Provisional Space Science Board

RAE   Royal Aircraft Establishment
RPE   Rocket Propulsion Establishment
RRS   Radio Research Station
RSRS   Radio and Space Research Station

SAS   Small Astronomical Satellite
SCOSTEP Special Committee on Solar and Terrestrial Physics
SERC   Science and Engineering Research Council
SMM   Solar Maximum Mission
SPC   Science Programme Committee
SPGC   Space Policy and Grants Committee
SPM   Solar Polar Mission
SRC   Science Research Council
SRMU   Space Research Management Unit
SSB   Space Science Board
SSC   Space Science Committee
ST   Space Telescope
STC   Scientific and Technical Committee
SUPARCO Pakistan Space and Upper Atmosphere Research Council

TADREC   Tracking and Data Recovery Sub-Committee
TD-1   Thor-Delta 1 (Satellite)
TD-2   Thor-Delta 2 (Satellite)
TMA   Trimethylaluminium

UKAEA   United Kingdom Atomic Energy Authority
UNESCO   United Nations Educational and Scientific Committee
URSI   International Union of Radio Science
UVAS   Ultra-violet Astronomy Satellite

WRE   Weapons Research Establishment
acu   attitude control unit
elf   extremely low frequency
rf   radio frequency
vhf   very high frequency
vlf   very low frequency

Glossary of abbreviations in the appendices

AFCRL	Air Force Cambridge Research Laboratories
AL	Appleton Laboratory of the Science and Engineering Research Council
ARD	Astrophysics Research Division of the AL
ARU	Astrophysics Research Unit of the Culham Laboratory
ASE	American Science and Engineering Inc.
BAC	British Aircraft Corporation
Bel	Queen's University, Belfast
Bir	University of Birmingham
Bkb	Birkbeck College University of London
Bn	University of Bergen
Bri	University of Bristol
BTI	Braunschweig Technical Institute
Camb	University of Cambridge
CNIE	Comisión Naciónal de Investigaciónes Espaciales, Argentina
CRC	Communications Research Centre, Ottawa
CRESS	Center for Research in Experimental Space Science, University of York, Toronto, Canada
Cul	Culham Laboratory of the UK Atomic Energy Authority
DFVLR	Deutsche Forschungs- und Versuchsanstalt für Luft- und Raumfahrt EV
EMIE	EMI Electronics Ltd London
ESTEC	European Space Technology Centre
Gr	Technical University Graz
GSFC	Goddard Space Flight Center of NASA
HCO	Harvard College Observatory
Hd	University of Heidelberg
IC	Imperial College of Science and Technology, London
ISRO	Indian Space Research Organization
JB	Jodrell Bank, Nuffield Radio Astronomy Laboratories
KTH	Royal Institute of Technology Stockholm
Lei	University of Leicester

MIT	Massachusetts Institute of Technology
MO	Meteorological Office
MPG	Max Planck Institute Garching
MPI	Max Planck Institute
MSSL	Mullard Space Science Laboratory, University College London
NASA	National Aeronautics and Space Administration
NDRE	Norwegian Defence Research Establishment
NRDE	Norwegian Research and Development Establishment
Oxf	Oxford University
PRL	Physical Research Laboratory, Ahmedabad
RAE	Royal Aircraft Establishment, Farnborough
ROE	Royal Observatory Edinburgh
RRS	Radio Research Station, later RSRS
RSRS	Radio and Space Research Station, later Appleton Laboratory
Shf	University of Sheffield
SSC	Swedish Space Corporation
Sth	University of Southampton
Sx	University of Sussex
Tb	University of Tübingen
UA	University of Adelaide
UCB	University of California, Berkeley
UCL	University College London
UCW	University College of Wales, Aberystwyth
UIO	Uppsala Ionospheric Observatory
WRE	Weapons Research Establishment, Australia

PREFACE

Space Science in Britain was initiated, and the foundations for its development were laid, very largely by one man, the late Sir Harrie Massey. Sadly, his untimely death in November 1983 occurred before this Preface could be written, but the main text of the History was complete. It bears witness to the enormous contribution which he made through his vision, foresight and determination. With the close co-operation of the late Sir David Martin, then Executive Secretary of the Royal Society, and the late Roger Quirk, then a senior member of the Ministry for Science, Massey as Chairman of the British National Committee on Space Research, took the lead in harnessing the essential components for a British Space Science programme.

The resulting combination of science, mainly from university departments of physics, and technology, mainly from government research establishments, supported by government funds and backed by electronic and aerospace firms in British industry, proved to be very successful and more than able to hold its own with the tightly organized Space Agencies of other countries. Massey always believed that his first duty lay with British Universities, but his vision extended far beyond those boundaries. From the mid-1950s to the end of his life he was tireless in stimulating and encouraging international co-operation in the furtherance of space science. The extensive and highly successful joint programme between Britain and the USA, the genesis and development of the European Space Research Organization, later to become the European Space Agency, the Commonwealth Collaborative Programme and many aspects of the work of COSPAR and of the European Science Foundation, all bear the marks of Massey's genius for leadership in co-operative programmes of science.

The concept of this History arose during discussions between the two authors in 1978. We realized that the scale of global space science had already outstripped the scope of a concise History. However, we thought that the time was ripe for the compilation of a History of the British contribution to the first 25 years or so of the subject, whilst memories were still relatively fresh and many of the leading participants still available for consultation. We believed also that there would be merit in putting on record how a new branch of science had actually been started and developed; how the resources of scientific manpower, the technological skills and not least the financial support, could be co-ordinated and brought together to achieve very complex and precise objectives.

We were very fortunate in enlisting the encouragement and support of Sir Geoffrey Allen, the Chairman of the then Science Research Council. This culminated in a Commission from the Science and Engineering Research Council in 1981 to prepare the present History. Whilst the activities of the first few years can readily be described in chronological order, the many parallel strands into which the programmes have developed necessitate a corresponding subject division in later chapters. Throughout, we have tried to give sufficient of the scientific and technological background to enable the reader to appreciate the major British contributions which have been made to this global subject.

We are indebted to very many individuals and organizations for information, comment and permission to include photographs and diagrams.

We have relied heavily on the excellent documentation held in the archives of the Royal Society, and on the help we were given by its Executive Secretary Dr R.W.J. Keay, and his staff. A particular acknowledgment is due to Mr P. Wigley, who was tireless in his search for information and never failed us in following up the most obscure references. Similarly, we were greatly helped by records held in the Rutherford Appleton Laboratory of the SERC and by the advice of Mr J. Delury and Mr J. Reed. Dr M.A.R. Kemp of the SERC Swindon Office has assisted us in many administrative matters.

Professor F. Heymann and his staff of the Physics and Astronomy Department of University College London have supported us in a variety of ways, and our thanks are especially due to Mrs M. Burton, for many years Secretary to Sir Harrie Massey.

The help of the following individuals is gratefully acknowledged: Mr R.W. Bain, Sir David Bates, Sir Granville Beynon, Professor A. Boksenberg, Sir Robert Boyd, Professor J.L.C. Culhane, Dr E.B. Dorling, Professor H. Elliot, Dr R. Frith, Professor G.V. Groves, Professor P.C. Hedgecock, Dr J.T. Houghton, Dr C.M. Humphries, Professor T.R. Kaiser, Dr J.W. King, Dr D.G. King-Hele,

Mr D.E. Miller, Professor K.A. Pounds, Dr J.J. Quenby, Mr J.A. Ratcliffe, Dr D. Rees, Dr W.H. Stephens, Dr K.H. Stewart, Professor A.P. Willmore, Professor R. Wilson.

Finally our thanks are due to the following organizations for permission to publish photographs, diagrams or documents:

The Royal Society
Planetary and Space Science
Royal Astronomical Society
Journal of Atmospheric and Terrestrial Physics
Astrophysical Journal
Royal Meteorological Society
COSPAR – Space Research
Nature
Solar Physics
NASA
ESRO (now ESA)
Royal Aircraft Establishment, Farnborough
Royal Observatory, Edinburgh
Macmillan London
Pergamon Press
Gale & Polden Ltd, Aldershot (now Aldershot News Ltd Aldershot)
Weapons Research Establishment, Australia
Fotopersbureau Lindeman
General Electric Co. Ltd
British Aerospace plc
B.A.J. Vickers Ltd
University of Leicester (Professor K.A. Pounds)
University College London (Professor G. Groves)
University of Sheffield
Hansard
HMSO
US National Academy of Sciences (a press release)

M.O.R.
1983

1

The scientific background

The growth of space science in the United Kingdom naturally depended very much on the scientific and technological background in the country in the years just after the Second World War. It was a fortunate fact that, by 1953, while there were a number of scientists whose research work would be greatly expanded if space research techniques became available, technological progress through defence requirements had proceeded to a stage where it could be utilized successfully. Once these possibilities became apparent to the scientists and arrangements made so that they could be realized, space science developed rapidly. The story of the way this occurred, involving many fortuitous circumstances, and of how British space science has developed to the time of writing, forms the subject matter of this account.

1 Ionospheric research in Britain

We begin by describing the scientific and technological background, the former in this chapter and the latter in the following chapter. Perhaps the most important early scientific discoveries in the present context were those of the E region of the ionosphere made by E.V. Appleton and M.A.F. Barnett in 1925 and the F region by Appleton in 1927. These confirmed the speculation of Kennelly and Heaviside that an ionized region in the high atmosphere was responsible for the long-distance transmission of radio waves demonstrated by Marconi. Appleton enthusiastically expanded this work to study the properties of the ionosphere and soon there was a vigorous school of British scientists interested in research in this subject. The discovery of the ionosphere was important not only for radio transmission but also for geomagnetism, of which the leading exponent at the time was S. Chapman then Professor of Mathematics at Imperial College London. He took an immediate interest in the ionosphere and in the earth's atmosphere generally and soon became the leading authority on the subject.

In the present context these developments are of special importance because they were concerned with the earth's atmosphere at altitudes far above those accessible by balloon. Thus the lowest layer of the ionosphere useful for radio transmission, the E region, is at an altitude of 100–120 km and the main, F_2, region extends upwards from 200 km or more. The maximum height which could be reached by balloon was about 32 km. While astronomers are used to observing regions which are permanently inaccessible, the ionosphere seemed to be tantalizingly just out of reach.

The fact that the ionosphere is subject to solar control was established at an early stage. Atmospheric ionization could be produced both by solar electromagnetic and corpuscular radiation. It was some time before it was established from eclipse observations that the ionizing agents responsible for the normal mid-latitude ionosphere travelled with the speed of light and hence must be solar ultra-violet or X-radiation. At higher latitudes, on the other hand, solar particles contribute to the ionization under conditions of great variability. It soon became clear that, for a proper understanding of the ionosphere, it would be necessary to know the intensity and spectral distribution of solar ultra-violet and X-radiation. Since this radiation is absorbed in the atmosphere at ionospheric heights in ionizing atmospheric atoms and molecules, no information could be obtained from ground-based or balloon observations. To attempt any theoretical discussion of the formation of the ionospheric layers before the days of space research it was necessary to make crude assumptions about the sun as a black-body radiator in the far ultra-violet.

It is obvious also that knowledge of the pressure, density, temperature and composition of the atmosphere is required as a function of altitude. Very slender information was available in pre-rocket days about these quantities at ionospheric altitudes.

Figure 1.1 shows a schematic representation of the ionosphere in which the regions where the electron concentration could be measured by contemporary ground-based sounding methods are distinguished from those which are inaccessible in this way. Even for this basic quantity it will be seen that ground-based sounding could only give a partial picture. It was not for many years (the 1960s) that the introduction of radar scatter techniques extended greatly the range of ground-based methods.

The problems of ionospheric theory were compounded by the need for knowledge of a wide variety of rates of reaction between electrons, atmospheric ions and neutral atmospheric atoms and molecules. In particular, the key question of determining the concentration n_e of electrons in the ionosphere at a particular altitude can be expressed in terms of the equation

$$\frac{dn_e}{dt} = q - \alpha \, n_e^2$$

where q is the rate of electron production by solar radiation at the altitude concerned and α is the effective coefficient for recombination of electrons and positive ions. In equilibrium

$$n_e = (q/\alpha)^{\frac{1}{2}}$$

Values for n_e could be derived from ground-based ionospheric observations. The interpretation of these results presented a major theoretical problem.

Despite the daunting lack of precise information about the solar ionizing radiation, the atmospheric structure at high altitudes and the rates of key ionic reactions, ionospheric workers in Britain began to tackle the wide-ranging problems involved. They included, in addition to Appleton and Chapman, J.A. Ratcliffe, W.G. Beynon and K. Weekes. In 1936, following

Fig. 1.1. Schematic representation of the electron concentration with altitude in the atmosphere under average conditions.
—— obtained directly from contemporary ground-based sounding.
- - - interpolated.

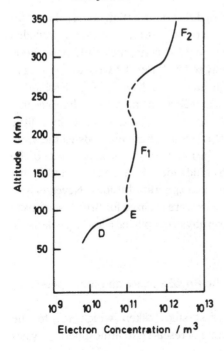

Electron Concentration / m³

some leading suggestions by D.F. Martyn, the first attempt was made by H.S.W. Massey to formulate an atomic theory of the ionosphere. This was devoted particularly towards the determination of the effective recombination coefficient, a subject which was followed up in succeeding papers. The problem proved very difficult, but already by 1947 D.R. Bates and Massey proposed that the basic process is one of dissociative recombination, as for example,

$$O_2^+ + e \rightarrow O' + O''.$$

This can only occur with molecular ions so that the effective recombination coefficient would depend on the fraction of molecular ions present. On this basis a consistent description of the variation of n_e with height in the ionosphere could be given but at the time could not be confirmed, partly through lack of knowledge of the reaction rates and partly through an equally profound ignorance of the molecular composition at ionospheric altitudes.

The reason for the layered structure of the ionosphere presented a baffling problem because of complete lack of knowledge of the short-wave solar spectrum. The high temperature of the solar corona was discovered in 1939 but attempts to link it with formation of one or more of the layers proved inconclusive.

It is no wonder that the possibility of direct observation of the solar ultra-violet and X-radiation, and of atmospheric structure, from rocket-propelled vehicles was a matter of great interest to those concerned with ionospheric research in Britain. This was especially so for those who were attempting to construct an ionospheric model at the atomic level. Some of these scientists played a considerable part in the war effort during which they became familiar with the existing methods of propulsion and deduced, by a little simple arithmetic, that while not yet adequate these methods would seem not so far short of being able to reach the ionosphere. The appearance of the V2 rocket showed at once that much could indeed be done to provide means of making *in situ* measurements at ionospheric altitudes. Nevertheless, much was to happen before these hopes were realized for British scientists; and in these developments ionospheric physicists played a leading role as we shall see.

2   The Gassiot Committee of the Royal Society and atmospheric research

Remarkably enough, the first steps taken which led to the establishment of a scientific programme of research in atmospheric physics using rocket propelled vehicles were taken during the Second World War. In 1941 the Meteorological Research Committee of the Air Ministry recommended that the Royal Society be asked to mount a programme of research in meteorology on a wider basis than hitherto, dealing in particular with the conditions of radiative equilibrium in the earth's atmosphere. The Air Ministry, in the very middle of the war, accepted this recommendation. A letter was sent to the Secretary of the Royal Society on behalf of the Air Council, enquiring whether the Society would be prepared to undertake this task.

The Royal Society referred the matter to its Gassiot Committee for consideration. This committee was set up as long ago as 1871 to supervise the management of the Kew Observatory when this responsibility was taken over by the Royal Society from the British Association for the Advancement of Science. The name of Gassiot was associated with it because Gassiot who, while an FRS through his contribution to experimental electrical science,[1] was also a wealthy wine merchant, had made possible the transfer to the Royal Society by establishing a trust fund of £10,000 to be used for the operation of the Observatory. He was naturally the first chairman of the committee which in the course of time became the Royal Society committee dealing with meteorological research.

In 1942 its terms of reference were as follows: 'To recommend as to the work of the meteorological and magnetic observatories with which the Society is connected and to administer the Gassiot and other trust funds applicable to their maintenance.' A further reference to Indian observatories was included which need not detain us here. G.M.B. Dobson was the chairman and apart from the ex officio members (the Astronomer Royal, the President of the Royal Astronomical Society and the Director of the Meteorological Office) the members were Sir Edward Appleton, D. Brunt, S. Chapman, A.E.G. Egerton, Sir Henry Lyons, Sir George Simpson, G.I. Taylor and Sir Gilbert Walker.

The Committee recommended that the proposal from the Air Ministry be accepted and the Council of the Royal Society then agreed in principle, pending a more detailed report. This was submitted on 8 March 1943. The chief problems to be investigated were summarized as the following.

(a) What gases are of primary importance in determining the radiation balance?

(b) What are the concentrations of different species at different altitudes?

(c) What are the absorption coefficients of different wavelengths of electromagnetic radiation under atmospheric conditions?

(d) What photochemical reactions lead to ozone formation in the atmosphere?

(e) What is the solar spectrum for wavelengths < 3000 Å?

The report was published in the Report of Progress in Physics in 1943.

2.1   The sub-committees of the Gassiot Committee

To implement the programme it was proposed that three sub-committees of the Gassiot Committee be set up. The first, Sub-committee A, was concerned with atmospheric composition, B with photochemistry of the atmosphere, and C with atmospheric temperature and radiation. Dobson, Massey and T.G. Gowling were named as the respective chairmen.

During wartime these sub-committees did not function at all regularly as many of the members were engaged in war work of some kind. However, soon after the war ended they became very active. From the present point of view, it was Sub-committee B which was most concerned with the possibility of *in situ* observations in the high atmosphere from rocket vehicles. Already at a meeting of the Sub-committee in 1946, A. Hunter had drawn attention to the possibility of observing short wave solar radiation in this way and at the same meeting the following recommendation was made. 'Contact should be made with departments developing rockets with a view to designing apparatus to be carried up by rockets and to measure the solar spectrum at heights above the absorbing atmosphere.' Bates, who had become chairman of the Sub-committee at the end of 1946, spent nine months in the USA in 1950. During this period he was in contact with developments there in the use of sounding rockets and suggested the possibility of seeding the high atmosphere with sodium evaporated from a rocket to produce a strong yellow glow in the sky – the first suggestion of an active experiment from a rocket (see Chapter 3, p. 30). It was not at all clear to the scientists of the Gassiot Committee and sub-committees in 1946 and for several subsequent years how the Sub-committee B recommendation could be implemented. Work on rocket development concerned with military applications was known to be going on vigorously in government establishments but security was so strict that little was known about what possibilities existed for the use of military rockets for scientific purposes.

3   The American Upper Atmosphere Research Panel

Meanwhile, important developments[2] had been taking place in the United States arising again through interest in the exploration of the upper atmosphere. It happened that, about the same time as a number of scientists and engineers at the Communications Security Section of the Naval

Research Laboratory in Washington were deciding in favour of rocket exploration of the upper atmosphere as a post-war research programme in which to engage, the United States Army offered space for such experiments in captured V2 rockets which were to be launched from the White Sands Proving Ground in New Mexico. As the rockets, of necessity, had to be launched nearly vertically for safety reasons, this was an excellent opportunity to introduce the research programme without delay. The interested scientists spontaneously formed in early 1946 a panel known at first as the V-2 Upper Atmosphere Research Panel, under the initial chairmanship of E.H. Krause who was succeeded in that post by J.A. van Allen in 1947.

While having no formal charter the Panel played from the outset a very influential role in the development of space science. Under its auspices the basis for a very sound and extensive research programme was established. By 1952 when all the V-2s had been used up, much information had already been obtained about the upper atmosphere, the solar radiation and cosmic rays. Well before this time a rocket to take over from the V-2, known as the Aerobee, was successfully developed locally and was in full operation. In 1948 the panel had assumed the title of Upper Atmosphere Rocket Research Panel.

We shall say more about the American rocket programme in Chapter 3. Here we are concerned with the part it played in bringing together the British scientists anxious to explore the upper atmosphere with instruments transported by rocket and the British rocket engineers who could provide the vehicles.

4 Proposal for a conference on rocket exploration of the upper air

The breakthrough came from a proposal by Chapman in 1951, that the Gassiot Committee invite the American Upper Atmosphere Rocket Research Panel to take part in a conference on rocket exploration of the upper air to be organized by the Committee in Britain. Chapman, by this time, had been in close touch with the Panel and had discussed the proposal with them. The Committee had no hesitation in accepting the proposal. In 1951 Massey had become Chairman of the Gassiot Committee in succession to Brunt who had become Physical Secretary of the Royal Society in 1948. A sub-committee to organize the Conference was set up consisting of Massey as Chairman with Chapman, W.S. Normand and F.A. Paneth.

In the event the Conference took place in 1953 at Oxford. In the course of preparation for the Conference, contact was established with work going on in the Ministry of Supply which proved to be very appropriate and at such a stage of development as to be very adaptable to sounding rocket research. However, before describing in Chapter 3 the sequence of events which led directly to the establishment of a sounding rocket programme we shall pause to describe in Chapter 2 what had been going on in rocket development and associated matters in the Ministry of Supply.

It is of interest to note in passing that the Gassiot Committee in 1952 made a grant to Paneth of £2000 to enable him to analyse samples of stratospheric air obtained from rockets, in collaboration with the Engineering Research Institute of the University of Michigan, the first example of Anglo–American collaboration in sounding rocket studies!

2

The technological background

The background of technology which was available in Britain in the mid-1950s and which was relevant, indeed was essential, for the development of a significant UK space science programme in the succeeding years, was generated largely during and shortly after the Second World War. Rocket technology was at the heart of the matter, and this had a long history at least back to the Chinese of the 13th century. We have no technical details of the rocket weapons used by the Chinese although it is reasonable to assume that the basic ingredient was black gunpowder.[1] Rockets were used intermittently in Europe either as weapons or in firework displays throughout the 16th, 17th and 18th centuries, although there appears to have been no systematic development of the techniques used.

One of the first major engagements in which Europeans were subjected to rocket attacks occurred during the invasion of the Indian state of Mysore by British forces under Wellesley (later the Duke of Wellington). In 1799 enemy rockets from the forces of Tipoo Sultan fell on the British encampment outside Seringapatam. Perhaps it was this first-hand experience which stimulated the British to take serious steps themselves in the development of rockets. At the Royal Laboratory of Woolwich Arsenal, Colonel (later Sir William) Congreve developed a 32 lb rocket with a range of 2000 to 3000 yards. Many thousands of these were produced, but without conspicuous military success. For instance, in the war of 1812 between Britain and the USA, Baltimore was bombarded by British rockets, and in the Peninsular war, Congreve rockets were given at least two trials by Wellington. In the second of these, in 1813 against cavalry, Wellington reported that 'they would have scared the horses stiff if only they had gone near them.'[2] The Congreve rockets were stabilized by a 15 foot long stick, which must have posed problems in handling on the battle field. However, in

9

the second half of the nineteenth century, an English inventor named William Hale developed and patented a method of spin stabilization, which was a significant advance in design. In the same period, Lt. Col. R.A. Boxer, a British officer at the Royal Laboratory, Woolwich, developed a two-stage tandem rocket, giving the possibility of much longer ranges.[3] Despite these improvements, only limited use was made of rockets in the period up to the end of the First World War. This was possibly due to the problems caused by propellant instability in storage, and to the superior accuracy, for military purposes, of conventional artillery.

It was the perseverance and enthusiasm of mainly non-government groups and individuals which led to the development by government agencies of the modern reliable and efficient rocket motor, the key to all operations in space. The theoretical studies of Tsiolkovsky in Russia and Oberth in Germany, and the experimental work of Goddard in the USA were pre-eminent in the years before 1939. In Britain, the subject was comparatively dormant until the War Office, in 1934, began to take serious notice of the activities of the German military authorities, who were embarking on an extensive rocket research and development programme. In 1935 the Research Department of Woolwich Arsenal was asked to propose a development programme for military rockets using cordite as the propellant. From 1936 onwards, Sir Alwyn Crow was in general control of an intensive programme which included studies of long-range (900-mile) rockets.[4] These might have been relevant to the design of vertical sounding rockets, but in the event the threat of war caused all efforts to be concentrated on the more pressing problems of rockets for anti-aircraft defence. For the immediate purposes of the war, only solid fuel propellant systems were considered, for reasons of logistics and economy, although some effort was devoted to liquid fuel systems on a longer term basis. Much technical progress was to be made on the design of solid fuel rockets with diameters of 2, 3 and 5 inches. Difficult problems relating to propellant burning rates, the ratio of propellant weight to overall weight, the manufacture of sufficiently accurate steel tubing and the stability of performance at extremes of temperature, were all mastered. The foundations were laid for the post-war development of propulsion systems for guided weapons, for upper atmosphere research rockets, and for satellite launching rockets. The wartime rockets were used almost exclusively for the carriage of explosive warheads but some experience was also gained in the carriage of relatively fragile items such as specially rugged thermionic valves and photo-cells. Advances in rocket technology during the war years may have been the most directly relevant contribution to the future space technology, but other essential techniques were advancing in parallel. Centimetric radio, advances in aerodynamics and structural engineering, and in instrumentation, all contributed. But in 1945 two major items still lay in the future; the solid state electronic device typified by the transistor, and the digital computer.

The rocket developments in Britain up to 1945 were overshadowed by the great technical achievements of German rocket engineers which resulted in the V-2 weapon. This long-range (200 miles) ballistic rocket with its sophisticated control system, convinced any remaining sceptics that the large controllable rocket, capable of mass production, was indeed a practical proposition. At the end of the war in Europe in May 1945, there was something of a scramble between the Western Allies and the Russians to capture the material and the leading German designers concerned with rockets. In the event, it seems that the major share of the material and the development engineers was taken by the Americans; this included parts for the assembly of over 60 rockets, and the leading German rocket engineer, Wernher von Braun, together with many of his colleagues. The Russians transferred a large contingent of the production staff to the USSR.

The Americans moved quickly to exploit the captured V-2 rockets, most of which were converted to carry scientific equipment into the upper atmosphere.[5] The first such flight took place in April 1946, the rocket being launched from the White Sands proving ground in New Mexico. Some 60 others followed in the next six years, covering an altitude range from 70 to 130 miles; a boosted version, code-named 'Bumper' reached 244 miles in 1949. The scientific measurements made possible at this time, covering phenomena of the ionosphere, cosmic rays, solar radiation and properties of the atmosphere, can fairly be said to mark the beginning of systematic space science based on *in situ* observations. Nearly half of the converted V-2s launched in America were classed as failures, but the experience gained by both the rocket technologists and the scientific experimenters must have been invaluable. It greatly enlarged the base on which the expanding programme of upper atmosphere sounding rockets could be built. The fortuitous supply of V-2 rockets was of course strictly limited, and in the late 1940s several rockets specifically designed for scientific research were coming into use in the USA. Amongst them, the WAC Corporal, the Viking and the Aerobee were outstanding, and the last, with variations, was to become the mainstay of the NASA sounding rocket programme for many years. As we shall see, it was to influence the first British vertical sounding rocket in due course.

The Russians are believed to have launched some 20 V-2 rockets with

German help in 1947, some carrying scientific measuring equipment.

The British, with German help, launched three V-2 rockets in October 1945, from a site near Cuxhaven, in an operation code-named BACKFIRE. However, the British were in no position, at the end of the war in Europe, to engage in rocket experiments on this scale, and a start was made in more modest ways.

To follow in outline the gradual development of the technological base in Britain during the 12 years from the end of the war until the first test launch of a British sounding rocket in 1957, we need to refer briefly to organization matters. The concepts of guided missiles which emerged as the war ended demanded a unified systems approach to design which did not fit easily into the Government's research and development organizations. The techniques of aerodynamics and of structural design for high-speed flight lay largely with the Royal Aircraft Establishment (RAE) at Farnborough. Radio and radar techniques required for guidance and tracking were concentrated in the main in specialist departments and establishments of the Ministry of Supply, and in establishments of the Admiralty. The design of rocket motors and the general handling of propellants and explosives lay within organizations such as Woolwich Arsenal and the Armament Research and Development Establishment. All were served by a broad cross-section of British industry.

Arrangements were very considerably rationalized by the formation of a Controlled Weapons Department at the RAE Farnborough, under G.W.H. (later Sir George) Gardner, in 1946. At about the same time the Rocket Propulsion Establishment was created at Westcott near Aylesbury. These two new organizations, together with the substantial activities in various specialist laboratories, formed the nucleus around which a new guided weapons industry was to be built, and the whole technical complex, required for safe firing trials and the assessment of results, developed. One of the first tasks to be undertaken, one which was to have a significant effect on the future activities of British space scientists, was the establishment of a major trials range. The requirements of public safety, freedom to instal the necessary tracking and monitoring devices, and the sheer size of the area required, could not be met within the British Isles. An Agreement between the governments of the UK and Australia, which specified the sharing of costs and responsibilities, enabled a range to be established in South Australia, based on Woomera, for the joint use of the two countries. When the first British sounding rocket, the Skylark, was ready for operation in the late 1950s, Woomera was to be the launch site. It remained the only available site until launch sites in Europe came into operation under the auspices of the European Space Research Organization many years later. Thus it was that all the early British experiments carried out by Skylark rockets measured conditions in the southern hemisphere, and observed the southern sky.

This is not the place in which to enlarge on the developing guided weapons programme in the early 1950s. We only need to identify a few of the strands which were soon to diverge from the military programme and serve the technological needs of the emerging civil space science programme. The small group of staff forming the Controlled Weapons Department of RAE Farnborough gained rapidly in experience, grew in size, and became the Guided Weapons (GW) Department as it assumed more responsibility in system assessment and design.

One of the relevant studies in the GW Department and the Rocket Propulsion Establishment concerned the general topic of ballistic missiles, the practicability of which had been proved beyond doubt by the German V-2 rocket of 1945. However, for various reasons, there was particular interest in the possible performance to be obtained from solid fuel rockets, as distinct from the liquid fuel system used in the V-2. There has been earlier reference to the relatively small solid fuel rockets of the wartime years. We now move to a stage where not only are the rocket motors to be much larger and more efficient in terms of total impulse for a given propellant weight, but also the predictability and reproducibility of performance are required to be of a higher order.

As part of the RAE studies in 1953–54, D.G. King-Hele and others assessed the performance of a variety of single and two-stage hypothetical solid propellant rockets, the general scale being such that payloads of the order of 100 lb could be carried to altitudes of 100 miles or thereabouts.[6] It was recognized that such rockets were potentially of value to upper atmosphere research, as well as to the research programme of RAE, and more detailed studies of likely performance were made by D.I. Dawton. These suggested that an existing solid fuel motor, if developed in ways which seemed feasible, including increasing the length and decreasing the burning rate of the propellant, might form the basis of an attractive single-stage vehicle capable of carring 45 kg of payload to altitudes of 150 km or more.

Concurrently with the discussions of University scientists about the possible exploitation of high altitude rockets for scientific research, to which we refer in Chapter 3, the RAE proceeded with its plans to extend a series of experimental rocket-propelled vehicles which had been developed through stages for various specific purposes. The latest in the series, to be designated CTV5 Series 3, was designed around the concept studied by King-Hele and

Dawton for the single-stage vehicle described above. At about the same time (June 1955) T.L. Smith and W.G. Parker of RAE visited the USA to discuss the design of rockets for upper atmosphere research. Of particular interest was the comparison between the American Aerobee to which we have referred earlier, and the CTV5 Series 3, which were not dissimilar in size. It was realized that it would be possible without much difficulty to design the British rocket in such a way that it could be launched from the tower at the White Sands proving ground should the occasion arise. The main impact on the design was the requirement that the stabilizing fins at the rear of the rocket which had to clear the launcher rails, should number 3, and not 2 or 4. This design feature was incorporated, although in fact no launching of the British rocket has ever taken place at White Sands. So we come to the genesis of the British upper atmosphere research rocket to become known as SKYLARK. In the next chapter the account of its development and harnessing to the needs of space science is carried further. Figure 2.1 shows some of the RAE team responsible for the development of Skylark.

One further facet of the technological background at RAE in the mid-1950s should be mentioned if only because it gives a foretaste of the

Fig. 2.1. Some of The Skylark team at RAE Farnborough 1955. *(Left to right):* F. Hazell, W.H. Stephens, E.B. Dorling, T. Moss, M.O. Robins, E.C. Cornford.

immense contribution which King-Hele and his co-workers in GW Department and eventually in Space Department RAE were to make to our understanding of the orbits of earth satellites. Studies completed in January 1956 by King-Hele and Miss D.M.C. Gilmore assessed the feasibility of launching into an earth orbit of altitude about 200 miles, a satellite of weight about 2000 lb. It was concluded, in a survey which took account of all the more obvious problems, that no insuperable difficulties appeared to stand in the way of such an achievement using the liquid fuel rocket technology which at that time seemed practicable.[7] The launch of Sputnik 1 by the Soviet Union in the following year was dramatic confirmation of the validity of the RAE assessments. King-Hele rightly identified the survival on the re-entry of such a satellite into the atmosphere as one of the severest problems likely to be encountered, if recovery were to be attempted. In a further study, completed in 1956, the dynamics of re-entry were analysed, and the basis laid down for the great advances in the deduction of atmospheric densities and motions from the observations of satellite orbits and re-entry trajectories.[8]

This chapter has been very largely about the background of rocket propulsion technology, but it is salutary to recall other rather more mundane, but essential, technology on which space science was dependent in the mid-1950s. In general, the available electrical equipment and instrumentation had been designed for use in aircraft, as being the nearest equivalent to spacecraft, and having a similar requirement for 'lightness and strength'. But rocket launches often imposed far more severe linear acceleration and vibrational loads than any aircraft, and caused many casualties amongst plugs and sockets, soldered joints, gyroscopes and the like. The use of thermionic valves was unavoidable until transistors became readily available in the early 1960s, and the problems of mechanical protection, heat dissipation and general volume requirements which arose in consequence were a severe limitation. However, the great pressure for reliable technology in the space environment ensured that this was eventually achieved, with corresponding benefits in the less demanding environment of earth-bound equipment.

3

The initiation of the Skylark rocket programme – the IGY and artificial satellites

In Chapter 1 we described the way in which the Gassiot Committee of the Royal Society had been taking the lead in developing a research programme in atmospheric (including upper atmospheric) physics, through its three sub-committees, since 1941. It was natural that the Committee would take a close interest in the development going on in the USA in the use of rockets as vehicles to transport scientific instruments to high altitudes in the atmosphere. The chapter concluded with an account of the arrangements made by the Committee in conjunction with the US Rocket Research Panel to hold an international conference in Oxford in 1953 on rocket exploration of the upper atmosphere.

1 Prelude to the conference – an offer of rockets

The atmospheric scientists up to this time were quite unaware of the work proceeding in the Ministry of Supply on the design and development of rockets in connection with ballistic missile development, described in Chapter 2. It so happened, however, that the American Scientific liaison officer in London, F. Singer, had a wide interest and involvement in both the scientific and technological aspects of space exploration. He knew the key figures in the Ministry of Supply concerned with rocket development and suggested to the Chairman of the Gassiot Committee that invitations to attend the Conference be sent to a number of these people. This was done immediately and was much appreciated as it opened to them a window on the outside scientific world interested in using their products.

On the morning of 13 May 1953 the Chairman of the Gassiot Committee (H.S.W. Massey) was just preparing to depart from his office in the physics department of University College London to take part in the annual departmental cricket match between the staff and students, when he received a telephone call from an official at the Ministry of Supply asking whether he would be interested in using rockets available from the ministry for scientific research. Without hesitation he said 'yes' and this really marked the beginning of the British scientific rocket programme.

The chairman was able to say 'yes' so readily because as head of a large and expanding physics department, strongly supported by the college and university, he was sure that the resources and the enthusiasm for a rocket programme would be forthcoming. Without delay he discussed the matter with R.L.F. Boyd, a lecturer in the department, who was carrying out experimental research work on rates of electronic and ionic collision processes in gases. Part of this work depended on the analysis of the processes occurring in electric discharges and for this he had developed new probe methods which were to stand him in good stead later on. Boyd, whose first degree was in engineering and who had spent the last two years of the war at the Mine Design Department of the Admiralty as assistant to D.R. Bates, was very well equipped to lead a scientific research programme to be carried out ultimately in the field, and which involved large-scale hardware. He reacted very favourably and plans were made for him to gain experience of the technical side by spending some weeks as a vacation consultant at the Royal Aircraft Establishment (RAE). It was during this period that Boyd first met M.O. Robins who at RAE was concerned with the control systems of rocket-propelled vehicles. The two were to work in close collaboration in the years ahead.

At the same time Massey realized that a research programme using rockets would certainly be very expensive relative to research in the laboratory. It would clearly not be possible for the necessary funds to be made available for one University department only so he approached a number of colleagues who might be interested to take part. These included D.R. Bates and K.G. Emeleus of Queen's University, Belfast, interested especially in optical phenomena in the high atmosphere, W.G. Beynon of University College of Wales, Aberystwyth, interested in carrying out radio propagation studies between a rocket and the ground, J. Sayers of the University of Birmingham, interested in measurement of the composition of positive ions in the ionosphere and P.A. Sheppard of Imperial College London, interested in measuring wind speed and direction at high altitudes. All of these University scientists indicated that they would be prepared to participate in a significant way if funds became available. This was despite the fact that at first it appeared from conversations which followed the telephone call of 13 May that the vehicles immediately available would hardly be able to achieve an altitude greater than 30 km!

2 The Oxford Conference

All of these discussions took place before the Conference which began on 24 August and continued to 26 August. The American Rocket Research Panel was well represented by 15 delegates including its chairman, J. van Allen, W.H. Pickering, who was to play a major part in many advanced space projects in the future, H.G. Newell and F.S. Johnson who were key figures in the development of space science in the USA. Apart from members of the Gassiot Committee and Sub-committees, the UK delegation included the Deputy Director of RAE (F.E. Jones) and two other representatives of that establishment, one (R.C. Knight) from the Ministry of Supply headquarters, three from industry, the director and two other representatives from the Meteorological Office, one (T. Gold) from the Royal Greenwich Observatory and one from the British Interplanetary Society. In addition there were delegates from Australia, France, West Germany, Sweden and Norway.

The meeting was a very successful one. The American delegates gave very thorough accounts of the work which had been carried out via their V-2 and Aerobee rocket programmes (see Chapter 1, p. 7). It is interesting to note in relation to future events that one paper presented by Singer was entitled 'Astrophysical Measurements and Artificial Earth Satellites'. This was thought of by most of the UK and European delegates as somewhat avant-garde!!

3 Initiation of a rocket programme

The American presentations were listened to most attentively by the British scientists present and provided a purposeful additional stimulus towards the establishment of a British programme. During the time in Oxford, opportunity was taken to arrange a meeting between the prospective British users that confirmed their desire to embark on a rocket programme.

A major problem still remained – where were the not inconsiderable funds to come from to finance such a programme? Massey first discussed the matter with the then Paymaster-General, Lord Cherwell on 14 August 1953. He was very sympathetic but could only suggest that the Meteorological Office might be prepared to take the initiative. This did not prove a very useful suggestion but a much more promising possibility was soon followed up.

Both the Physical Secretary (Sir David Brunt) and the Assistant Secretary (D.C. Martin) of the Royal Society were very interested in the Gassiot

Committee's research activities and were most helpful. They thought that a direct approach to the Treasury for funds to initiate the programme could be made and Brunt was prepared to make such an approach.

At a meeting of the Gassiot Committee on 20 October 1953 the chairman reported the approach made to him from the Ministry of Supply involving the possible use of facilities at Woomera (see Chapter 2, p. 12) or Aberporth for rocket launching. The Deputy Director of Farnborough mentioned the high cost of a rocket programme while Boyd, based on his experience during two months' vacation work at the RAE, reported on the possibilities as he saw them. In his report the chairman envisaged a three-year programme to be carried out, initially with a special grant from the Treasury.

The Committee resolved that the chairman approach the director of RAE Farnborough (Sir Arnold Hall) to arrange a meeting to discuss the possibilities with a representative group consisting of the chairman, G.M.B. Dobson, D.R. Bates, P.A. Sheppard, F.E. Jones, the director of the Meteorological Office or his representative, R.L.F. Boyd and J.A. Ratcliffe.

This meeting took place at RAE in February 1954 and was very satisfying for the scientists. They had understood that vehicles attaining only low altitudes would be available but Sir Arnold Hall revealed to their delight the proposed development of the CTV5 Series 3 rocket (see Chapter 2, p. 13) which would be capable of taking a payload of 100 lb to an altitude of 200 km or so, well beyond the E region of the ionosphere which is located between 100 and 120 km. The prospects for collaboration seemed very good. This was reported to the Committee at its meeting on 29 June 1954.

On the basis of these discussions, Sir David Brunt made a direct application to the Treasury on behalf of the Royal Society for a sum of £100,000 for rocket research in the upper atmosphere over a four-year period. Of this, half would be paid direct to the Royal Society to provide grants to the University scientists to enable them to develop and construct the necessary instruments. The remainder would be paid to the Ministry of Supply for the cost of rocket and other facilities which they would contribute.

There was considerable delay before any decision was made by the Treasury but during this time the prospective investigators began planning the scientific experiments which they proposed to carry out. In March 1954 Massey was awarded a grant to enable him to employ G.V. Groves, for one year in the first instance, to undertake preliminary work on rocket instrumentation in the Physics Department at University College London. This proved to be very important, for Groves had been on the staff of the Ministry of Supply for years and was very familiar with the rocket side of the work. This experience combined with his enthusiasm and ability both in theory and practice helped very much to get a good programme going.

4   Advice on the rocket programme – Gassiot Sub-committee D

Early in 1955 it seemed likely that the Treasury would look favourably on the application from the Royal Society. Accordingly, in April 1955 the Gassiot Committee recommended to the Council of the Royal Society that a further Sub-committee be set up to advise on the rocket programme. The terms of reference of the Sub-committee, referred to either as the Rocket Sub-committee or Sub-committee D, were as follows: 'To advise the Gassiot Committee on the programme of upper atmospheric research using rockets, and that its work in detail would consist in approving and co-ordinating the programmes of the different research centres, making recommendations to the Gassiot Committee for grants for research and establishing priorities of proposals for research. Copies of minutes of the meetings would be sent to Sir Owen Wansbrough-Jones the Chief Scientist, Ministry of Supply.'

In many respects this Sub-committee was the nearest analogue to the American Rocket Research Panel. The foundation members were: H.S.W. Massey (Chairman), D.R. Bates, W.G. Benyon, R.L.F. Boyd, Sir David Brunt, G.M.B. Dobson, K.G. Emeleus, F.E. Jones, W.G. Parker, J. Sayers, P.A. Sheppard and the Director of the Meteorological Office or representative. It first met on 3 May 1955. The Chairman stated that RAE expected to have within two years a single-stage rocket capable of carrying 100 lb to 210 km, which would cost about £2000. It was hoped that the development would make it possible to use the rockets during the International Geophysical Year about which more will be said below (p. 34). F.E. Jones, the Deputy Director of Farnborough, nominated W.G. Parker of the Rocket Propulsion Department, RAE Westcott (see Chapter 2, p. 12) as liaison officer between RAE and the university scientists.

5   The first grant applications

While still no official statement had been received from the Treasury it was considered worthwhile to discuss grant applications from the user groups which had been planning their programmes. These had been crystallized as follows:

Queen's University, Belfast (Bates and Emeleus): Photoelectric study of night and day airglow in the optical band as well as a search for micrometeoritic material by the microphone method

Birmingham University. (Sayers): Measurement of ion density in the ionosphere

University College London. (Boyd and Groves): Determination of atmospheric pressure, temperature and winds by the Grenade method; (Boyd): Ionic composition of the ionized layer

University College Swansea. (Beynon): Determination of the ionization profile in the ionosphere

Imperial College, London. (Sheppard). Determination of wind structure in the atmosphere by optical and radar methods

Grants were recommended to the respective groups of £2250, £1500, £2500, £900 and £1400 respectively.

6   Arrangements at Woomera

While it was still believed very optimistically that launching of the research rocket would be possible from Aberporth in Wales, it was clear from the outset that most of the vehicles would be launched from the rocket range at Woomera in Australia. As in the defence agreement, the responsibility for the launching would rest with the Australian Weapon Research Establishment (WRE). The idea of carrying out research in which your instruments would be launched 11,000 miles from the home base and your success would depend on co-operation between three separate bodies – the Royal Society, RAE and WRE – seemed to be ridiculous to many universities and other scientists. To the physics department at University College London, which had already operated a cosmic ray station in the Dolomites and carried out research in nuclear particle physics at accelerators outside the university, it did not seem to be that difficult and considerable research effort was directed to the programme.

The key role to be played by WRE made it important that good relations be established as quickly as possible with the Australians concerned. In this respect it was a great advantage that R.W. Boswell, who was in charge of the operations, had attended the same High School in Melbourne as had the Chairman of the Gassiot Committee and both were physics graduates of Melbourne University. A good relationship already existed between them and there was no difficulty at all in securing very close collaboration with the Australians that become even closer as the programme developed.

It was clear that a visit to Australia from a representative of the Gassiot Committee would be very important. Accordingly, the Chairman paid a visit to WRE at its Salisbury headquarters near Adelaide as well as to Woomera. He travelled by the regular service maintained by the Transport Command of the RAF between Lyneham, Wiltshire and Salisbury, South Australia, to transport men and equipment. The aircraft was a Hastings in the full string and sealing-wax tradition, coming down for the night at Tripoli, Habbaniya, Karachi, Singapore and Darwin. This service, which in later years replaced the Hastings by Comet IIs, was made great use of as the scientific work developed. During the visit, Massey was driven over the site for the rocket launching by Boswell, in a Land Rover, and was able to visit all relevant aspects of the work, meeting those who would be concerned with the detailed arrangements.

He also talked with D.F. Martyn, L. Huxley and other Australian scientists interested in atmospheric physics. Huxley, who was then head of the physics department at the University of Adelaide, offered a home from home in his department to any UK scientists spending time at Salisbury and Woomera in connection with the programme. This offer was gratefully accepted and was taken up on many occasions.

Shortly after Massey's return, the agreement of the Treasury was finally obtained for a grant, to be funnelled through the Air Ministry, to the Royal Society of the amount requested for the four-year period 1955–59. A formal announcement of the programme and an account of its general character was published in a Note to *Nature* submitted by Jones and Massey.[1]

As soon as the programme got underway, a further organizational arrangement was made. Regular technical meetings of those concerned with actual experiments and representatives of RAE, usually the liaison officer W.G. Parker, and J.F. Hazell who was in charge of the development of the rocket, took place in Massey's room in the physics department at University College London. These were roughly at two-monthly intervals, the first being on 27 September 1955. Reports of each meeting were presented to the Rocket Sub-committee and formed an important part of the decision-making as well as ensuring very close personal relations between all sides.

In November 1955, Robins visited Australia to discuss with WRE and the Woomera range authorities the problems of the near-vertical launching of rockets to a height of a hundred miles or so. The Australians were understandably concerned about the safety aspects of the proposed British project, but were very co-operative in finding satisfactory solutions.

In October 1956, Boyd and Hazell paid a joint visit to WRE to discuss the work necessary for the grenade experiments designed by Boyd and Groves, as well as many other matters such as data recovery. Matters were by then well under way but before describing the interaction with the International Geophysical Year we must pay some attention to the characteristics of the rocket that proved so effective a vehicle as the programme expanded.

7   The Skylark rocket and its development
In 1955 when it was decided to go ahead with the development of the CTV5 Series 3 rocket referred to in Chapter 2, overall responsibility for the project was given to Hazell who had been in charge of an earlier project in the series. This is the rocket which later (in 1956) was dubbed Skylark following a suggestion first made in RAE's house magazine in September 1956. An excellent account of the problems encountered in the early stages was presented to the XXVIth Congress of the International Astronautical Federation (IAF) at Lisbon on 21–27 September 1975 by E.B. Dorling in a paper entitled 'The Early History of the Skylark'. After referring to Hazell's appointment he says 'Much of the ultimate success of the new venture was due to him. The nominal characteristics of the motor were by then already decided, but the motor had still to be designed and made as had the rocket head and the launcher. Steel was still in short supply ten years after the end of World War II and finding the right material for motor-case and launcher posed problems. Digital computers were still some months away and performance calculations were done slowly and laboriously by hand. Little was known about the dynamic behaviour of fin-stabilized rockets at high speeds and low dynamic pressures; roll-yaw resonance[2] was known to be plaguing the rocket programme but no adequate theoretical treatment of the aerodynamics of fin-stabilized vehicles existed to give a guide as to what might be expected with the CTV5 Series 3. But estimates had to be made for the actual weight of the motor case, propellant, unburned propellant, structure and payload. Dispersion of the rocket's impact point was of primary concern and much of Hazell's time was taken up, together with that of his assistants, in assessing the likely behaviour of the rocket once it left the launcher rails. Meanwhile, the design and engineering team got to work on the rocket head and launcher, whilst at Westcott work began on the motor.'

Much effort was devoted by Hazell and his assistants towards the calculation of wind corrections. Quoting once again from Dorling's paper – 'Studies such as these were aided immeasurably when, in 1956, the first digital computer, the Ferranti Pegasus, became available. Prior to Pegasus, calculations were laborious and slow; hair-raising extrapolations had to be made for want of calculated results. Analogue computers were already in use in other departments and it seemed reasonable to enquire what help they might be. None it was learned . . .!' All in all it was remarkable that the teams succeeded in producing rockets for the test series of flights at Woomera beginning on 13 February 1957. Moreover, already on the fourth test flight University experiments were carried out and operated successfully in advance of the first flight specifically for the scientific programme on 17 April 1958. More will be said of these early flights on p. 26, below.

As mentioned in Chapter 2, p. 13, the Skylark differed from the V-2 and the Aerobee rockets in using solid propellant. While this meant a somewhat heavier engine there was a considerable gain for flexibility as well as in the avoidance of a complex fuel handling system. If, for example, a launch had to be postponed a Skylark could be left in the launcher for days while awaiting a second attempt, whereas a liquid propellant in such circumstances would have to be pumped out to be replenished later.

Figure 3.1 shows a sectional drawing of the rocket. Its overall length, 25 feet, is made up of the solid propellant Raven motor, the head section and the tail fins.

The Raven motor was developed by RPE (later RAE) Westcott. The initial weight of the propellant, a plastic formulation of ammonium perchlorate oxidizer and ammonium picrate coolant in a polyisobutylene binder, was close to 840 kg. It was enclosed in a tube of closely circular cross-section and was burned radially along its length. This was achieved by shaping the charge (star-shaped section until Raven IX) and including an igniter, which ran along the length. To ensure stable combustion 5% aluminium powder was added. This also had the effect of reducing the unburnt fraction after firing to only about 4%. The first motor, Raven I, gave a total impulse of 1450 kN at sea level. This was soon superseded by Raven II which increased the total impulse to 1539 kN. Gradual development has today led to Raven XI, with a slotted charge, giving 2230 kN.

The head section consisted of a nose cone and a cylindrical instrument bay. This was designed to be flexible so as to accommodate a variety of instruments apart from standard devices such as tracking beacons and telemetry. Arrangements could be made to pressurize either or both nose

Fig. 3.1. Sectional drawing of a Skylark rocket.

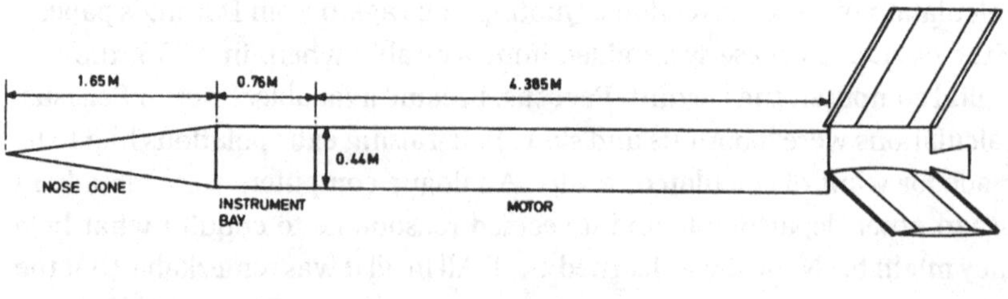

cone and bay. A photograph of the rocket is shown in Figure 3.2.

The launcher used was a metal structure 31 m high. The launching tower and supporting bays were fabricated from Bailey bridge panels at the Royal Ordnance Factory, Woolwich, to a design by RAE (Figure 3.3). A photograph of a Skylark launching from Woomera is shown in Figure 3.4.

It was hoped at first that launching could take place from Aberporth for which a similar launcher was built. However, safety considerations ruled out this possibility, much to the chagrin of some of the University experimental groups.

Accurate tracking of the rocket throughout its flight was essential. Each rocket included a standard Doppler transponder[3] and a 5 GHz beacon to operate with the Missile Tracking System (MTS)[4] used by the range. In addition, supplementary information was obtained from tracking by radar and by optical methods using kinetheodolites.

Telemetry compatible with the Woomera range system was built in. For many flights this was a standard RAE system.[5]

In the early stages of the programme, transistors had not yet been introduced. Miniature, ruggedized valves were used, mounted in aluminium blocks, both to protect them and to act as heat sinks.

Fig. 3.2. Photograph of a Skylark rocket.

8   The first Skylark flights

The first six flights of Skylark at Woomera were primarily to test the performance of the vehicle and the standard tracking and telemetry facilities. Nevertheless, advantage was taken on the last three of these flights to test out some of the scientific experiments and in fact some good measurements were made as we shall see. The first test flight at Woomera on 17 February 1957 was carried out with the launcher set at an elevation of

Fig. 3.3. Photograph of the Skylark launcher.

Fig. 3.4. Photograph of the launching of Skylark.

75°. The rocket returned to ground 43 km from the launch site after attaining a peak altitude of 12 km. All in all it was a most successful flight, the rocket being stable throughout the flight and the motor (a Raven I) behaving as expected.

The following four test flights were also successful from the same point of view so that the way was clear for the launch of the first Skylark specifically for the scientific programme. However, in fact, the first experiments were successfully carried out with the fourth test Skylark, equipment from three University groups being given a ride on what was still primarily a proving flight. The launch date, 13 November 1957 at 21.22 hours, therefore marked the actual beginning of the scientific programme even though it began officially with the launch of the 7th Skylark on 17 April 1958. Before describing the results obtained in the first few 'scientific' Skylarks we shall give a brief account of the design of the experiments carried out at this stage.

9   First experiments carried out in the Skylark programme

9.1   The grenade experiment

This experiment, requiring the most elaborate mechanism in the rocket, was designed to measure the temperature, density and wind distribution in the atmosphere up to an altitude of around 80 km. It is essentially a sound ranging method based on the production of a series of explosions from grenades ejected at regular intervals during the upward trajectory of the rocket.

Initially, the experiments were carried out at night using, on the ground, ballistic cameras, photoelectric flash detectors and an array of sounding microphones. The ballistic cameras photographed the flashes against the star background so that they could be located accurately in position. The flash detectors recorded the arrival of light from each grenade flash from which it is possible to determine the times of each explosion. Finally, the microphone array recorded the time of arrival of the sound pulse from each explosion at a number of known locations. An array rather than a single microphone must be used because of the tilting of the wave front of the sound by horizontal winds and refraction. Using all this information it is possible to determine not only the mean value of the speed of sound between each explosion and hence the mean atmospheric temperature but also the mean horizontal wind speed. Of course this involves a great deal of analysis but an elegant and effective procedure was worked out by Groves.[6]

In a typical experiment, such as the first on 13 November 1957, eighteen grenades were carried. A considerable amount of experimentation on the effectiveness of the grenade and detection equipment had already been carried out by Boyd and Groves at various suitable locations in the UK including Shoeburyness. The grenades were carried in the instrument bay as shown in Figure 3.5 and could be ejected horizontally.

Figure 3.6 shows a plan of the location of the ground-based detecting equipment at Woomera. It was originally planned in some detail to carry out this experiment in Wales with a launching at Aberporth, but it was not to be.

Although background noise in the microphones makes it impossible to observe acoustic signals at altitudes beyond 80 km it is still possible to obtain useful information from observations of the internal and external motions of the glow produced by grenade explosion at an altitude beyond 100 km. The glow arises from photochemical reactions between the explosion products and atomic oxygen which is a major atmospheric constituent at these altitudes. It lasts for 20 min or so and observations may be both of the mean motion of the cloud through the air and of its rate of expansion. From the former, the wind speed may be obtained and, from the latter, the diffusion coefficient of the glow gases through the local atmosphere which provides an estimate of the local air density. The occurrence of the glow was an unexpected bonus which was put to very good use.

9.2   The window experiment

A quite distinct method of measuring wind speeds was used by Sheppard and R.M. Goody of the meteorological department of Imperial College London. This involved the ejection from the rocket at selected

Fig. 3.5. The Grenade Bay for use with Skylark.

heights of clusters of metallic strips ('chaff') which are effectively radar reflectors. They were used during the Second World War under the code name 'Window' to spoil radar detection by giving rise to spurious signals. Once the chaff was ejected its motion could be followed by radar reflection and hence the wind speed obtained.

9.3   The sodium glow experiment

It was mentioned in Chapter 1, p. 6 that Bates, in a visit to the United States, had suggested that if vapour from a few kilograms of sodium is ejected from a rocket at an altitude of 60 km when it is twilight on the ground it should produce a bright yellow glow due to fluorescence in sunlight at the high altitude. Once such a glow is obtained it can be made to provide information about atmospheric density and wind speeds just as with a grenade glow. In addition it is possible under certain conditions to determine the atmospheric temperature at the glow height from measure-

Fig. 3.6. The location of cameras, flash detectors and microphones for the grenade experiment at Woomera. The flash detectors were operated at the camera sites.

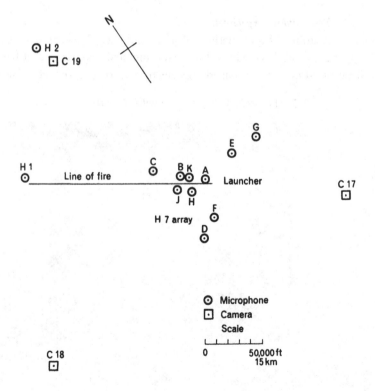

ment of the Doppler width of the Sodium D lines which are the source of the yellow colour.

Naturally Bates was anxious to carry out experiments of this kind at Woomera and he was assisted in this by E.B. Armstrong who had already carried out distinguished ground-based studies of the night airglow.

9.4 Optical measurement of the intensities of night and day airglow lines as functions of altitudes

Bates and Armstrong were also concerned with making these observations which were particularly important because the results of ground-based observations of the variation of the intensities of the lines of the night airglow were completely incompatible with the theory to which Bates had made major contributions. In retrospect the theory turned out to be correct.

9.5 Measurement of electron concentration as a function of altitude by the dielectric method

Sayers introduced an ingenious method for measuring electron concentration n_e, based on the fact that the dielectric constant K, for electromagnetic waves of angular frequency ω, of a medium containing n_e free electrons per unit volume, is given by

$$K = 1 - 4\pi \, n_e \, e^2/m\omega^2$$

where e is the charge and m the mass of an electron. This ignores dissipative effects due to collisions but these are unimportant at altitudes above 90 km. It follows that the capacity of an electrode system enclosing a gas containing free electrons will be a known function of n_e.

Applying this in the first instance to Skylark, Sayers arranged for the nose cone to be insulated from the remainder of the rocket. The capacity between the two insulated fractions was measured by monitoring the frequency of oscillation of a circuit in which they were included.

As in many other rocket experiments, care must be taken that the immediate ambience of the rocket is not disturbed by the latter's presence. Thus if the insulated nose cone is not at space potential, electrostatic fields will be present in the close neighbourhood which will modify the electron concentration from that which would be present in the absence of the rocket. To avoid errors arising from this source the potential of the nose cone was varied over a range ± 3 V at 25 Hz frequency. From the variation of the signal with potential, that of space could be identified.

9.6   *Mass spectrometers to study ion composition in the ionosphere*

Just as with the ion probe discussed in 9.5 above, it is important in making measurements from a rocket of the ion composition in the ambient ionosphere to avoid observing in a region which is affected by the presence and/or motion of the rocket. Sayers therefore developed equipment in which a time-of-flight mass spectrometer is ejected from the rocket on 100 feet of flex so as to observe in an undisturbed region. The spectrometer also had a high acceptance aperture much greater than for standard time-of-flight designs at the time.

Boyd and Willmore at University College London also prepared to make direct observations of the ionosphere. They designed a Langmuir probe mounted ahead of the rocket on a four-foot long spur, again to diminish effect of disturbance by the rocket. From the current–voltage characteristic of this probe when operated in the ionosphere not only the electron concentration but also the electron temperature could be determined. In this work they had had the benefit of experience[7] in the use of Langmuir probes for similar purposes in electric discharges.

9.7   *Observation of short-wave solar radiation*

The importance of measuring the intensities of the short-wave radiation from the sun in order to provide basic data for the interpretation of ionospheric structure and behaviour has been stressed in Chapter 1. Indeed it was one of the major reasons why British ionospheric scientists were interested at an early stage in the use of rocket vehicles.

The D region of the ionosphere, below the main reflecting layers at an altitude of 90 km or less, in contrast to the ionization above, is a nuisance for radio transmission as it absorbs radio waves. Considerable importance therefore attaches to obtaining a thorough understanding of this region as well as of the reflecting, non-absorbing, regions above. At the time, in the early 1950s, it was fairly well established that the ionization in the D region is produced through ionization of nitric oxide by solar Lyman α radiation at a wavelength of 121.6 nm and by ionization of molecular oxygen by soft X-rays. The relative importance of these two sources varied considerably from time to time. At times of strong solar disturbance the intensity of penetrating X-rays can increase to such a magnitude as to produce great enhancement of the ionization and hence the absorption at D region heights. This may become so strong as to produce a complete blackout in communication which depends on passage of radio waves through the region. For these reasons it is of special value to be able to measure the intensity of the Lyα and

X-radiation from the sun during rocket flights. Such measurements are also of importance for solar physics. Boyd and Willmore therefore used the technique described by H. Friedman of the Naval Research Laboratory in the USA for measuring the intensity of Ly-α radiation by means of a tubular counter containing nitric oxide which can only be ionized by radiations with wavelengths shorter than 130.0 nm. By requiring the radiation to enter the counter through a lithium fluoride window it is possible to block off all wavelengths shorter than Lyα so that the counter responds to radiation in a narrow band round 121.6 nm.

With hindsight, some of the most significant further experiments which were initiated at University College London were those directed to the measurement of solar X-radiation. In the first instance they used a photographic technique but were soon directing their attention to the use of photon counters as detectors. These may be used to give special information over a broad range by using different foil thickness for the counter window. Attention was soon being turned towards obtaining data with higher resolution by pulse height analysis. At the same time a further group concerned with X-ray observations from Skylark was established at Leicester University under E.A. Stewardson and K. Pounds, who had worked as a postgraduate student with Boyd and Willmore on these detectors. This group was also to play an important role in the later development of the subject.

By the end of 1958 several of these experiments were ready for flight in Skylark rockets and the remainder were in an advanced state of development. We shall describe the results obtained during this period which spans the International Geophysical Year, 1957–58. The value of the experience gained from preparation of the instrumentation for the early Skylark flights for the development of British experiments flown in artificial satellites cannot be exaggerated.

10   Later development of the Skylark

Earlier in this chapter, we referred to the development of the original Raven motor, obtaining thereby a substantial increase in the total impulse. It is a tribute to the soundness of the original concepts and the skill with which individual rocket motors have been developed and combined in stages, that over the years the performance of Skylark has been improved almost beyond recognition. This has been due both to the improvements in the Raven motor, and to the use of short-burning booster rockets as additional stages of propulsion.

The first booster was known as the Cuckoo, and burnt for four seconds, giving a total impulse of 364 kN. Later versions were developed, up to Cuckoo V. A larger booster, the Goldfinch, gave a total impulse of 699 kN s. A still larger booster rocket, the Rook III is now available. With this selection of sustainer motors and boosters, British Aerospace have been able to market a range of Skylark sounding rockets of which the following are the most important:

SKYLARK 5   A single-stage rocket using the Raven XI motor.
               A payload of 250 kg can be carried to an altitude of 200 km.
SKYLARK 7   A two-stage system using the Raven XI motor and a Goldfinch booster.
               A payload of 250 kg can be carried to an altitude of 300 km.
SKYLARK 12   A three-stage system using the Raven motor, the Goldfinch booster, and Cuckoo IV as a third stage.
               A payload of 200 kg can be carried to an altitude of 570 km; alternatively, 100 kg can be carried to 1000 km.
SKYLARK 14   A two-stage system using a Rook III booster and a Raven XI motor.
               A payload of 250 kg can be carried to 450 km.
SKYLARK 15   A three-stage system using the Raven XI motor, with the Rook III as first stage and Cuckoo V as third stage.
               A payload of 100 kg can be carried to about 1500 km.

The financial constraints placed upon the UK space science programme in the late 1970s meant that it has not been possible for the full potential of this range of Skylark rockets to be exploited by UK scientists, and the regular use of Skylark as part of the national space programme came to an end in 1978.

Developments which led to the production of stabilized Skylarks for scientific purposes are discussed in Chapter 9.

11   The International Geophysical Year

While the development of the Skylark programme was proceeding, quite independently preparations were going on for the International Geophysical Year (IGY) 1957–58, the largest international co-operative scientific enterprise ever undertaken. It had a profound influence on the future direction of space science, as we shall see.

As long ago as 1875 an Austrian naval lieutenant, K. Neypracht, returning from an Arctic expedition, suggested that such scientific expeditions would be much more effective if they were co-ordinated in time with other similar ventures. Eventually this idea was taken up and in 1882–83 the first Polar Year took place. During this period 12 stations in the Arctic were operated by different countries, including a joint Anglo–Canadian station at Fort Rae in Canada. Meteorological, geomagnetic and polar auroral observations were made. Following the success of these operations, a proposal for a Second Polar Year, to follow 50 years after the first, was made at a meeting in Hamburg in 1927. An international commission to administer the programme was set up. No less than 44 countries participated, half organizing special expeditions during the assigned year, 1932–33. In addition to the subjects studied in the First Polar Year, co-ordinated ionospheric observations were also made.

In 1950 L. Berkner, who was well known in international scientific circles for his work both as a radio scientist and as an organizer of large-scale scientific projects, suggested to some colleagues that, with the great increase in the pace of scientific research it would be appropriate to organize a Third Polar Year 25 rather than 50 years after the Second, i.e. in 1957–58.

This suggestion was very well received in the USA and, at a meeting of the Mixed Commission on the Ionosphere in Brussels in June 1951, Berkner explained his idea to an international audience. Again this was received very favourably, especially strong support coming from the International Unions[8] of Geodesy and Geophysics and of Astronomy. The parent body of the scientific unions, the International Council of Scientific Unions (ICSU), not only adopted the proposal but extended it in a way which greatly increased its general appeal. They realized that to understand the behaviour of the polar regions it is necessary to make a wide variety of observations at other latitudes. Why therefore should not the Third Polar Year be expanded to become a Geophysical Year during which co-ordinated studies of the earth at all latitudes and longitudes, including the solid earth, the oceans and the atmosphere, would be made? This was agreed and so the IGY was born. It was chosen to take place from 1 July 1957 to 31 December 1958. These times were specially suitable, for they covered a period of maximum solar activity in contrast to a solar minimum during the Second Polar Year.

To organize and supervise this vast enterprise ICSU set up a Special Committee[9] for the IGY known as CSAGI (Conseil Scientifique Annuaire Geophysique International). Chapman was appointed President, Berkner, Vice-President and M. Nicolet, a leading Belgian atmospheric physicist, Secretary General. The committee was made up of representatives from all the scientific unions concerned[10] as well as of the World Meteorological Organization, an international association at government level.

The IGY enlisted the interest not only of scientists but of the general public the world over and strong financial support was forthcoming from government sources. Among the twelve sections[11] into which the work was divided by CSAGI there was one labelled Rockets and Satellites. It was by good luck that the Skylark programme, which had developed quite independently, came to fruition during the IGY and, in all, seven flights with scientific instruments aboard took place in that period. The IGY influenced British space science through its involvement with earth satellites as will shortly be described. British participation in the IGY was organized and supervised by a National Committee for the IGY set up by the Royal Society in 1953, first under Chapman's Chairmanship but, because of his responsibilities in CSAGI, he was soon succeeded by J.M. Wordie. This was very appropriate as Wordie had much experience of Antarctic exploration in which area it was expected that there would be strong British participation during the IGY. At the time the Skylark programme was still in an early stage but it seemed likely that it could make a significant contribution during the IGY. Accordingly, Massey, as Chairman of the Gassiot Committee, joined the IGY Committee as representative for Rockets and Satellites.

CSAGI held its first meeting in Rome in 1954 and, at that meeting, the Soviet Academy of Sciences adhered to the IGY programme. At the same meeting it was recommended that 'thought should be given to the launching of small satellite vehicles, to their scientific instrumentation, and to the new problems associated with satellite experiments, such as power supply, telemetering and orientation of the vehicle'.

Little notice was taken of this recommendation in the UK until on 29 July 1955 President Eisenhower announced that the USA planned to launch an earth satellite during the IGY. It happened that the Chairman of the Gassiot Committee was due to leave for Woomera two days after this announcement. He was considerably embarrassed by newspaper interviews about what Britain was doing because at the time the possibility of a Skylark programme was still confidential, the Treasury having not yet agreed to fund it. In the event, some papers imagined more going on than actually was.[12] Nothing happened in Britain until the next meeting of the IGY National Committee on 23 April 1956. Many members of the Committee were very sceptical about earth satellites ever appearing and if they did, that anything of real interest would result. However, Massey was asked to advise at the next meeting about what action, if any, the Committee should take as a result of the American announcement. It was already realized that from accurate determination of the rate of change of the orbital elements of the satellites with time, information could be derived about the air density at satellite altitudes as well as about the departure of the gravitational field of the earth from spherical. There was in particular much scorn thrown on the latter possibility by some geodesists who mentioned that, if given the financial resources required for satellite launching, they could do much better by spending money on gravity surveys!

12 The Artificial Satellite Sub-committee

Massey discussed the matter at an informal meeting at the Royal Society on 2 July 1956. Those present were W.G. Beynon, J.G. Davies (Jodrell Bank), W.A. Johnson (RAE), F.E. Jones (RAE), C.M. Minnis (Radio Research Station, RRS) and J.G. Porter (Nautical Almanac Office). By this time, a quite comprehensive discussion of the 'Scientific Uses of Earth Satellites' had been published as a book edited by J. van Allen. It was in effect a report of the American Upper Atmosphere Research Panel (see p. 7) and gave a very clear account of the potentialities from the American side including the kind of information available from analysis of accurate data on satellite orbits. The meeting was able to take these considerations into account as well as the work of D.G. King-Hele and his associates at RAE (see Chapter 2, p. 15).

A passive satellite may be tracked by optical or radar means even though this was thought to be rendered difficult by the high speed of the satellite. If, however, the satellite includes a radio beacon at a known frequency it is possible to use radio tracking methods also. Thus the angular co-ordinates of the satellite relative to a ground station may be determined from a radio interferometer located there. Again, from measurement of the Doppler Shift in the received signal as a function of time the satellite velocity V as well as its distance d from the receiving station may be obtained. It follows that with interferometer and Doppler measurement at one receiving site or by both one interferometer and one Doppler recording station at different sites all three coordinates of the satellite may be obtained as functions of time.

All this ignores the effect of the ionosphere on the transmission of the signals from the beacon to the ground. If a sufficiently high beacon frequency is chosen this effect can be very greatly reduced. On the other hand, by using a lower signal frequency it should be possible, by comparing satellite positions determined by accurate optical methods and by radio methods, to obtain information about the ionosphere. This aroused the interest of the ionospheric physicists in Britain who could see an opportunity for them to participate in ionospheric research using satellites simply as orbital radio beacons. The meeting had no hesitation in advising that the National IGY Committee should set up an Artificial Satellite Sub-committee composed of representatives from the same institutions as in the *ad hoc* meeting, together with one from the Cavendish Laboratory, Cambridge. It further recommended that the Sub-committee first meet on 5 September. Massey was designated Chairman and the other members were: Sir D. Brunt, W.T. Blackband (RAE), F.E. Jones (RAE), G. Sutton (Met. Office), R.D'E Atkinson (Royal Observatory, Greenwich), J.G. Davies (Jodrell Bank), C.M. Minnis (Radio Research Station) and J.A. Ratcliffe (Cavendish Lab., Cambridge).

12.1   What to do about tracking American satellites?

By this time the information was available about the planned orbit for the American satellite. The orbital inclination was to be close to 30° with a perigee distance of 200 km or so. Under these conditions it would not be visible from Britain so that optical tracking would have to be carried out in British dependencies closer to the equator. Furthermore, the frequency for the satellite beacon, 108 MHz, was selected so as to minimize effects on the transmission to ground due to the ionosphere. This made radio reception from the satellite of much less ionospheric interest. Nevertheless, it was realized that later satellites might well be directly observable from Britain and/or contain transmitters of somewhat lower frequency. The possibility of establishing an American Minitrack station, designed especially for radio tracking of American satellites operating at 108 MHz, at the Radio Research Station, Slough, was discussed. This is of interest in view of later developments.

The question whether additional facilities for satellite tracking should be set up and if so, on what scale, was discussed in some detail but in the absence of any likely immediate application because of the low inclination orbit of the planned satellite, it was difficult to clinch the matter.

13   The Russians steal the show

Meanwhile the Russian delegation to the CSAGI meeting at Barcelona in 1956 announced that the Soviet Union, in addition to carrying out a rocket research programme, intended to launch satellites during the IGY to measure atmospheric temperature and pressure, cosmic ray intensities, micrometeor fluxes and solar radiation. No details were given – but it seemed possible that the Soviet satellites might carry radio beacons of the same frequency as the American. It was in fact recommended that this should be so to ensure compatibility of radio tracking systems.

Western scientists were unable at the time to judge the scale and likely effectiveness of the Soviet programme. A thaw in relations with the West, following the death of Stalin, had not long begun. The first appearance in Britain of Russian scientists connected with rocket research was at a meeting on rocket technology held in July 1957 under the auspices of the College of Aeronautics at Cranfield. These scientists distributed copies of publications, in Russian, of research work on the upper atmosphere but gave no inkling of how far satellite launching plans had progressed in the Soviet Union. However, it was shortly reported that the Russian satellites would carry radio beacons operating at 40 and 20 MHz, despite the Barcelona discussions. As there was considerable scepticism about these satellites, no planning about observing them was initiated.

Throughout the West the matter rested until October 1957. From 30 September to 5 October a special CSAGI Conference on rockets and satellites was held at the National Academy of Sciences in Washington. Massey, Davies (J.G.), Blackband, and Anthony of the British Joint Scientific Mission in Washington, attended as the British delegation. They were the only representatives from Western Europe. The Soviet delegation was led by A. Blagonravov and included some of the scientists who had been present at the Cranfield Conference.

On 4 October the Soviet Embassy gave a reception for the Conference delegates. This was well under way when, amid much surprise, the chief American delegate, L. Berkner, the Vice-President of CSAGI, called for silence to make an announcement. To still greater amazement he then said that a Soviet satellite had been successfully launched into orbit. He had heard this from American reception of a BBC news broadcast following the announcement from Moscow Radio.

14 British observers take advantage of Soviet satellites

The consequences of the Soviet achievement were far-reaching, especially in stimulating American activities. For present purposes, however, there were two very favourable aspects of the Soviet satellite. Its orbital inclination near 65° carried it over the British Isles. The chosen frequencies of the satellite beacon, 40.002 and 20.001 MHz were about the best compromise between reasonably accurate radio tracking and ionospheric research. Moreover, not only the satellite but the rocket case went into orbit. The latter was a relatively bright object which could be tracked optically with much less difficulty than the smaller satellite itself.

Under these conditions British scientists exhibited the skill in improvization so widely regarded as a British characteristic. Very soon they were obtaining the most accurate observations of the tracks of the satellite.

At the Cavendish Laboratory, Cambridge, they rapidly modified equipment operating at 38 MHz to function at the satellite beacon frequency of 40 MHz. They were then able to make very accurate interferometric and

Doppler measurements. The Doppler shift was of order 1 kHz. From these observations the eccentricity of the orbit was found to be 0.053 ± 0.001. The orbital period at the beginning of 15 October 1957 was determined as 5750.0 ± 0.3 s decreasing at 2.2 ± 0.1 s daily, the orbital inclination to the equator as $64°40' \pm 10'$ and the perigee and apogee heights as 197 ± 10 km and 934 ± 10 km respectively. It was found that the orbit was precessing westwards by $3°10' \pm 5'$ daily.

The Royal Aircraft Establishment was also very active, carrying out Doppler observations at Bramshot Cove and Aberporth and radio-interferometric observations at Lasham. Their results agreed well with those from Cambridge and gave much more accurate orbits thanks to a programme for analysing the interferometer data by means of a Pegasus computer programme. This development, initiated by E.G.C. Burt and D.G. King-Hele, was the beginning of a most important series of investigations on atmospheric density and the earth's gravitational field which has remained, under King-Hele, in the forefront of this work right up to the present time.

The BBC Monitoring and Frequency Measuring Station at Tatsfield first detected the satellite transmission on 5 October 1957 at 12.15 am. They were able to make very precise Doppler shift measurements and even obtained evidence through polarization methods that the satellite was rotating slowly.

Radio observations were also made by the Radar Research Establishment, Malvern by the Radio Research Station, Slough, the Admiralty Signals Research Establishment and Members of the Radio Society of Great Britain.

Visually, the satellite was a faint object of magnitude $+6$ at best. On the other hand, the rocket case, which was separately in orbit, was quite bright in sunlight, of magnitude -1. While comparatively few optical observations of the satellite were made, the rocket case was frequently located optically with good precision.

Perhaps the most remarkable of all was the achievement of the Jodrell Bank laboratory. The 250-foot radio telescope was in an incomplete state and had never been used for radar detection at the time Sputnik I was launched. However, five days later (9 October) emergency arrangements were in operation which made it possible to use the telescope to obtain satisfactory echoes from the moon. On 11 October the first radar echoes were obtained from the rocket case, the beginning of many observations of space vehicles with the new radio telescope.

The Nautical Almanac Office, which holds itself responsible for providing information about the movement of all heavenly bodies, was steadily improving the accuracy of its orbital calculations for both the satellite and the rocket case.

15 Working groups of the Artificial Satellite Sub-committee

The Artificial Satellite Sub-committee met on 15 October 1957 and in a state of euphoria heard first-hand comments of the observations made by representatives of the different institutions concerned who attended the meeting by invitation if they were not members.

It was obvious that satellite tracking had attained a new dimension. The Chairman suggested that a number of working groups should be set up to deal with detailed arrangements for radio observations, visual observations, computation and the scientific evaluation of experiments. Pending the setting up of these groups, arrangements were made for a meeting to be held in which representatives of institutions carrying out satellite radio observations would discuss with representatives of the Radio Society of Great Britain and of the Radio and Electronics Section of the British Astronomical Association, who were concerned with radio observations made by amateurs. This meeting, under the Chairmanship of Blackband, was held on 21 October and was the first in which steps were taken to ensure effective contribution from the many enthusiastic amateurs.

Another important matter was raised at the Meeting of the Artificial Satellite Sub-committee by D.H. Sadler. He drew attention to the need for setting up a central facility for orbital prediction and analysis which would also have equipment for radio tracking at the Russian and American beacon frequencies and for telemetry reception. The Nautical Almanac Office could not undertake this task although it was dealing with orbital prediction in the initial phase and the prediction service was transferred to RAE early in 1958.

The Radio Methods, Optical Methods and Computing Methods working groups met in the first week of November 1957 under the Chairmanship of J.A. Ratcliffe, R. v. de R. Woolley and Sadler respectively. They addressed themselves particularly to the discussion of proposals for continuing the satellite tracking programme and the question of a central facility. The Optical Methods working group stressed the importance of using kinetheodolites for accurate optical tracking of satellites. Such instruments, capable of measurement of angular location of 20″ of arc and 10–20 ms in time for each of 20–40 points observed on a single transit, were in use in the Ministry of Supply. It was recommended that these instruments be operated when appropriate.

Before the reports of these groups were considered on 13 November 1957, the Russians launched their incredible second satellite (Sputnik 2) with the bitch Laika aboard. This represented a big advance on the first satellite as far as weight was concerned but the choice of payload amazed the West.[13]

Nevertheless, it all added to the excitement of those heady days – literally one did not know what would happen next. The radio transmitter of Sputnik I had already failed and it was encouraging to hear another beacon working aloft in Sputnik II. However, this also failed after two weeks but optical measurements of high precision were made using kinetheodolites operated by the Ministry of Supply at their Trials Outstations. This was an important step forward as precision optical tracking was still poorly organized internationally. In fact the most accurate observations of Sputnik II were made with those kinetheodolites and fed into the work of the orbital analysis groups at RAE under King-Hele and R.H. Merson. The results which were obtained about the polar flattening of the earth from these and other less accurate tracking data are given in Table 3.4.

On 29 November 1957 a discussion meeting on 'Observations of Russian Artificial Satellites and their Analysis' was held under Massey's leadership at the Royal Society. No less than 14 papers were presented by British scientists. The full programme is shown in Annex 1. The fact that such a substantial discussion meeting could have taken place less than 2 months from the launch of the first satellite is a testament to the range and vigour of the British investigations.

16   Central satellite facility

The Scientific Value Working Group (Chairman, H.S.W. Massey) at its meeting on 13 November, supporting the recommendation that a central facility for tracking, orbit prediction and analysis, and telemetry reception be set up, suggested that it could become IGY World Data Centre C on Rockets and Satellites and be located at the Radio Research Station Slough. This was favourably received by the Director of the Station, R.L. Smith-Rose. Financial estimates of the capital and running costs of the facility as well as recommendations for grants to university research groups for satellite tracking were made.

The Royal Society Council in April 1958, set up an *ad hoc* committee composed of W.D. Hodge (Chairman), D. Brunt, H. Massey, R. v. de R. Woolley and R.M. Wordie to discuss the whole question of the organization of the UK observations of artificial satellites. In the light of the recommendation of the Artificial Satellite Sub-committee and its working groups, the committee proposed that a central facility, including a World Data Centre C, be set up at Slough at an estimated capital cost of £33,500 and running costs of £41,000 including salaries. Application was duly made to the Treasury for these funds as well as for a sum of £18,000 for 1958–59 to enable university groups to participate in the programme. At the meeting of the

National IGY Committee on 30 April 1958 it was announced that the Treasury had approved expenditure by the Radio Research Station of £50,000 to establish a central facility as suggested. It was understood that this would be an activity continuing beyond the IGY. The recommended sum for university research was also approved.

The way ahead was now clear. On 26 June 1958 a proposal was made to CSAGI for establishment of a World Data Centre C and accepted at the Moscow meeting on 25 July 1958. B.G. Pressey was appointed to take charge of the work and gradually took over the prediction service from RAE. King-Hele and his colleagues, who had handled this task to everybody's satisfaction, continued their work on the analysis of satellite orbits about which more will be said shortly.

17 Further satellite developments – mainly but not exclusively American

Meanwhile much had been going on. The Americans, who had been having very little luck with their Vanguard satellite programme, decided to switch to a different vehicle and launching system. This was to be based on the Jupiter C test vehicle which used the Redstone missile developed for the US Army by a team under W. von Braun. The first launch of these so-called Explorer satellites was planned for early in 1958. R. Porter, the American IGY Correspondent for Rockets and Satellites was in touch with his UK counterpart (Massey) about possible contributions to tracking this and later satellites. Arrangements were made for telemetry signals to be recorded at the substation of the Radio Research Laboratory at Singapore, and at Ibadan by N.S. Alexander and his group at University College Ibadan, Nigeria. Alexander had been in touch with the Artificial Satellite Sub-committee from its formation. Through a grant from the government of Nigeria of £2000 he planned to install a radio interferometer for satellite observations at the frequency 108 MHz chosen for the US satellites. In both Singapore and Ibadan, American Microlock receiving equipment was provided by the US National Committee for the IGY and both were in operation following the launch of Explorer I on 31 January 1958 and of later US satellites, including the spherical Vanguard I on 17 March 1958.

On 15 May Sputnik 3 was launched by the Soviet Union. It was the heaviest satellite to date, weighing 2926 lb, as compared with 1120 lb of Sputnik 2. As with the latter it was accompanied in orbit by the rocket case. This was a bright object of magnitude -1 under favourable conditions in contrast to the satellite which was not only much fainter but also fluctuated greatly between invisibility and magnitude $+1$. British observers using radio and optical methods, including kinetheodolites, were very active in observing Sputnik 3 and its rocket case, while King-Hele and his associates at RAE lost no time in analysing the new data in terms of air density and gravitational field of the earth.

During May, a further exciting development began to unfold. This was the possibility of launching objects containing scientific equipment to great distances from the earth, even as far as the moon. Porter was again in touch with Massey about assistance in tracking such vehicles. Four tracking stations were considered necessary – at Hawaii, California and Singapore and, if available, the large telescope at Jodrell Bank. A whole week of telescope time would be taken up in tracking each probe so that Lovell agreed to collaborate in tracking the first three probes without making any promise beyond this. The USA would provide the necesary Microlock equipment in a van stationed at Jodrell Bank during the experiments. This equipment would be operated by US personnel whereas the telescope would be operated by the regular staff from Jodrell Bank.

It was also agreed that the Singapore station would be based on the Radio Research Station (RRS) substation there. A more directive aerial had to be created and telemetry recording equipment set up. US personnel would make their modifications and RRS would send out a senior scientific officer to help in the programmes. The Americans insisted that the planning of the project be carried out in great secrecy. Even a meeting held at the Royal Society on 6 June 1958 to discuss it had to be treated as confidential and raised some problems for a Society which stood for open discussion of scientific research.

On 11 October 1958 the first American space probe, Pioneer I, was launched and reached a distance from the earth of 75,000 miles. This was followed on 6 December by Pioneer III which reached a comparable distance of 68,399 miles. Useful results were obtained from onboard instruments about particle radiation out to great distances from the earth. A substantial contribution to the tracking of these probes was made by the stations at Jodrell Bank and Singapore.

18   Looking to the future

It was realized quite early during the IGY that the rocket and satellite programmes would continue indefinitely beyond 31 December 1958.

The amount of activity in Britain which needed to be co-ordinated and supervised grew very rapidly after the launching of the first satellite. Up to the middle of 1958 responsibility for the co-ordination of the programme fell mainly to Massey as Chairman of the Gassiot Committee and its Rocket Sub-committee as well as of the Artificial Satellite Committee of the National IGY Committee. Despite the devoted enthusiasm and skilful assistance of D.C. Martin, the Assistant (later Executive) Secretary of the Royal Society, it had become clear that appointment of a full-time assistant to Massey at senior level would be necessary. The scope of the programme work was growing steadily and policy issues relating to international as well as national activities were arising almost daily.

The Chief Scientist of the Ministry of Supply, Sir Owen Wansbrough-Jones, who had been most co-operative from the outset in making available facilities within the establishments under his control, was again very helpful. He was prepared to second a scientific officer to the Royal Society to assist Massey in all matters relating to rocket and satellite research. Moreover, he presented a panel of candidates for the post who were interviewed by Massey, the Physical Secretary of the Royal Society and a Senior Ministry official. M.O. Robins was chosen. He was first based in the physics department of University College London so as to be readily available for consultations with Massey on the wide range of policy, management and other questions which arose. From this time on, Robins was closely involved in the British space science programme in all its aspects.

At the same time, R.N. Quirk, an Under-Secretary in the office of the Lord President of the Council, was taking a close interest in British IGY activities especially in the area of rockets and satellites. Until his untimely death in 1964 he played an increasingly important part in the organization and administration of British space science (see Chapter 4, p. 63, Chapter 5, p. 86, and Chapter 6, p. 118).

As funds were available only up to 1959 for both the Skylark and artificial satellite programmes, consideration was already being given early in 1958 to future, post IGY, organization, and the resources which would be required to prosecute it. We shall take up the story of the post-IGY plans, and how they were realized, in the next chapter. We shall conclude here by giving a brief summary account of some of the work of British scientists in rocket and satellite research during the IGY.

19 Contribution by British scientists to IGY research using rockets and satellites

We have already described how, by good fortune, the first Skylark launchings carrying scientific equipment took place during the IGY. Of these, two were part of the trial programme and five of the research programme proper.

Table 3.1 summarizes the seven flights and the performance of the instruments carried.

The grenade and window experiments worked well in the first two flights and gave good results for the temperature and wind variation with altitude over Woomera, which are shown in Figures 3.7 and 3.8 respectively for the first flight. For the wind observations, comparison is made between the grenade and window results and also some observations made using meteorological balloons. After this initial success the grenade experiment went through a depressing sequence of three failures, due to failure to eject the grenades and in one case a rocket motor failure. In fact, two flights on 18 and 19 June were specially for the IGY programme and were the object of official interest in Australia, the Minister of Supply attending the launchings. Boyd was also present from the UK. As often happens under such circumstances, one Skylark motor chose to fail while no grenades were ejected from the second Skylark. However, the dielectric method for measuring electron concentration worked well for the first time in the latter flight. It was launched at night and clearly showed the presence of a narrow

Table 3.1

Flight	Firing date	Experiments	Performance of instruments
1	13.11.57	Grenades Window Dielectric method	Successful Successful Failure
2	17.4.58	Grenades Window Dielectric method	Successful Successful No results
3	20.5.58	Airglow photometer	Failure
4	18.6.58	Grenades Window Dielectric method	Rocket motor failure
5	19.6.58	Grenades Window Dielectric method	Failed Failed Successful
6	19.6.58	Grenades Window Dielectric method	Failed Failed No telemetry
7	3.12.58	Sodium vapour ejection	Successful

shelf of ionization at an altitude near 100 km due to the presence of sporadic E ionization. This observation came in good time for it to be reported by Massey at the last CSAGI conference in Moscow in July 1958.

While the first trial flights of the airglow photometer failed, probably because a shutter failed to open, the last flight during the IGY was very successful in producing a strong twilight glow of sodium vapour ejected from the rocket at an altitude of 100 km. Observations were made by Bates and Armstrong of the width of the sodium line, yielding data on the ambient atmospheric temperature, and of the motion of the cloud and hence the atmospheric wind by Groves.

On the whole, in view of the fact that the programme only got under way

Fig. 3.7. Atmospheric temperature profile derived from the first grenade experiment (on Skylark 4). Estimated errors are as indicated. Measurements made from balloons at altitudes below 30 km are shown for comparison.

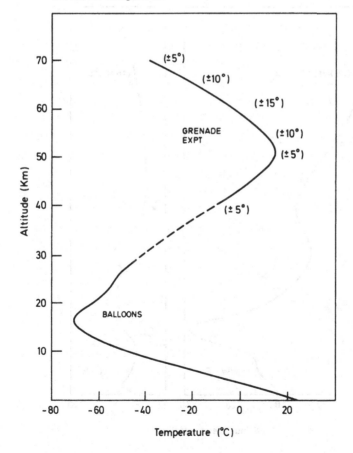

during the IGY, good results were obtained and much experience gained which proved invaluable not only in the later Skylark programme but in satellite experiments.

The grenade, dielectric and sodium vapour experiments which began in the IGY all had a fruitful future. An account of the experiments carried out in the post-IGY period is given in Chapters 13 and 14 and a full list of Skylark launchings with UK experiments on board in Appendix A1 and A2. In all, over 200 unstabilized skylarks with scientific instruments aboard were launched in this period.

Fig. 3.8. Wind profiles derived from the first grenade and window experiments (on Skylark 4). The full line curves above 40 km are derived from the grenade experiments with an estimated error of ± 6 m/s. Below 30 km the full line curves have been obtained from balloon measurements. The points ● refer to results from window.

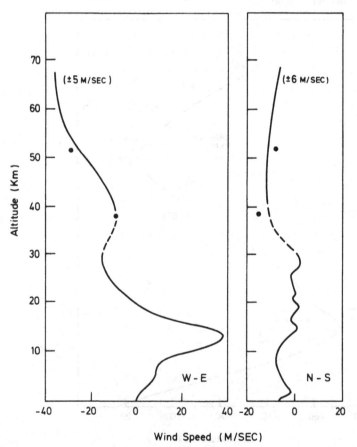

20 Results obtained from observations of earth satellites

We have described how observations of earth satellites were made by UK observers using radio, radar and optical methods. From precise measurement of satellite positions, accurate satellite orbits could be derived and from the temporal variations of these orbits information obtained about air density at satellite altitudes and about the gravitational field of the earth. Analytical procedures were set up at RAE to carry through such analyses. For these purposes precision optical tracking is required. Radio tracking of a satellite with a radiating beacon is less accurate than optical, especially because of the influence of the ionosphere. On the other hand, it is possible to deduce information about the ionosphere by comparing radio with optical satellite locations. Such studies were pursued vigorously by many groups in universities and other institutions. Tables 3.2 and 3.3 list optical, radar and

Table 3.2. *United Kingdom optical and radar satellite observing stations operating at the end of the IGY*

Station	Visual	Transit camera	Astronomical camera	Kinetheodolite	Radar
Thurso	×				
Edinburgh (Royal Observatory)	×	×			
Newton Stewart	×				
Armagh	×				
Harrogate	×				
Jodrell Bank	×				×
Cambridge	×	×			
Malvern	×		×		×
Orfordness	×			×	
Chelmsford	×	×			
Mill Hill (Univ. of London Observatory)	×	×			
Slough		×			
Tatsfield	×				
Farnborough (Bramshot)				×	
Farnham	×				
Herstmonceux (Royal Greenwich Observatory	×				
Singapore		×			
Aberporth				×	

radio tracking stations operating in the UK in the latter half of the IGY. Stations at Singapore operated by the Radio Research Station of DSIR and at Ibadan operated by the physics department of University College Ibadan, Nigeria, which formed part of the overall programme, are also included.

The early radio observations of Sputnik I at Cambridge and RAE showed that the orbit was precessing Westward at $3.17 \pm 0.03°$ per day. Ignoring for the moment any dependence of the earth's gravitational potential V on longitude we may write

$$V = \frac{GM}{R} \left\{ \frac{R}{r} - \sum_{n=2}^{\infty} J_n \left(\frac{R}{r}\right)^{n+1} P_n (\cos \psi) \right\}, \; r > R \qquad (1)$$

where ψ is the co-latitude angle and r the distance from the centre of the earth, G is the constant of gravitation, M the mass of the earth, R its radius and the J_n are numerical coefficients. In pre-satellite days J_2 alone was known from ground-based gravity surveys. The polar flattening of the earth is related to J_2, for if a and b are the equatorial and polar radii

$$\frac{a-b}{a} = \frac{3}{2} J_2 + \frac{\omega^2 R}{2g} + \text{terms of order } J_2^2$$

where ω is the angular velocity of the earth's rotation and g the acceleration of gravity at the equator. The value of J_2 obtained from gravity surveys gave $\frac{a-b}{a} = \frac{1}{297.1}$ J_4 was then chosen so that the potential was exactly of the correct form for a spheroidal earth. All higher terms in (1) were ignored.

Table 3.3. *United Kingdom radio satellite observing stations operating at the end of the IGY*

Station	Interferometer	Doppler	Direction finding	Variation in field strength	Telemetry reception
Banbury		ABC		ABC	
Cambridge		ABC		AB	
Chelmsford		A		A	
Farnborough (Lasham)	ABC			ABC	
Ibadan					C
Jodrell Bank				ABC	C
Malvern	B	B			
Portsdown		AB	AB	AB	
Singapore		C		C	
Slough		ABC	A C	ABC	ABC
Tatsfield		ABC	A	ABC	ABC

ABC distinguish equipment operating respectively at frequencies of 20, 40 and 108 MHz

Using these values of J_2 and J_4 the calculated rate of precession for Sputnik I came out to $3.176 \pm 0.024°$/day, quite close to the observational result. However, much greater accuracy in orbit determination was possible for Sputnik II using kinetheodolites operated by the RAE at Orfordness, Aberporth and Bramshot. In Table 3.4 we give the values of the orbital Westward precession rate derived from these observations and those calculated with the pre-satellite gravitational potential.

It will be seen that, with the increased accuracy, the calculated values of $d\Omega/dt$ are significantly larger than those derived from the observations. This could be due either to an error in J_2 or a relatively larger error in J_4. It was possible to separate these two possibilities by noting that from data for the American Explorer and Vanguard satellites, which followed orbits of smaller inclination, the calculated values of $d\Omega/dt$ were also greater than the observed. If the effect were mainly due to J_4 instead of J_2 it would have reversed in sense. The main error must then be attributed to J_2. The corrected J_2 which gives the observed $d\Omega/dt$ leads to a change in the polar flattening so

$$\frac{a-b}{a} = \frac{1}{298.21 \pm 0.03}$$

In the first year of satellite launching, King-Hele and his collaborators had been able to correct the only term in equation (1) from gravity surveys.

This was a good beginning of a remarkable programme in which King-Hele has continued to play a leading part. It has yielded information not only about many higher terms in equation (1) but also has been able to take into account the dependence on longitude. We shall say more about this work in Chapter 13.

King-Hele and his colleagues at RAE[14] and Groves[15] at University College London also analysed the orbit data to obtain information about air density.

Table 3.4. *Comparison of observed and calculated rates (°/day) of orbital precession $d\Omega/dt$ and of rotation of the perigee $d\omega/dt$ in the orbital plane for Sputnik 2 as analysed by King-Hele**

Date		25 November 1957	4 January 1958	13 February 1958
$\frac{d\Omega}{dt}$	observed	2.685 ± 0.003	2.814 ± 0.003	2.992 ± 0.003
	calculated	2.705 ± 0.003	2.832 ± 0.003	3.009 ± 0.003
$\frac{d\omega}{dt}$	observed	-0.403 ± 0.014	-0.422 ± 0.003	-0.447 ± 0.004
	calculated	-0.409 ± 0.017	-0.428 ± 0.017	-0.455 ± 0.017

*D.G. King-Hele, *Proc. Roy. Soc.* A., *247*, 49, 1958.

In this case it is the rate of change of orbital period which is the sensitive quantity. Figure 3.9 shows the variation of air density with altitude in daytime derived from the satellites launched during the IGY, a period of very high solar activity. The extension of the atmosphere out to great altitudes which these results exhibit was very important for atmospheric physicists, confirming the existence of high dynamic temperature at these altitudes. This was also a good beginning and it was soon found that with increasing numbers of satellites much could be found out about the great variability of high altitude atmospheric density. We shall say more about the post-IGY work in Chapter 13, p. 264.

King-Hele and Merson[16] found, from analysis of kinetheodolite observations of Sputnik 2, that the inclination of its orbital plane changed by 0.13° during its lifetime. This they attributed to rotation of the high atmosphere at a rate considerably faster than that of the earth on its axis. This was confirmed by later studies (see Chapter 13, p. 267). In addition much work began on analysing the immense number of radio observations taken by British observers, particularly at the lower frequencies of the Russian satellite beacons. Evidence was found that the rate of decrease of electron concentration in the first 50 km above the F_2 layer maximum is very slow indeed in accordance with the current theory of the origin of the layer as determined by dissociative recombination (Chapter 1, p. 4).

Considerable interest was attached to observations of the last stage in the life of a satellite when it burnt up in the lower atmosphere. Figure 3.10 shows the variation in the orbital period of the rocket case of Sputnik 1 with time observed through radar observations from Jodrell Bank and by optical

Fig. 3.9. Variation of air density with altitude for daytime in 1957–8 derived by analysis of satellite orbit data (from King-Hele and Walker).

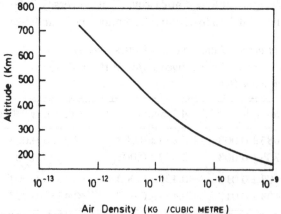

methods as indicated.[17] These data indicate that the rocket case descended through the atmosphere on orbit 879 to 880 on 1 December 1957. Jodrell Bank, being a centre for research in meteor physics, was well placed to detect any special features of the burn-up, but none appeared.

The IGY proved very exciting for those concerned with rockets and satellites. Although the later missions to the moon and planets were also exciting, they were isolated events whereas there was hardly a time during the IGY when something new was not turning up to keep us even busier than before.

Fig. 3.10. The decrease in orbital period with increasing number of revolutions for the rocket case of Sputnik I. The 'observed' points ○ were obtained according to the coding ■ combination of two radar observations from Jodrell Bank, ▲ a radar and a visual observation, ● two visual observations.

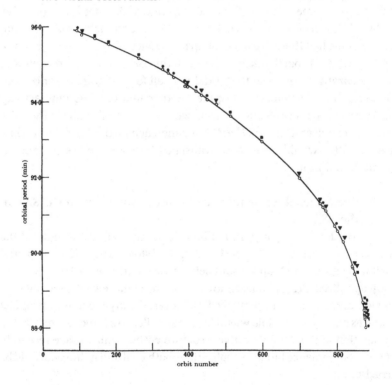

4

Post-IGY developments – NASA – COSPAR – British National Committee for Space Research – British satellite experiments

We have already pointed out that it was quite clear that the rocket and satellite programme in the IGY would continue and, indeed, expand after that 'year' had terminated (31 December 1958). Permanent organizational arrangements had therefore to be set up in advance of that date. Even in the USA the pre-IGY programme was based on the use of sounding rockets and was comparatively small so that it did not call for elaborate organization. The launching of satellites was quite another matter. This new activity called for a much more elaborate organization in the USA while in the UK it was obviously clear that the *ad hoc* IGY arrangement of the Artificial Satellite Sub-committee would need to be translated into some more permanent organizational form.

1 The National Aeronautics and Space Administration (NASA) in the USA

The launching of Sputnik I had a traumatic effect throughout the USA. To many, it seemed that the Russians had stolen a march on them and that it was vital to catch up without delay if the security of the USA were not to be jeopardized. President Eisenhower was one of the few who did not take this view and remained unperturbed. However, the urge to make up for lost ground as rapidly as possible was at first channelled to a large extent into the examination of the administrative structure within which space research, not only space science but also applications both civil and military, would be carried out.

All such work had been carried out hitherto in the USA under some military sponsorship and it might have been taken for granted that any new space agency would fall within some department of defence. There was, however, strong pressure from the scientific community expressed through

54

the National Academy of Sciences, which established a Space Science Board in June 1958, the Rocket and Satellite Research Panel and the President's Scientific Advisory Committee (PSAC), that any new space agency should be a civil one. Through the publicity and success of the IGY the advice of the scientists carried great weight. The view that an open organization for US space research was desirable both nationally and for the international image of the USA persisted despite the strong involvement of the Pentagon in the past and in the foreseeable future. This was a remarkable decision, particularly in that it was the bearing of Sputnik I on military matters which excited much of the interest in the American public generally, including many politicians. It is likely that, under other circumstances, when the prestige of pure scientific research was not so high, no major civilian agency would have been set up. This must be said with some reservation as President Eisenhower, who regarded large military–industrial establishments with grave concern, certainly favoured very firmly the civil choice.

The next question was how to establish such an agency. To build one from scratch would be fraught with great difficulties and in any case would probably not be able to begin seriously expanding the space programme quickly enough – the aim was to overhaul and pass the Soviet Union with minimum delay. It was therefore decided to build on an existing agency, the National Advisory Council for Aeronautics (NACA). This agency had gradually changed from an advisory role to one of research in aeronautics at three main research institutes (Langley, Lewis and Ames) but was on a much smaller scale than the proposed new space agency.

On 2 April 1958 Eisenhower submitted a bill to Congress proposing that a civilian space agency be established on the existing NACA structure. This was very thoroughly considered by both the Senate and House of Representatives who finally passed the bill with some very significant modifications. Thus no attempt was made to prescribe the space programme or to divide it between the military and the new space agency which was to be known as NASA (National Aeronautics and Space Administration). Furthermore, the administration of NASA was to be directly responsible to Congress for the programme rather than carrying out the advice of a supervisory board. This made for great flexibility and helped to get the new agency moving more effectively and rapidly than would otherwise have been the case.

J.K. Glennan, who was at the time President of Case Institute of Technology, was appointed the first administrator of NASA with H. Dryden, who had been Director of NACA, as his Deputy. It was formally opened in October 1958. The important role played by the scientific community through the National Academy and other bodies ensured that space science would form an important part of the activities of NASA. No time was lost, for example, in taking over the Vanguard and Pioneer programmes as well as other activities developed during the IGY. A fourth research centre, shortly to become known as the Goddard Space Flight Center, was set up. This was to be concerned especially with space science and it soon became well known to the world-wide space science community. The Jet Propulsion Laboratory (JPL) of the Californian Institute of Technology, which had previously been under contract to the US army, was transferred to undertake contractual work for NASA and soon began playing a major role in the development of large missions. Rocket launching facilities were shared with the military at Cape Canaveral while a much smaller launching range was established at Wallops Island which had previously been part of the Langley Research Center. However, a major contentious issue soon arose which led from time to time to active disquiet in the scientific community. This was the question – who controls the space science programme?

It was only natural that the scientists who had played an important pioneering role and whose influence had been very significant in the choice of a civil agency, should feel that they should play a dominant part even though they did not propose to join NASA. The Space Science Board (SSB) of the National Academy, soon after its establishment, drew up a detailed space science programme which they forwarded to NASA in the expectation that this would be accepted by the agency. In fact, while NASA did follow through most of these proposals, the NASA management considered that they alone had the responsibility for the detailed choice and execution of the space science programme. At the same time they were taking steps to obtain the views of the scientific community but pointing out that, in carrying out space operations, so many factors were involved in addition to scientific priorities that NASA staff alone were in a position to take these factors into account.

However, in 1959, Dryden arranged for NASA to join the Department of Defence and the National Science Foundation in financing the work of the SSB on the assumption that it would provide advice to NASA on long-range planning of the Space Science programme, including planning for specific scientific disciplines, as well as on international programmes and on the way in which scientific data and results obtained through space experiments should be handled. In doing this, emphasis was placed on planning as against execution. Relations between NASA and the Board were rather stormy at first but despite many vicissitudes the relationship has been maintained and is now very well established (see Chapter 12).

While arrogating to themselves such matters as the choice of experiments and experimenters, NASA established a Space Sciences Steering Committee with a number of sub-committees concerned with particular scientific disciplines. Among their tasks was to recommend which experiments and experimenters should be selected for particular missions. The committees and sub-committees included outside consultants as well as liaison representatives from the SSB – a reciprocal arrangement was made also whereby the Deputy Administrator and the Director of the Space Science Division of NASA were invited to attend meetings of the Board.

A large part of the space science carried out would be done in the universities and NASA not only provided grants to university scientists working directly in space science but also supported laboratory research related to the subject. Allowance was made for the need for continuity as most scientific programmes envisaged an indefinite series of observations of increasing accuracy and/or significance. For some years NASA also provided funds for universities to build laboratory facilities.

An important question at the outset was the extent to which research in space science was confined to universities. Should there be also some space science carried out within NASA itself? Despite considerable opposition from many sides of the scientific community, NASA decided that it would establish what was called an in-house capability in this area. Partly with hindsight, this was justified on two main grounds.

The first is the need for the presence, within the organization, of active scientists who are able to represent the scientist's viewpoint in discussions with engineers and administrators about many questions.

The second and closely related consideration is that the contribution to the research projects of outside scientists made by the agency would only be first class if first class scientists were included within it.

For both reasons the quality of the space scientists within the organization should be high. To recruit such scientists they must be allowed research opportunities, at least for part of their time.

It was also pointed out that the space scientists within NASA were exposed to the thinking of the engineers and technologists within the agency and therefore had the opportunity to propose wider ranging projects which benefited the whole community.

These problems are common to any large-scale scientific organization set up to serve a large outside user community as we will see when discussing ESRO in Chapter 6. It is generally agreed now that under these conditions it is desirable to allow in-house research on a reasonable scale in order that the in-house scientific staff should be highly competent. Nevertheless, there is always much argument about the amount of research carried out in-house if it is competitive with the outside users, the in-house workers having the advantage of 'knowing the ropes'. By encouraging co-operative research between the internal and external scientists this problem is minimized. The organizational problems we have discussed can never be completely solved. What is required is to minimize the irritation while maintaining a high level programme. Whether this has been secured at any time is a matter of opinion. While the scientific community often raised issues related to the above questions they also lost no time in expressing their views on two other major issues. The first concerned the rate at which space research facilities expanded. There is always a likelihood that an organization such as NASA, served by first-class engineers, will tend to take the bit between its teeth as far as technological developments are concerned. Many members of the scientific community argued that this was so in NASA. On the whole they preferred a large number of relatively small projects to a few large ones. The 'Man in Space' programme aroused particular venom in scientific circles. It was judged to be of small value scientifically and yet so very expensive.

All of these matters are highly subjective but it was of special interest, and still is, to see them argued in the open as has been the case since the establishment of NASA. It grew up in a period of great expansiveness which gave it a very good start and it was of much benefit to international as well as national space science as we shall see in Chapter 5.

2   The Committee on Space Research (COSPAR)

It was obvious by the beginning of 1958 that scientific research using rockets and satellites would continue after the end of the IGY. The Bureau of ICSU therefore proposed to set up a Special Committee on Space Research to deal with the international aspects of these research activities, again as a non-governmental organization. The suggested composition of the Committee was as follows.

(a) Representatives from each of the countries which are actually launching earth satellites and of those which have major research programmes using sounding rockets.

(b) Three representatives designated on an agreed system of rotation from among those countries actively participating in tracking and other aspects of space research.

(c) One representative from each of the following scientific unions – IAU (Astronomy), IUGG (Geodesy and Geophysics), IUPAC (Pure and Applied Chemistry), URSI (Scientific Radio), IUPAP (Pure and Applied Physics), IUBS (Biological Sciences), IUTAM (Theoretical and Applied Mechanics), IUPS (Physiological Sciences) and IUB (Biochemistry).

This proposal was circulated to National IGY Committees by the Secretary General of ICSU and at its meeting on 30 April 1958 the British IGY Committee agreed to support the proposal. It was approved by the General Assembly of ICSU in Washington in October 1958. A convenor, designated by the Bureau of ICSU, of which L. Berkner was Chairman, was to call the first meeting on 14–15 November 1958 in London. This was to formulate a charter of the responsibilities of the new Committee and to prepare a detailed organization for conduct of its affairs for submission to the ICSU Bureau.

Britain with its well-established Skylark programme was already able to nominate a representative under (a) above. The British National IGY Committee nominated H. Massey. He also attended as the representative from IUPAP. H.E. Newell was chosen as convenor on behalf of ICSU and the meeting took place as planned, in the rooms of the Royal Society in London immediately after a two-day discussion meeting on Space Research which the Society had organized.

At the meeting it was agreed to set up an Executive Council of the Committee of five members as follows:
President, H.C. van de Hulst (IAU); Vice-Presidents, E.K. Fedorov (USSR) and W.A. Noyes Jun. (USA); Members, H.S.W. Massey (IUPAP) and M. Roy (IUTAM). Three working groups were also established to deal respectively with (a) Tracking and Transmission of Information; (b) Scientific Experiments; (c) Data and Publication. The annual subscription was fixed at 15,000$ (US) for the USA and USSR and 5000$ for the other adherents. South Africa was elected under the rotating membership rule (b).

The next meeting was fixed to take place on 12–14 March 1959 at The Hague. Some time before this (18 December 1958) the Royal Society had established a British National Committee for Space Research (BNCSR) to deal with the British connection with the Special Committee. In fact as we shall see on p. 62 BNCSR almost immediately assumed much wider responsibilities but for the moment all we need to note is its existence. As Massey would be representing IUPAP, M.O. Robins was chosen by the BNCSR as the British delegate.

The American Vice-President, Noyes, was forced to resign before the meeting owing to illness and he was replaced by R. Porter who had been very active during the IGY.

Figure 4.1 is a photograph of those attending the meeting.

Soon after the meeting began on 12 March, the Russian Vice-President, Fedorov, raised a major issue which he considered would have to be resolved before the Soviet Union would be prepared to join. The composition of COSPAR which had emerged from the London meeting was heavily weighted in favour of the West. In fact, as it followed the Discussion Meeting at the Royal Society it was not surprising that many of the representatives were from Britain and the Commonwealth whereas there was only one from the Soviet Union and none from any other member of the Eastern bloc. This was a very serious matter because it was hard to see how COSPAR would function without Soviet adherence – in fact, one of the main aims was to provide a forum in which scientists from both East and West could talk about their scientific work. An *ad hoc* committee was then set up to consider the matter. It was composed of Massey (Chairman), van de Hulst (Holland), Fedorov (USSR), Porter (US), Roy (France), Nicolet (Belgium), Maeda (Japan) and Fraser (ICSU Secretariat).

This committee soon agreed unanimously that the present membership was unsatisfactory. It first received a proposal from Fedorov. Following what had been agreed in the early days of the United Nations he wished to include as permanent members, representatives from the Academies of Science of the Ukraine and of Byelo-Russia, and the list of rotating national members to

Fig. 4.1. Participants at the second COSPAR meeting at The Hague, March 1959. *(Front row left to right):* H. Massey, M. Roy, E.K. Federov, H. Kallman-Bijl (sec.), H.C. Van de Hulst, R.W. Porter, K. Maeda. *(Second row):* –, –, J. Bartels, D. Rose, H.E. Newell, A. Wexler, A.C.B. Lovell, F.A. Kitchen, A. Dollfus. *(Third row):* –, –, A.W. Frutkin, J. Kaplan, –, –, M. Nicolet, M. Florkin. H. Odishaw, M.O. Robins. *(Back row):* –, L.G.H. Huxley, –, F. Hewitt, –.

include Czechoslovakia, Poland, Rumania, Albania, Bulgaria and Hungary. Furthermore, he suggested that no country (or more properly, academy) should have more than one vote in COSPAR, even though one or more of its nationals represented scientific unions.

This proposal was not very well received. Porter and Massey discussed the matter together for some time and came to the conclusion that a more satisfactory solution might be based on the arrangement within ICSU in which the Executive Council reported to a Bureau and thence to the General Assembly. Porter then worked out a constitution for COSPAR in which the additional flexibility arising from the existence of a Bureau as well as a Council could be put to good use.

In Porter's proposed constitution all countries wishing to join COSPAR would be able to do so provided they were prepared to pay the appropriate subscription. This national membership together with representatives of the scientific unions would form the main committee. The chief executive organ would be a bureau consisting of the President of COSPAR, two Vice-Presidents, one from the United States and one from the Soviet Union, and four other members, two of whom would be elected by the main Committee from a slate of names presented by the American Vice-President and two from a slate presented by the Russian Vice-President. In addition, there would be an Executive Council, consisting of the Bureau members and the representatives of the Scientific Unions, which would report to the Bureau. During one half of the year one Vice-President would be the senior and for the second half the other.

This nicely balanced proposal was accepted by the Committee as a basis for resolution of the problems raised by the Soviet Academy and it was forwarded to ICSU with a strong recommendation that it should be accepted. However, it was not accepted by the ICSU Bureau who made the alternative proposal that the constitution should remain as before but with the restriction that no one country should have more than two representatives through Union representation. The Executive Council of COSPAR met on 8–10 June 1959 in Paris, rejected the ICSU proposal and reiterated their support for that of Porter. This was at last accepted by ICSU at the end of September 1959. It remained then to work out the details at a final meeting of the Executive Council of COSPAR in Amsterdam on 10 November 1959. There proved now to be little difficulty in satisfying the Soviet Union which then formally joined the committee. Fedorov, the tough negotiator, had completed his task and was replaced by A. Blagonravov, who had led the Russian delegation at the famous meeting in Washington in 1957, as Russian Vice-President. The final form of the agreed constitution of COSPAR is given in Annex 2.

The next meeting of COSPAR at Nice in January 1960 was the first with the new constitution. The new Bureau was composed of van de Hulst, President; Blagonravov and Porter, Vice Presidents; Members, Massey (UK), Roy (France), E. Buchar (Czechoslovakia), W. Zonn (Poland).

The early COSPAR meetings were valuable in bringing together space scientists from many countries though, perhaps, as time went on this became less important. However, both the Hague and Nice meetings were of special interest to UK scientists as it was at these meetings that two major developments for them first began – the Ariel satellite programme (Chapter 5) and the European Space Research Organization (Chapter 6).

COSPAR was set up in the first instance for one year only but its life was soon extended indefinitely. It still continues, the 24th meeting being held at Ottawa in June 1982.

3 The British National Committee for Space Research

Soon after the proposal for establishment of a Special Committee on Space Research had been approved by the ICSU meeting in October 1958, a meeting of a number of Fellows of the Royal Society was called to consider the establishment of a British National Committee for Space Research. While the primary motivation for this proposal was the need for a liaison committee with COSPAR, it was realized that the co-ordination and management of British post-IGY activities in space science called for a central committee, bringing together the rocket, tracking and telemetry activities at present under the Sub-committee D of the Gassiot Committee, and the Artificial Satellite Sub-committee of the National IGY Committee. Also British scientists were naturally anxious to have opportunities to fly experiments in satellites and a central committee would provide a base from which various possibilities could be explored.

Provided it was agreed that the Royal Society should be responsible for the consideration and management of the National space science programme then the British National Committee for Space Research (BNCSR) would be the most appropriate one to assume these additional major responsibilities. Its constitution and terms of reference were framed with this in mind. The Royal Society Council agreed to set up the Committee on 18 December 1958 and agreed its membership on 12 February 1959, which was as follows.

Chairman: H.S.W. Massey; Ex-officio members, The Astronomer Royal (R. v. de R. Woolley), The Astronomer Royal for Scotland (H.A. Brück). The Director General of the Meteorological Office (Sir Graham Sutton), The Biological Secretary of the Royal Society (Sir Lindor Brown), the Physical Secretary of the Royal Society (W.V. Hodge) and the UK delegate to COSPAR (M.O. Robins); and representatives of: The Atomic Energy Authority (I. Maddock), The Department of Scientific and Industrial Research (R.C. Smith-Rose and P.D. Greenall), Institution of Electrical Engineers (J.S. McPetrie), Ministry of Supply (Sir Owen Wansbrough-Jones, Sir George Gardner, R. Cockburn and A.W. Lines), Office of the Lord President (R.N. Quirk), The Physical Society (W.J.G. Beynon), Physiological Society (Air Commodore W.K. Stewart), Royal Aeronautical Society (Sir Arnold Hall), Royal Astronomical Society (A.H. Cook), Royal Meteorological Society (P.A. Sheppard); Royal Society, (D.R. Bates, Sir Edward Bullard, W.R. Hawthorn, F. Hoyle, A.C.B. Lovell, J.A. Ratcliffe and L.R. Shepherd).

It looked a very grand affair but the range of activities covered was so wide that many institutions and societies had to be represented. Indeed, at the first meeting, it was decided to add a still further ex-officio member – The Director of the National Physical Laboratory. The terms of reference were 'to co-ordinate UK activities in connection with space research, particularly with regard to COSPAR'.

At its first meeting on 4 March 1959 the Committee received the final reports from the Artificial Satellite Sub-committee and from the Rocket Sub-committee of the Gassiot Committee. Henceforth, the activities covered by these two bodies would fall within the responsibilities of BNCSR. To carry out these and other responsibilities a number of sub-committees and working groups were set up.

The first sub-committee concerned with tracking, orbit analysis and data recovery was known as TADREC. J.A. Ratcliffe was appointed Chairman. Working groups on tracking (Chairman C.W. Allen), ionospheric studies (Chairman J.A. Ratcliffe) and orbit analysis (Chairman A.H. Cook) were set up. This was a natural succession to the Artificial Satellite Sub-committee and its working groups and there was no lack of matters to be dealt with.

The second Sub-committee known as the Design of Experiments Sub-committee (DOE) had a clear remit to continue the work previously done by the Rocket Sub-committee in choosing experiments to be flown in sounding rockets. Hopefully, however, DOE was given the responsibility for studying experiments which might be flown in artificial satellites comparable to those already launched. Because of the wide range of physical science, including astronomy, which had to be covered the membership of DOE was large. Initially it consisted of H.S.W. Massey (Chairman), D.R. Bates, W.G. Beynon, R.L.F. Boyd, Sir David Brunt, Sir Edward Bullard, H.E. Butler, J.G. Davies, J.S. Hey, F. Hoyle, F.E. Jones, R.J. Lees, A.W. Lines, A.C.B. Lovell, B.W. Lythall, I. Maddock, R.O. Redman, M.O. Robins, G.D. Robinson, J. Sawyer, L.R.

Shepherd, P.A. Sheppard, W.H. Stephens and G.B.B.M. Sutherland. A working group on rocket experiments, replacing the old Rocket Sub-committee, and a Study group on satellite experiments were set up under Massey's chairmanship.

The third sub-committee, under the Chairmanship of Sir E. Bullard, was set up to maintain liaison with the World Data Centre at RRS Slough.

3.1   Activities of TADREC

The TADREC continued the work of the old artificial satellite sub-committee without any loss of continuity. The central facility and World Data Centre had been established at the Radio Research Station, Slough, and work was proceeding on the design and construction of a 60 ft diameter radio telescope at Chilbolton, specifically for the tracking of artificial satellites and deep space probes. Considerable progress was soon made as far as optical tracking was concerned through an offer made by Chief Scientist, Ministry of Supply (Sir Owen Wansbrough-Jones) to loan kinetheodolites to appropriate institutions who would provide the staff to operate them for precision satellite tracking. In view of the successful use of these instruments for tracking, in particular, the Sputnik III rocket case, this offer was taken up without delay. Eventually, after careful consideration, kinetheodolites were located at the Royal Greenwich Observatory, Herstmonceux; The Royal Observatory for Scotland, Edinburgh; The Royal Military College, Shrivenham and at Diendi in Malta, operated by the Meteorological Office. The latter station was a site with much better seeing and at a considerably lower latitude so was a very useful addition to the observing network. Staff were also provided so that the Ministry of Supply stations at Bramshot and Aberporth could continue to make observations of artificial satellites.

Increasing attention was paid to the analysis of orbit data which yielded very much of interest about the figure of the earth and the properties of the high atmosphere. An account of this work is given in Chapter 13.

4   The key question – British satellites?

The rocket working group of the Design of Experiments Sub-committee continued the work of the original Gassiot Sub-committee and was soon considering new developments in this area, such as the use of smaller, cheaper rockets and the need for stabilized Skylarks. Much progress was made with both these developments which are described in Chapters 8 and 9. At the same time many new rocket experiments were introduced. These are summarized in Appendix A1 and many are referred to in Chapters 13, 14 and 15.

Initially, the satellite instrument working group had nothing to bite on. It was necessary first to resolve the big issue as to how British instruments were to find space in artificial satellites. This was the key question that was faced by the National Committee. We now give an account of the way in which this question was answered, in the first instance through Anglo–American co-operation. The later negotiations which led to European co-operation will be discussed in Chapter 6.

5 A national launching system or international collaboration?

The essential groundwork which led to government support for a scientific programme using satellites was carried out in 1958. Within government, the lead appears to have been taken by Sir Owen Wansbrough-Jones, Chief Scientist of the then Ministry of Supply. This Ministry had a wide responsibility for scientific and technological developments aimed primarily at the support of the Defence Services, and included the technology of rockets and guided weapons, both very relevant to the technologies required to support scientific space research. Outside government, the widespread scientific interest in using satellites for scientific observations in the fields of geophysics and astronomy was channelled through the Royal Society where, in addition to Massey, it also received strong support from the President (Sir Cyril Hinshelwood) the Physical Secretary (Sir David Brunt) and Assistant Secretary (D.C. Martin) of the Society. Quite early in 1958 Massey was discussing British launching possibilities with members of the RAE staff. At about the same time Lovell addressed the Parliamentary and Scientific Committee. He made a strong plea for an all British satellite for scientific purposes.

In June 1958, the Chief Scientist of the Ministry of Supply wrote to the Executive Secretary of the Royal Society supporting the preparation of a full scientific case for British satellites. Shortly afterwards, the Physical Secretary of the Royal Society (Sir W. Hodge) suggested that DSIR should include £250,000 in estimates for the next five years to cover space research activities by DSIR mainly for tracking satellites and space probes.

Meanwhile Martin summarized the state of co-operation between Britain and USA in space research. The scientific aspects were under the care respectively of the Royal Society and the National Academy of Sciences. Co-operation was through the IGY organization with a pattern of visits, international meetings and information exchanges. So far, American satellites had been out of range of observation from the UK, and observations under Royal Society auspices had been restricted to sites at Singapore and Ibadan. When future American satellites came within range of UK they would be observed in the same way as were satellites launched by the USSR. The World Data Centres for rockets and satellites in the USA and USSR were being complemented by a third centre to be set up by the Radio Research Station at Slough.

The stage was thus set for the next moves towards greater British participation. During the summer of 1958, Massey assembled the scientific case, whilst the Chief Scientist of the Ministry of Supply prepared his views on other aspects of the subject. These included a preliminary assessment of the time scale and cost of adapting the two British rockets code named Blue Streak (an intermediate range ballistic missile) and Black Knight (a smaller experimental rocket), to launch satellites of a few hundred to 2000 lb weight (depending on the orbit) from Woomera. It was thought unlikely that worthwhile satellites could be launched in this way in less than five years and the cost of adapting the two rockets to launch five satellites of about 1000 lb each would be about £9 million. The essential tracking facilities might cost about £200,000. If no more than two satellites were launched in any one year, the annual costs might be about £100,000. Thus total costs of some £10 million to £11 million might suffice to cover the launch of four or five British satellites, with British rockets, by the mid-1960s.

Lord Hailsham, the Lord President of the Council, was informed of the initiatives being taken in the Royal Society and the Ministry of Supply. He asked that the Advisory Council on Scientific Policy (ACSP) chaired by Lord Todd, should be consulted on the scientific case when the advice from the Royal Society was available.

The Memorandum prepared by Massey is reproduced in Annex 3. It outlined those topics in the scientific disciplines of geophysics and astronomy for which earth satellites offered unique opportunities for advance.

The paper recommended that a programme of all British satellites should be developed concurrently with that of a British rocket launching system as described by the Chief Scientist, Ministry of Supply. Although the estimated costs would be high for pure scientific research, they were comparable with the costs for research in nuclear physics. The five-year time scale seemed long but there would be no possibility that all discoveries would have been made in that time. The paper referred to the possibility of launching British satellites by foreign rockets, and the inclusion of British instruments in foreign satellites. Whilst offering methods of obtaining early experience, such arrangements were regarded as second best options, much less attractive than the development of an all-British system.

The Memorandum was considered on 23 October by a group of Fellows of the Royal Society, all of considerable experience and standing in the relevant fields of science. With the exception of one member the group was emphatically of the opinion that the proposed satellite project was of very great scientific importance, with possibilities far beyond those immediately predictable. It would give the UK a major new method of physical, meteorological and astronomical research, and failure to proceed would incur grave handicaps in the future. Only the Astronomer Royal dissented from these views of the majority, believing that any available resources could be more effectively deployed in supporting more conventional astronomical facilities.

At about the same time that these considerations were taking place in London, an authoritative report published in Washington indicated the direction in which the US government policy might move. This report, dated 15 October 1958 and entitled 'International Co-operation in the Exploration of Space', a staff report of the 85th Congress Select Committee on Astronautics and Space Exploration, gave very favourable views on the value of international co-operation in science. The report doubted that Western Europe could launch any type of space vehicle for some time, but if European scientists could share the 'know how' with American scientists, substantial benefits could occur for both parties and unnecessary expense to the USA could be avoided. The report continued that in Western Europe the UK was visibly the most advanced in space technology, with the development of the military rocket 'Blue Streak' and the research rocket 'Black Knight'. In the scientific field the work of the Jodrell Bank radio telescope was favourably mentioned. This and other less formal remarks by American officials were pointers to the eventual offer of co-operation made to the members of COSPAR about six months later.

The Memorandum urging British participation in scientific research using British satellites launched by British rockets was circulated under cover of a supporting letter from the President of the Royal Society to, amongst others, the Lord President, the Minister of Supply and the Chairman of the ACSP. Lord Hailsham requested speedy consideration by the ACSP.

In the background, although not yet in the open, was the question mark hanging over the future development of the military missile 'Blue Streak'. The possible courses of action for British participation in satellite research would be very greatly influenced by whether or not 'Blue Streak' was maintained and became available as the basis for a launch vehicle. This crucial decision would be taken largely on the advice of Defence Chiefs and the Government was bound to take this into account in formulating its policy towards space research.

The ACSP considered the matter in a preliminary manner at its meeting on 10 December 1958. It gave particular attention to three important aspects, viz.

(a) the strength of the case for a British scientific satellite programme,
(b) the likely long-term demands on money and manpower,
(c) the prospects for, and advantage of international co-operation, e.g. with the USA, the Commonwealth, NATO.

One outcome of this preliminary discussion was that the author of the Memorandum (Massey) was invited to prepare a further note giving more details of the programme costs and manpower requirements. This was done and submitted to the ACSP in time for its meeting on 11 February 1959. The subsequent recommendations from the Advisory Council to the Lord President fell short of full support for the programme proposed in the Royal Society Memorandum, but were sufficiently positive to give an initial impetus just when it was needed. The most important recommendations were as follows.

(a) The scientific component of an artificial earth satellite programme should be initiated, the level of costs being about £100,000 per annum.
(b) Whilst the simplest course would be to launch the satellites with a British rocket system, it would be costly and tied to the military programme. It was recommended therefore as a first step that an immediate approach should be made to the authorities in the USA to ascertain the terms under which suitable rockets could be supplied.
(c) If satisfactory arrangements for US rockets could not be made, preliminary steps in an adaptation programme for 'Blue Streak' should be considered. This would defer a decision on the launching vehicle for about two years.
(d) In the international field, it would be most desirable to explore the possibilities of collaboration with the Commonwealth and the USA. COSPAR should be encouraged but problems were foreseen if NATO was brought in. This, it was thought, could have the scientifically disastrous result of a total withdrawal of the USSR from COSPAR.
(e) It was believed that the high cost of lunar, planetary and solar probes could not be justified.

Events moved quickly in the weeks following the submission of the Advisory Council's recommendations

6 The US offer

At the Hague meeting of COSPAR on 14 March 1959 (see p. 59) R. Porter, the American delegate, announced that the United States, through NASA, would be prepared to launch without charge, scientific equipment for scientists of other countries, subject to certain provisos including mutual interest in the experiments and compatibility with the launching rocket system. Either individual experiments or complete scientific payloads could be considered. NASA proposed to use a newly developed relatively cheap launching system to be known as SCOUT.

The formal offer was made by letter from the National Academy of Sciences to the President of COSPAR. A copy is reproduced in Annex 4. It was clear from early discussions with the USA that whilst COSPAR would be informed about collaboration, it was NASA's intention to make bilateral arrangements for the execution of joint programmes. The American offer came at a most opportune time for the five groups of British scientists who had been active for the previous four or five years in the preparation and use of the Skylark rocket for upper atmosphere experiments. It seemed to promise much earlier access to scientific satellite experiments, with the added bonus of close working arrangements with engineers in NASA, then at the forefront of space technology, than any existing alternative.

The BNCSR formally noted the NASA offer and moved quickly so that proposals for a scientific programme of experiments could be in an advanced stage of preparation should it become possible to proceed. The Design of Experiments Sub-committee of the BNCSR, at its meeting on 23 April 1959, established *ad hoc* working groups on seven topics. These groups were to formulate lists of experiments under three headings:

(a) Experiments which could be prepared quickly for early launch.
(b) Any necessary programmes of laboratory work.
(c) Possible experiments suitable for a longer-term programme of space research.

The topics covered by the Working Groups were:

Radio transmissions and galactic noise
Cosmic rays and corpuscular radiation
Meteorology
Interplanetary dust
Astronomy
Magnetic and gravitational fields of the earth
General relativity

A Technical Co-ordinating Group was set up to assess the compatibility of the experiments which were proposed. It was envisaged that as far as possible experiments would be associated in groups which interlocked, so that an integrated scientific payload resulted, rather than a set of independent unrelated measurements. Soon after this preparatory move by the BNCSR, the way was cleared for further progress by an announcement of Government support. Meanwhile, immediately after the Royal Society delegation returned from the COSPAR meeting to London, the Lord President's Office was told of the American offer. At the same time Lord Todd the Chairman of the ACSP was informed, just before leaving on a visit to Washington. Although he could make no official response, he was, fortuitously, very well placed to explore informally the possible opportunities. This he did, and was able to report that the US authorities were indeed very willing to discuss close co-operation with the British.

Meanwhile, within Whitehall, consideration was being given as to how a British programme could be organized and financed should any or all of the recommendations of the ACSP be accepted by the government. It was clear that a number of departments would have an interest in the content of the programme and the deployment of resources, and would be in a position to support the activities of the scientists. It would be necessary to discuss with the American authorities the precise terms under which co-operation might be possible. The government decision was made known in a statement by the Prime Minister in the House of Commons on 12 May 1959, (see Annex 5) in which he said that a programme for the design and construction of instruments had Government approval. As to the means of launching into space, there appeared to be scope for joint action with the USA, or possibly with some Commonwealth countries. A team of experts would shortly be going to Washington for detailed discussions on the matter. Meanwhile, continued the Prime Minister, design studies were being put in hand for the adaptation of British military rockets which were then under development. This would put the UK in a position, should it be so decided, to make an all-British effort.

Encouraged by this announcement, the British team under the leadership of the Chairman of the BNCSR (Massey) visited the USA in late June, taking details of eleven experiments which might be suitable for launching by NASA within about two years (see Figure 4.2).

The NASA authorities and American space scientists were very forthcoming both in discussing their own plans for future experiments and also in describing the possible technical facilities which might be available for joint Anglo–American projects should a programme be formally agreed. The

Administrator of NASA (K. Glennan) and his colleagues had discussed the conditions and outlined the facilities which might be involved in such a programme.

With the valuable help of H. Billingsley, the NASA Foreign Relations Officer, a provisional agreement was reached, during the visit, by Dryden and Massey for the launching by the Scout Rocket system of three satellites containing British experiments, at intervals of approximately one year, in the first instance. In the first satellite NASA would provide also the shell of the satellite and the auxiliary services such as power supplies and telemetry. The NASA world-wide tracking and telemetry network would be available but might need augmentation and this could perhaps be undertaken jointly. The UK would progressively take over responsibility for the satellite engineering in the later launchings.

At a meeting of the BNCSR held after the return of the party from the USA it was agreed that the proposed co-operation gave an excellent opportunity for British scientists to gain early experience, but it was equally concerned that it should not be allowed to affect adversely the UK launcher design studies. The future of these studies was in fact bound up with certain aspects

Fig. 4.2. The British party at Wallops Island, June 1959 *(Front row left to right):* Walker, −, H. Dryden, H. Billingsley, −, J. Sayers. *(Back row):* R.L.F. Boyd, D.G. Robinson, J. Gait, M.O. Robins, H. Massey, R.L. Smith-Rose, R. Quirk, H. Bourne, -

of European co-operation as is discussed in Chapter 6, p. 113.

The provisional agreement was made formal at government level and one way found for providing space for British experiments in satellites. A detailed account of the way this programme was implemented will be given in the following chapter. Meanwhile, we shall return to discuss the organization set up to deal with all British space science activities including the new Anglo–American co-operation.

7 The overall organization for space science

It has already been pointed out that with the end of the IGY and of the 4th year of the Skylark programme there was no regular procedure for financing space science. With the advent of the Anglo–American programme a Steering Group for Space Research was set up within the Lord President's Office. Its Chairman was Sir E. Bullard, the Royal Society was represented by the Physical Secretary (Sir W. Hodge), the Chairman of BNCSR and the Astronomer Royal; the Secretary of DSIR (Sir H. Melville) and the Director of the Meteorological Office were ex officio members while a number of officials represented other interested government departments. It would be responsible within government for matters of policy (including financial matters) in space research and would receive recommendations on scientific aspects of the programme from BNCSR which would continue its supervisory and co-ordinating role. The financial accounting officer was the Secretary of DSIR. In order that the grant recommendations made by BNCSR should be in accordance with DSIR practice a further sub-committee of BNCSR was set up, known as the University Projects Expenditure Sub-committee, to consider and advise on the financial aspects of applications from university research groups. Scientific proposals from government departments, while financed separately, competed with proposals from universities for selection by the Design of Experiments Sub-committee.

This rather unusual arrangement worked very well. The BNCSR occupied a central role and was concerned with an increasing variety of activities – the establishment of a co-operative European programme, the development of small research rockets and of stabilized Skylarks, all either began under the initiative of the BNCSR or were involved with it closely from the outset. These developments are discussed in Chapters 6, 8 and 9. In addition, the realization of a means for providing satellite space for British scientists meant a rapid increase in the number of interested university scientists with a corresponding increase in administration and a greater need for co-ordination and supervision.

Thus on 2 March 1961 the Steering Group formally agreed the

recommendations for research grants made by the BNCSR since the last months of 1959 to no less than fifteen groups for the development of instrumentation in sounding rockets and/or earth satellites. These included, in addition to the five groups who were involved from the outset in the Skylark programme, groups from the Universities of Cambridge (F.G. Smith), Leeds (P.L. Marsden), Leicester (E.A. Stewardson), London, Imperial College (H. Elliot, S. Hall and D. McGee), Manchester (R.C. Jennison), Oxford (J.T. Houghton), Reading (S.D. Smith), Sheffield (T.R. Kaiser) and South-ampton (G.W. Hutchinson). The total sum awarded was £247,000 but this covered different periods of award. In fact, for the year 1960/61, the total expenditure in grants to universities on rocket and satellite instrumentation was £123,000. Government departments spent, from the financial votes, £26,500 over the same period. The corresponding expenditures on satellite tracking were £6500 and £20,000. The cost of the rocket vehicles, which was covered by a vote to the Ministry of Supply, is not included.

This was the last year in which no provision had to be made for participation in a European programme. We shall discuss the financial situation in later years in Chapter 6 which deals particularly with the setting up of the European collaboration.

8   Further development of launching systems

The opportunities provided by the American offer to launch UK experiments and satellites reduced the emphasis from the point of view of space science on the provision of an all-British launching system. The situation was further modified when Blue Streak was cancelled as a military rocket and the possibility arose of its use as the main booster in a co-operative European launching system. An account of the subsequent history is given in Chapters 6 and 11. In the event the only all-British launching system to be used was that based on the Black Arrow rocket, which placed the Prospero satellite in orbit on 28 October 1971 (see Chapter 15).

5

The Ariel programme

In the previous chapter an account was given of the events which led to the establishment of a co-operative programme in which, in the first instance, the National Aeronautics and Space Administration (NASA) would launch three satellites at roughly yearly intervals with British scientific instruments aboard. We now describe in more detail the nature of the programme and how it worked out in practice. Most attention will be concentrated on the initial three satellites but something will also be said of the three further satellites whose launching was arranged at a later stage.

1 General co-operative arrangements

It was likely that a first launch for UK experiments would take place from the launch site on Wallops Island on the east coast of the USA. The arrangements with experimenters would be those normal in the USA. Each experimenter would have first call on his own data, and if the first launch attempt failed every experimenter could expect a second launch for his equipment. In return, NASA would require the equipment to have satisfied the stringent environmental tests appropriate to the launch and orbit conditions. In addition, evidence that the scientific instruments had operated satisfactorily in vertical sounding rockets would be required. If this was not possible in all cases using British rockets NASA would consider offering test flights in American rockets. Financial arrangements would be on a 'no billing' basis, each party paying for those items for which it was responsible.

It was possible at this stage to outline the procedures which would be involved in the preparation of experiments. These procedures would, with appropriate modifications, establish the methods to be used for many years ahead.

(a) The experiments would be selected.
(b) Prototypes would be constructed in UK.
(c) Prototypes would be taken to NASA for compatibility tests.
(d) Designs would be finalized.
(e) Construction of several flight models in UK would be completed.
(f) Flight models would be subjected to environmental tests in UK.
(g) Acceptable flight models would be taken to NASA for installation, accompanied by members of the appropriate scientific groups.

For further consideration, the Chairman of the British National Committee for Space Research (BNCSR) proposed that of the original list of eleven possible experiments discussed with NASA, the first satellite should contain, as the core of the instrumentation, equipment to measure electron concentration, electron temperature, ion concentration, solar Lyman α radiation, solar X-radiation and a cosmic ray counter experiment. The second or later satellite should be considered for the galactic radio noise, meteorological and micro-meteorite measurements.

It had transpired during the discussions with NASA that joint experiments with the Canadians were being planned which would measure characteristics of the ionosphere from a satellite above that region, the so-called 'top-side sounder' experiments. No further consideration was given therefore to direct British activity of this type. The Director of the Radio Research Station, where there was much interest in this work, accepted an offer for a British scientist to sit with the appropriate NASA/Canadian Working Party (see also Chapter 7, p. 165).

A further area of co-operation emerged from the NASA wish to see a ground station for the tracking of satellites located in Britain. These early discussions led eventually to the establishment of a NASA Minitrack station on a site near Winkfield in Berkshire, operated on behalf of the Radio and Space Research Station (RSRS) as part of the NASA network.

The stage was therefore set for the first of the joint Anglo–American satellites and in due course formal agreement was reached between the two Governments through an exchange of Notes on co-operation in space research (Annex 6). This embodied the 'no-billing' arrangement to which earlier reference has been made.

2 The first payload
In December 1959 the Design of Experiments Sub-committee confirmed to the BNCSR the first payload to be carried in a NASA built satellite. Their choice had been partly dictated by the desire, at least in the

first satellite, to take advantage of experience gained by research groups in the Skylark rocket programme. This suggested that the payload should be largely concerned with observations of the high altitude ionosphere and of solar radiations important in producing the ionosphere. The particular instruments which it was proposed should form the payload and the institutions and principal scientists involved were as follows.

1. A radio frequency plasma probe to measure electron density: University of Birmingham. J. Sayers and J.H. Wager.
2. Langmuir probes to measure electron density and temperature in the satellite neighbourhood: University College London, R.L.F. Boyd and A.P. Willmore.
3. Positive ion spectrometer to determine ion composition and temperature: University College London, R.L.F. Boyd and A.P. Willmore.
4. X-ray spectrometer to measure solar radiation in a particular wavelength band: University College London, R.L.F. Boyd and A.P. Willmore in association with the University of Leicester, E.A. Stewardson and K. Pounds.
5. Detection of solar Lyα radiation: University College London. J.A. Bowles and A.P. Willmore.
6. Detector of solar aspect angle: University College London, J. Alexander and R.J. Bowen.
7. Cosmic ray detector: Imperial College London, H. Elliot, R.J. Hynds, J.J Quenby and A.C. Durney.

Of these 1, 2, 4 and 5 followed on very similar lines to the corresponding instruments developed for flights in the early Skylark programme (see Chapter 3, pp. 31–33). The positive ion spectrometer referred to in 3 was of an unusual type depending on measurement of the energy distribution of the positive ions relative to the satellite. It took account of the fact that the kinetic energy of an ion of mass m relative to the satellite, due to the satellite's velocity v, is close to $\frac{1}{2} mv^2$.

An additional experiment 7 was also included to be carried out by a group with wide experience in ground-based cosmic ray observations and some limited use of the Black Knight rocket for experiments in the upper atmosphere.

It was a great advantage that it was possible to propose a payload which was scientifically coherent and designed by research groups familiar through the Skylark programme with the problems encountered in this type of scientific research.

Following agreement by the BNCSR and The Steering Group, the payload was subject to further detailed scientific and technical discussions in Washington. Particular attention was given to compatibility with the specified performance of the SCOUT rocket system, orbit requirements and the demands for electric power, data transmission, and the many new environmental factors inherent in designing for operations in space. The Committee on Space Research (COSPAR) was also informed of the proposed payload and commented favourably on it. Agreement was reached with the Americans in 1960.

3 Meeting the satellite requirements

The joint project now about to be undertaken by NASA in the USA and groups of scientists based in three universities in the UK posed difficult and unusual co-ordination problems of which neither side could claim much, if any, experience. The overall shape, size and weight of the satellite would be tightly constrained by the mechanical performance characteristics of the rocket launching system, which in 1960 had yet to be finalized. The data transmission and electrical power systems could within limits be designed to meet the needs of the experimenters. At the same time they had to be compatible with the world-wide network of telemetry receiving stations and the capabilities of miniature batteries and solar cells. The dynamic stability and heat balance criteria were stringent. Good management at all levels would be essential for the success of the project, and a simple framework was established. A project manager was appointed on each side of the Atlantic, and all substantive communications about the project passed between these two managers. NASA appointed Robert C. Baumann, an engineer at the Goddard Space Flight Center (GSFC) near Washington, as the US Project Manager and R.E. Bourdeau (GSFC) was appointed the US Project Scientist. M.O. Robins became the UK Project Manager and E.B. Dorling UK Co-ordinator based at University College London (UCL). A Joint US/UK Working Group was established, meeting at intervals of about three months and overseeing the design, construction and testing of the satellite. A direct TELEX link between GSFC and UCL via the Radio Research Station provided by NASA was invaluable in keeping the two national activities in close day-to-day contact on scientific, technical and programme details as the work progressed.

However, at the beginning of 1960, the problems loomed large, and a high degree of optimism was an essential requirement in those working on the project. The SCOUT launching system was still in the design and development phase; predictions of payload capacity and orbit capability had yet to be verified by trials, and reliability had yet to be proven. The environmental conditions such as temperature and vibration, to which the satellite equipment would be subjected during launch, could only be specified in general terms as no data from the rocket itself were available.

Although individual components such as tape recorders, solar cells and batteries existed as building blocks, there was no ready-made design suitably tailored for the planned, but yet to be defined, payload shape and volume. The instrument sensors needed to carry out the proposed British experiments, although proved in principle, did not exist in the form suitable for long-life operation in a space environment.

The arrangements for this first satellite involved NASA in designing the satellite structure, power supplies, data storage devices and encoders, and telemetry transmitters. Groups in Britain would design the instrument sensors which would interlock and interact with practically every aspect of the NASA design activities. The scope for mishaps was immense. The fact that relatively there were so few was a tribute to the skill and devotion of the NASA scientists, technologists and engineers, and the enthusiasm and inventiveness of the British scientists involved, very ably supported by various British industrial firms and Government Laboratories.

At the first Joint Working Group meeting in March 1960, it was proposed that the final design be started in October 1960, with a launch date in late 1961 or early in 1962. At the fourth meeting of the Group in December 1960, detailed layout designs were being considered and the many problems of packaging the required equipment into the defined volume were being tackled. The data encoding and transmission requirements were proving to be much more complex than anything hitherto undertaken. The electrical power demands led to a reluctant acceptance of the need for four external paddle structures to carry the necessary solar cells. This introduced considerable mechanical complexity. The safety requirements for a SCOUT rocket launch from the NASA range on Wallops Island, off the coast of the USA, appeared to restrict the maximum orbit inclination to 52 degrees, which was less than the UK scientists had hoped. Thus, at the turn of the year, much had been accomplished, but the many detailed and difficult problems still to be solved left no one in any doubt about the formidable amount of work which lay ahead if a successful launch was to be achieved in early 1962.

4   Tracking and data acquisition

During late 1960 and early 1961, aside from the mainstream design and development of the satellite and rocket system, much consideration was being given in the UK to the problems of data acquisition, processing and analysis which would follow immediately upon a successful launch into orbit. A leading part in this was taken by the Tracking Analysis and Data Recovery (TADREC) Sub-Committee of the Royal Society, chaired by J.A. Ratcliffe, at that time director of the Radio Research Station. At the time we are considering, NASA had in operation about eleven radio stations at widely dispersed sites throughout the world equipped for the tracking of satellites by the Minitrack system, and for the reception of radio telemetry signals. This array of stations constituted the STADAN network and was under the control of the Goddard Space Flight Center. The availability of this network for the joint project was a prerequisite for the efficient use of the satellite. In the event, it was operated to meet both the British requirements and those of other American satellites, without any distinction between them. As will be recounted later, the STADAN network was augmented by additional British facilities in time for the launch of the first satellite.

However, the reception of the telemetry signals is only the beginning of a complex chain of processing and computation which eventually leads to comprehensible data. In 1961, it was necessary to work out in detail with NASA at which stage in the chain the data should be transferred across the Atlantic and to ensure the compatibility of the procedures in the two countries. The generosity of NASA at this time in preparing to undertake substantial data handling for the project was greatly appreciated, indeed, it was essential to avoid serious delays. However, it was fully realized that plans in the UK must be sufficiently far-sighted and coherent to enable all routine processing to proceed in Britain in due course, particularly if further satellites should follow the first one. Here we only give the minimum of technical detail to enable the main stages in the process to be followed. The signals received at each ground station were to be recorded on magnetic tape. This record, in analogue form, in a frequency modulated code, would contain the intermingled outputs of the different experiments, timing information, and data about the satellite performance. It would be necessary to transport all these tapes, which might number up to 40 per day, to a central point for cataloguing and editing so that faulty sections could be eliminated. Next, the analogue data on the tapes would, by a complex device known colloquially as a 'digitizer' be converted and compressed into a digital formation on another series of tapes, less in number by a factor of about ten.

These digitized tapes would then need further processing in specialized equipment in which the data for the various experiments would be separated out, and corrections made for scale factors and non-linearities. The data output from this process would be in a chosen format suitable for further detailed analysis by the individual experimenters. It could either be printed out in tabular form as a series of numbers or transcribed into graphical form, or subjected to further analysis so that each experiment was separately examined and the results presented in a consolidated form.

NASA offered, for the first satellite only, to receive the telemetry tapes from the various reception stations, process them into digital format and send these to UK for further analysis. They also provided equipment to give a 'quick look' at the tapes so that the general performance of the experiments could be rapidly assessed without the delays associated with the full analysis.

It was recognized that for any joint satellites beyond the first, the data processing, including digitization and all subsequent operations, should be carried out in Britain. These proposed arrangements did, in the event, form the basis of all data processing for the series of satellites to follow. In Britain it was fortunately possible to engage the willing assistance of several groups of experienced workers in the data processing field. The Radio Research Station prepared to install and operate the equipment for digitization of the telemetry tapes. Specialized advice was obtained from staff experienced with IBM computers at the Atomic Weapons Research Establishment (AWRE) at Aldermaston, whilst staff in Space Department at the Royal Aircraft Establishment (RAE) helped in the compilation of programmes for the analysis and separation of experimental data.

5   Preparation of scientific equipment

Meanwhile the preparation of the equipment being made for the satellite in Britain proceeded well, and in May 1961 the first prototypes were taken to GSFC (see Chapter 4, p. 56) for exhaustive testing. They were accompanied by Willmore (UCL), Durney (Imperial College) and Wager (University of Birmingham), who were to remain in the USA[1] effectively until launch. The prototypes were followed by two sets of equipment suitable for launch into space, and the three British scientists were, from then on, very fully occupied in a thorough testing programme and in solving the many problems arising in the integration of their instruments into the satellite system.

Figure 5.1 is a photograph of the completed satellite.

6   The planned orbit – observation of the first phase

Another part of the programme had been proceeding more slowly than predicted. The development of the SCOUT rocket fell behind schedule, and it became apparent that the planned launch date of early 1962 could not be met with SCOUT. NASA, at considerable cost, decided that plans should be changed and the satellite launched by a THOR–DELTA rocket.

This liquid-fuelled rocket was already well developed with a proven record of success. The launch would be transferred from Wallops Island to the much larger and more sophisticated site at Cape Canaveral on the Atlantic coast of Florida. The change, with its greater assurance of an early and successful launch, was welcomed by the British participants. Some additional work arose from the different mechanical features of the new rocket system, but this was relatively minor in nature. However, the greater attention now given to the orbit possible from a Cape Canaveral site showed

Fig. 5.1. The Ariel 1 satellite.

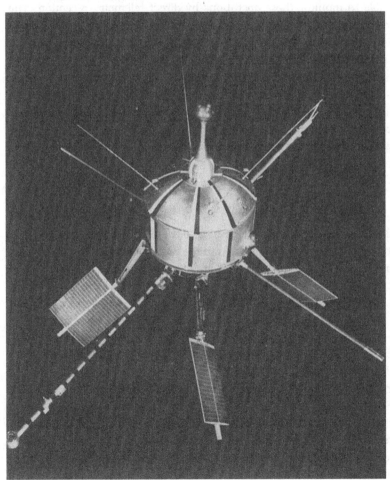

that much enhanced telemetry coverage could be obtained by taking advantage of the Department of Scientific and Industrial Research (DSIR) radio stations at Singapore and at Port Stanley on the Falkland Islands in the South Atlantic. The latter location was particularly well placed for the expected orbit, covering a gap in the STADAN network. Accordingly, arrangements were made for telemetry reception equipment to be installed at the two sites. With the co-operation of the Royal Air Force, service personnel were posted to augment the resident DSIR staff. A very important although shortlived phase of the satellite's life would occur very soon after injection into orbit. At this time various mechanical operations should take place, including deployment of several booms and solar panels. The spin rate of the satellite should also decrease markedly. It would be extremely desirable to monitor these operations by direct telemetry reception. The optimum place at which to receive the telemetry signals appeared to be on the island of Tristan da Cunha, in the South Atlantic. Location of the staff and equipment on the island itself posed big problems because of a recent volcanic eruption. However, the British Admiralty responded to a call for assistance made by the Chairman of BNCSR to the Deputy Controller Sir John Carroll and the Royal Navy agreed that the necessary equipment should be installed in HMS 'Protector'. It was planned that she should take up station near Tristan da Cunha and receive the signals during the initial phase of the first orbit.

7 The launching of S-51

All was ready for the launch on 10 April 1962. The combined British–American teams were in position at the site.[2] HMS 'Protector' was in radio communication at her station in the South Atlantic. The network of telemetry and tracking stations was standing by and all the diverse range safety precautions were in a state of readiness. Even the weather promised well, so the ten-hour count down began in the small hours of the morning. It was an exciting time for the British party. These were early days in the development of satellite launchings and the extensive public television coverage which has made the procedures relatively well known still lay in the future. The individual scientists monitored the performance of their instruments in the satellite as the various tests were carried out. The two project managers shared responsibility at the control console (Figure 5.2) for clearing the satellite for launch. A party of some eighty visitors from the United Nations Committee on the Peaceful Uses of Outer Space came to inspect the proceedings, and then retired to a safe distance to observe the launch.

Unfortunately during the countdown a fault in the fuel system of the rocket was revealed by the test procedures. This could not be corrected immediately and the launch attempt was perforce abandoned.

The sense of anticlimax was overtaken by the immediate action of the launching team in bringing the abort procedure into play and making the rocket safe to approach.

Some two weeks were to elapse before the rocket had been cleaned, refuelled and the launch rescheduled. This allowed some of the British team to relax at nearby Coco Beach after many strenuous months.

HMS Protector returned to base at Simonstown and the Royal Navy arranged a rapid transfer of the telemetry receiving equipment to the frigate HMS Jaguar. This was done in time for her to be back on station near Tristan da Cunha for the next launch attempt on 26 April. This time the launch was completely successful to the relief of the British experimenters who not long before had witnessed (see Figure 5.3) the awesome sight of an Atlas rocket blowing up on its launch pad. They were delighted to see the Thor–Delta (Figure 5.4) lifting off majestically with gathering speed into the blue carrying British equipment into orbit for the first time. The launch, a

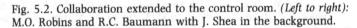

Fig. 5.2. Collaboration extended to the control room. *(Left to right):* M.O. Robins and R.C. Baumann with J. Shea in the background.

photograph of which is shown in Figure 5.5, took place near 1 pm. It was not until 2.30 pm, after the globe had been encircled once, that information received was sufficient to show that the satellite was actually in orbit very close to that planned. There was a further wait for some hours until telemetry signals from the satellite could confirm that the experiments were working. When these signals were at last received it appeared that all was well except for the solar Lyα emission detector.

The successful launch called for a celebration party and this was given by the British at nearby Patrick Air Force Base. The party grew larger as the night wore on and the British excused themselves near midnight as they were due to leave by plane for Washington at 5 am to attend a press conference. On that plane, Boyd, Willmore and Massey, using telemetry data already received, were able with the help of a 12-inch slide rule to analyse the results obtained with the ion mass spectrometer. They found that, even at this stage, the data showed that while O^+ was the major ion present in the high atmosphere up to an altitude of about 900 km it was replaced by He^+ at higher altitudes, a most gratifying result, not least

Fig. 5.3. Waiting to see the Atlas launch *(Left to right)*: E.B. Dorling, H. Elliot, R.C. Baumann talking to M.O. Robins, H. Massey and J. Sayers.

because no high technology was used in the analysis. Preliminary assessments of data from the other experiments were also made in time for the COSPAR meeting in Washington a few days later. There was a brief alarm at this time when it appeared that the telemetry signals from the satellite had ceased. However, it transpired that this was due to faulty ground equipment, soon corrected, and the British escaped what would have been a devastating blow.

Fig. 5.4. The Thor–Delta launcher just before take-off. *(Left to right):* E.B. Dorling, M.O. Robins, H. Massey, J. Sayers, P. Bowen, H. Elliot, A.C. Durney and R.L.F. Boyd.

The signals received by HMS 'Jaguar', which were not available for analysis until some time later, eventually showed that while the deployment of aerials and solar paddles, etc. was satisfactorily achieved it was largely by good luck because it occurred through an unplanned sequence of events which might well have been disastrous.

On successfully entering orbit the satellite, previously known as S-51, became known as Ariel 1. The name Ariel, with its Shakespearean connotations was first suggested by Mrs Quirk, wife of R. Quirk at the Ministry for Science (formerly at the Office of the Lord President).

Fig. 5.5. The launching of Ariel 1 from Cape Canaveral.

8 Subsequent life of Ariel 1

On 9 July 1962 Ariel encountered an unexpected hazard. A high altitude hydrogen bomb explosion, code named 'Starfish', took place off Johnston Island in the central Pacific Ocean, at a time when Ariel 1 was about 7400 km away. Nearly all the instruments carried in the satellite were initially saturated by the radiation effects, but most recovered and continued to operate satisfactorily. More seriously, the solar cells providing electric power, and some other components, suffered permanent radiation damage. This resulted in intermittent and erratic operation for the remainder of the useful life of the satellite. A failure which was rather fortuitous but not unusual was that of a timer which should have cut off the satellite transmissions after one year of operation. The fact that it did not do so enabled further data to be obtained until, in November 1964, it was judged that the performance no longer justified the efforts required for data reception and analysis. The orbit of Ariel 1 finally decayed and the satellite was destroyed on entering the atmosphere on 24 May 1976. This first international satellite was to be followed by many more, but the pattern was set by Ariel 1, and as a pilot exercise it was an unqualified success. Most, if not all, the scientific aims were achieved, and the techiques of design and testing established for this satellite to ensure operation of equipment in the space environment were to become standard practice for many years. It also answered those who doubted whether university scientific groups could design and construct experiments capable of operating for long periods without maintenance. The experiment of Boyd and Willmore to measure electron concentration and temperature was still operating two years after launch until it became too expensive to continue to tape the data from it, while other experiments operated for satisfactorily long periods.

9 Scientific results from Ariel 1

It is not our intention here to give a detailed account of the scientific results obtained with the instruments carried in Ariel 1. A much more detailed discussion is given in Chapter 14. Nevertheless, some indication of the observations made is appropriate here. On 2 and 3 May 1963, about one year after the launch, a discussion meeting[3] on the results obtained was held at the Royal Society under Massey's leadership. All the scientific groups concerned reported the results of analysis of their data to date.

Sayers described measurements of electron concentration obtained with his capacity probe. He found extensive regions of enhanced ionization in the topside ionosphere (that above the F_2 maximum, see Figure 1.1 Chapter 1) related to the earth's magnetic field. In addition to these systematic observations, Sayers also obtained records of the increased ionization density over Johnston Island soon after the Starfish explosion. This is shown in Figure 5.6.

Bowen, Boyd, W.J. Raitt and Willmore described their observations of ion composition. These were the first to make systematic observations of both the light ions H^+ and He^+ and the heavier ions such as O^+, though earlier observations of He^+ and O^+ had been made on separate occasions.[4] Figure 5.7 shows typical results obtained for the so-called transition altitudes as a function of the solar zenith angle. For $O^+ - He^+$, for example, this is the altitude at which the O^+ and He^+ concentrations are equal. At higher altitudes He^+ is dominant until the $H^+ - He^+$ transition altitude is marked. A great number of systematic observations of this kind were made and have proved very useful for the interpretation of the behaviour of the high atmosphere.

Bowen, Boyd, C.L. Henderson and Willmore gave an account of measurements of electron temperature T_e obtained with their Langmuir probe technique. Figure 5.8 shows the observed variation of T_e with latitude

Fig. 5.6. Comparison of the ionization density recorded by Sayers over Johnston Island a few hours after the explosion with that recorded on the previous and following days.

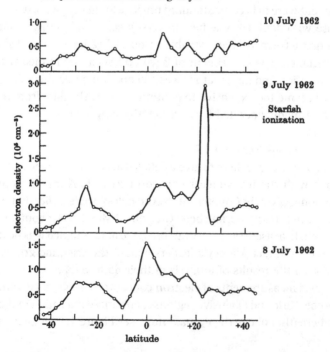

Fig. 5.7. Positive ion transition altitudes as a function of solar zenith angle $O^+ - He^+$ transition _ _ _, ○, 35°N–45°N; _ _ _, ●, 10°S–10°N. $He^+ - H^+$ transition _ _ _, △, 35°N–45°N; _ _ _, □, 10°S–10°N, observed from Ariel 1.

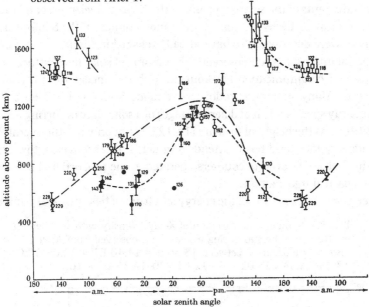

Fig. 5.8. Variation of electron temperature with latitude at midnight at 1000 km altitude observed from Ariel 1 on magnetically quiet days.

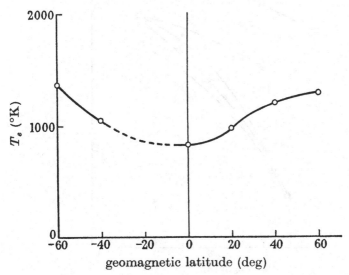

at an altitude of 1000 km on magnetically quiet days. Data of this kind were obtained systematically and are important for the analysis of the thermal balance in the ionosphere.

Measurements of the solar spectrum in the X-ray wavelength band from 0.4 to 1.4 nm made by Bowen, K. Norman, Pounds, P.W. Sanford and Willmore were described. We have already referred in Chapter 3, p. 32 to the special interest of such observations at a time of solar flare disturbance leading to a communication blackout through enhancement of D region ionization. Many such observations were made from Ariel 1. Figure 5.9 shows X-ray spectra taken at different stages of a solar flare occurring on 27 April 1962. At the height of the flare (14.12 UT) not only has the overall X-ray intensity increased but it extends to much shorter wavelengths. This was the first use in orbit of counters operated in the proportional mode to obtain spectral information.

Accurate measurements of the energy spectrum of heavy (Z > 6) primary

Fig. 5.9. Showing the variation of X-ray intensity with wavelength emitted from the sun during a solar flare, observed from Ariel 1. The visible flare occurred between 13.50 and 14.40 UT × 12.05–12.08 ○ 13.48–13.53 ● 14.02 □ 14.12 + 14.30–14.35 ● 14.10.

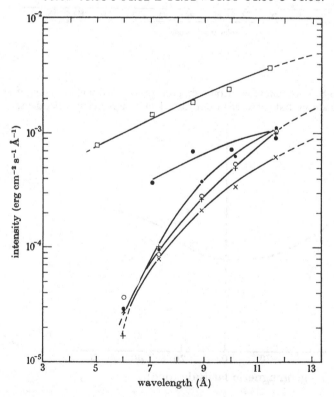

cosmic rays were successfully made by Durney, Elliot, Hynds and Quenby. An example is shown in Figure 5.10. These results are of importance in providing a means of studying the large scale properties of the interplanetary magnetic field. The same investigators were also able to use their equipment to observe many features of the energetic particle flux distribution arising from the Starfish explosion.

These early results show that the instruments on Ariel 1 provided data of much interest for atmospheric, solar and cosmic ray physics.

10   The second Ariel satellite

We can now turn, in rather less detail, to Ariel 2, which began life under the NASA code number S-52. The choice of British experiments as potential candidates for inclusion in the satellite did not present a big problem

Fig. 5.10. Integral primary spectrum of cosmic rays, observed from Ariel 1.

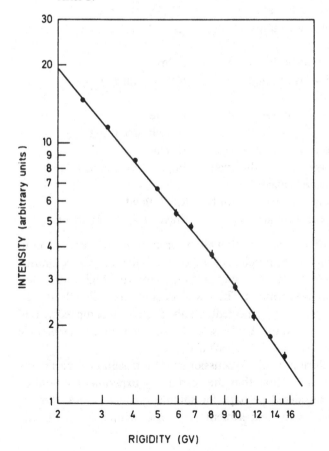

once the scientific payload for Ariel 1 had been selected from an initial list of eleven experiments drawn up by the Design of Experiments Sub-committee of the BNCSR. The experiments concerned with galactic radio noise, meteorological measurements, and measurement of the micrometeorites flux formed the core of the proposals not accommodated in Ariel 1. To these were added proposals for observations of very low energy cosmic rays and solar X-rays.

A working group to study and consolidate proposals for this second satellite payload was set up in June 1960, as part of the Royal Society's BNCSR structure. Preliminary discussions with NASA about the proposals took place in Washington soon afterwards. The opportunity was taken to have further informal talks with NASA scientists and engineers concerned with S-51 during their visit to London in September 1960.

At this stage the proposals under consideration were:

1. Measurements of galactic and extra-galactic radiation in the radio frequency bands around 1–2 Mc/s
F.G. Smith
Cambridge University
(later Manchester University)

2. Measurement of the ozone distribution in the high atmosphere
R. Frith
Meteorological Office

3. Detection of micrometeorites
R.C. Jennison
Manchester University

4. Detection of very low energy cosmic rays (only in the event of unallocated space)
H. Elliot
Imperial College London

5. Solar X-ray observations (only if the opportunity arose)
R.L.F. Boyd
University College London

The general view at this time was that all the proposals had scientific merit, but that orbit compatibility problems would arise if all were to be contained in the same satellite. Elliot, for instance, would require a high inclination orbit for his proposed experiment. There was pressure to simplify the design of S-52 in view of the unexpectedly high degree of complexity and sophistication inherent in the Ariel 1 scientific payload and the formidable data encoding and transmission problems.

It was decided during further discussions in the Design of Experiments Sub-committee in early 1961 that the cosmic ray experiment should be withdrawn and considered perhaps for submission to a NASA Nimbus satellite. The solar X-ray experiment was also withdrawn with the suggestion that it might possibly be considered for a NASA Orbiting Solar Observatory satellite.

Thus the three experiments concerned with the galactic radio astronomy, ozone distribution and the detection of micrometeorites were approved in Britain and accepted by NASA in due course. Launch was to be by a SCOUT rocket from the Wallops Island launch site.

Whilst the general arrangements for the design, construction and operation of S-52 were broadly similar to those employed in the Ariel 1 programme, there were important differences. In the USA, NASA placed more of the work out to industrial contractors and passed responsibility for the handling of the data tapes, including the digitization process, to the UK. In Britain, the RRS assumed a central role in this phase of data processing which was to continue for the remainder of the Ariel series. With advice from NASA, equipment for the analogue to digital conversion process, and necessary auxiliary equipment, was procured, and continuing assistance was received from the experienced computer staff in Space Department RAE and in the AWRE.

The three groups of University scientists could, in the design and preparation of their equipment, benefit from the experience of their colleagues with Ariel 1. However, two of them, Smith and Jennison, were breaking new ground in having for the first time to operate equipment in orbit having moving parts. In addition, Smith required long antennae to be extended from the satellite. Jennison's micrometeorite penetration detectors required the winding on from time to time of thin aluminium foils. Novel problems were overcome in the successful designs which were evolved. The equipment was made by British contractors.

11 The Space Research Management Unit

It has been described in Chapter 3, p. 45, how Robins was seconded from the Ministry of Supply to the Royal Society to assist at a senior level in the managerial and technical problems inescapable in the growing space research programme. With the advent of S-52 there was need for further managerial support and Dorling, a physicist experienced in the Skylark programme (see Chapter 3, p. 23), who had been seconded from RAE, became the UK project manager to work with E. Hymowitz of NASA as his US colleague. Shortly afterwards, J.F. Smith, an engineer from the UKAEA, joined Robins and Dorling to help on the managerial side. The expanding activities on this side gradually outgrew the space available at University College London and led to a rather more formal relationship with the Ministry for Science. In the early part of 1963 the group under Robins moved to the office premises in Chester Gate, Regent's Park, London. It had been joined by B.G. Pressey, a radio engineer from the Radio Research Station who had been concerned with the World Data Centre at Slough, and by A.C. Ladd, a scientist from UKAEA.

This small but well-balanced team, with supporting staff, became known as the Space Research Management Unit (SRMU). It was for administrative purposes responsible to the Ministry for Science, but in scientific and technical matters it responded to the needs of the British space scientists.

Fig. 5.11. The launching of Ariel 2 from Wallops Island.

12 The launching of Ariel 2

The relatively uneventful launch of the satellite, henceforth known as Ariel 2, took place on 27 March 1964 (Figure 5.11). It was possible quickly to check the successful operation of the equipment measuring ozone concentration and galactic radio noise. The nature of the micrometeorite detectors precluded any early conclusions about their operation in orbit, but subsequent analysis confirmed that this equipment also was capable of successful detection although no conclusive evidence of micrometeorites was obtained.

The ozone experiment provided useful data until the end of September 1964 when the tape recorder failed and the rotation of the satellite had become so slow and irregular that the method was unworkable. Figure 5.12 shows the altitude profile of ozone concentration obtained by D.E. Miller and K.H. Stewart[5] from the satellite data. It agrees well with results of early rocket experiments except above 60 km where it is significantly lower. The satellite data also confirmed that the attenuation of 380 nm was substantially larger than expected from ozone alone, in agreement with results from a calibrating Skylark flight in April 1964. This increased attenuation could be ascribed to dust from a volcanic eruption in Bali in March 1963.

Interesting results were obtained from the cosmic radio noise experiment.

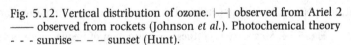

Fig. 5.12. Vertical distribution of ozone. |—| observed from Ariel 2
—— observed from rockets (Johnson *et al.*). Photochemical theory
- - - sunrise – – – sunset (Hunt).

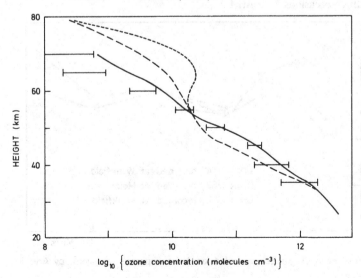

Figure 5.13 shows the mean sky brightness as a function of frequency between 1.25 and 3.4 MHz obtained for three orbital passages (J. Hugill and Smith[6] 1965). This differs from that obtained by extrapolating the data available at higher frequencies in being of greater intensity near 3 to 5 MHz and also in not continuing to increase below 3 MHz. Possible interpretations have been discussed by Smith.[7]

Some further account of these experiments is given in Chapter 13.

13   Ariel 3

There was a major change in the national sharing of responsibilities for the third satellite of the Ariel series. The UK undertook all the engineering design, construction and testing of the satellite itself, in addition to the provision of the scientific experiments. This progressive step in the arrangements was foreshadowed in June 1961 during a visit to Washington by R. Quirk, of the Office of the Lord President, and the two present authors. In a general review of the joint activities, H. Dryden, Associate Administrator of NASA, explained that whilst his organization had been able to accept responsibility for designing and constructing the first two satellites, this could not be a continuing activity for NASA. He had no doubts about the capabilities of British engineers to carry out such work, and must ask that the UK undertake the design and construction to an acceptable flight standard, of the third and subsequent satellites in the series. Under these conditions NASA would be prepared to provide launch and range facilities

Fig. 5.13. Mean sky brightness as a function of frequency recorded on three occasions from Ariel 2.

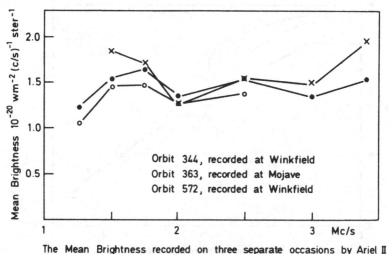

Orbit 344, recorded at Winkfield
Orbit 363, recorded at Mojave
Orbit 572, recorded at Winkfield

The Mean Brightness recorded on three separate occasions by Ariel II

for further Ariel satellites beyond the three originally agreed. NASA would continue to be available for consultation and advice.

On the British side, this was seen as both inevitable and welcome to engineers and technologists in the UK, who were just as enthusiastic to gain practical experience in the art of designing for the space environment as were the scientists.

New organizational and financial arrangements within the UK were necessary to meet the changed situation. Coincidentally between the launching of Ariel 1 and Ariel 2, a major change in responsibilities followed the abolition of DSIR and the formation of the Science Research Council (SRC) in 1965. Reference is made to these events in Chapter 10.

Here we need only record that responsibilities for the satellite, at this time beginning to be known as the spacecraft, were undertaken by the Ministry of Technology, acting through RAE and British industry. Staff of the RAE had been close observers of the technological development of the Ariel 1 and Ariel 2 satellites, and were in a very good position to play a leading part in the subsequent programme.

Although the SRC in 1965 assumed some of the space science policy responsibilities hitherto carried by the BNCSR, at the inception of the S-53 programme (in NASA terminology) the scientific experiments were selected and recommended by the BNCSR.

Detailed considerations of a possible scientific payload began after the policy discussion in Washington in June 1961. It soon became clear that at least one proposal would require S-53 to be launched in a near polar orbit. This could not take place from either Wallops Island or Cape Canaveral. NASA willingly agreed that for planning purposes it could be assumed that launch into a near polar orbit from the Western Test Range in California would be possible.

In February 1962, the Design and Experiments Sub-Committee of the BNCSR gave preliminary consideration to eight proposals for a variety of experiments from groups in the Universities of Cambridge, Sheffield, Manchester, London, Birmingham, and in the RRS and the Meteorological Office (MO). The initial limits on the total weight of scientific equipment (35 lb) total power consumption (1.25 watt) and data rate (50 samples/s) were, by later standards, very tight. In addition to these engineering constraints, a number of scientific desiderata helped to narrow down the choice of experiments to five. They were:

1. Measurements of electron   J. Sayers
   density and temperature   Birmingham University

2. Measurement of galactic and ionospheric noise	F.G. Smith Cambridge University (later Manchester University)
3. Measurement of natural VLF radiation	T.R. Kaiser Sheffield University
4. Measurement of HF noise from thunderstorms	F. Horner Radio Research Station
5. Distribution of molecular oxygen	K. Stewart Meteorological Office

These proposals were accepted by the BNCSR and the Steering Group for Space Research, whilst agreement with NASA was reached in January 1963. Three of the experiments were linked with or extended experiments carried in Ariel 1 and Ariel 2. The two concerned with natural VLF radiation such as hiss and whistlers, and HF radio noise from lightning discharges, broke new ground. Setting up the working arrangements, obtaining the necessary financial approvals and letting the contracts to industrial firms, took time, and it was early in 1964 before all the necessary effort was brought to bear on the project in the UK. The management of the project in the USA was in the hands of the GSFC of NASA, whilst management in the UK was the responsibility of the SRMU, formal management control being achieved through Joint Working Group meetings. The UK Programme Manager was A. Ladd of the SRMU.

The Space Department of the RAE Farnborough played a major role in the design of the spacecraft, under the management of J.H. Sketch. The main industrial contractors were the British Aircraft Corporation and the General Electric Company. Whilst tracking and data acquisition were largely in the hands of NASA through the operations of the STADAN network, the Radio and Space Research Station (RSRS) originally the Radio Research Station, continued to operate stations in Singapore and the Falkland Islands. The RSRS developed further the data digitization facilities installed for Ariel 2. Processing the digital data was primarily the responsibility of the Data Processing Group of the AWRE and of the SRC Atlas Computer Laboratory.

Satellite S-53 (UK3 in British terminology), shown in Figure 5.14, was launched into an orbit very close to that specified, by a SCOUT rocket from the Western Test Range on 5 May 1967, and became Ariel 3. The scientific experiments were mostly very successful. Great care had been taken before launch to eliminate all causes of radio interference between the experiments. Unfortunately, the galactic noise experiment suffered seriously from an interaction with the Birmingham experiments on electron density and temperature. This occurred solely in the ionosphere and was not in any way a malfunction of the spacecraft. It did not prevent useful results being obtained.

An account of the results obtained was given by the different experimental groups at a discussion meeting[8] held at the Royal Society on 24 April 1968. We select here some typical examples. Further discussion is given in Chapters 13 and 14.

Fig. 5.14. The Ariel 3 satellite.

Figure 5.15 shows the variation of electron concentration and temperature along a satellite track obtained by Sayers, J.W.G. Wilson and B. Loftus.
Many results were obtained from this experiment.

Horner and R.B. Best described their measurements of terrestrial radio
noise at frequencies near 10 MHz from which they derived information
about the location of thunderstorms. Figure 5.16 shows results obtained on
a satellite pass over Europe and Africa on 31 May 1967. Peak signals are
associated with storms.

R. Bullough, A.R.W. Hughes and Kaiser presented a wide geographical
range of measurements of low frequency (3.2, 9.6 and 16 kHz) electromagnetic fields outside the ionosphere and discussed their origin (see Chapter 14,
p. 326).

The Meteorological Office equipment to measure O_2 concentration as a
function of altitude (Stewart and P.I.C. Weldman) worked well. Figure 5.17
illustrates a typical altitude profile.

Although at this stage much of the analysis of cosmic noise data (Smith
and R.B. Best) was concerned with obtaining means for disentangling the
interaction with the electron concentration experiment, this was satisfactorily achieved a little later.

The scientific success of Ariel 3 was important but of equal significance
was the fact that the UK had, for the first time, successfully designed and
developed a complete satellite. That the necessary technology and design

Fig. 5.15. Variation of electron density and temperature observed from
Ariel 3 for orbits 99/100, on 12 May 1967.

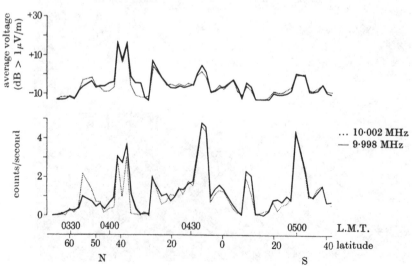

Fig. 5.16. Variation in radio noise intensity observed from a pass of Ariel 3 over Europe and Africa on 31 May 1967 05.00 UT.

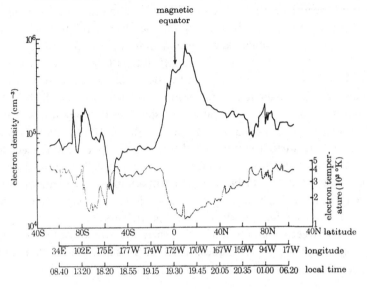

Fig. 5.17. Vertical distributions of molecular oxygen observed from Ariel 3.

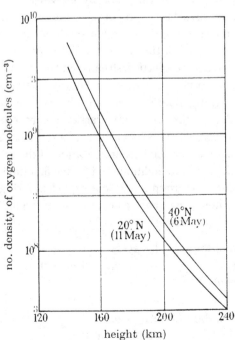

techniques had been available was very largely due to the research and development programme undertaken at RAE in the immediately preceding years and concerned with the many individual systems required by satellites – power, attitude control, telemetry, command, data handling and so on – and with their mechanical and thermal design. This programme was initially supported by the Ministry of Supply. It later (1964) developed with the Black Arrow programme, funded by the Ministry of Technology and involving notably RAE, RRE, and industrial firms. This, with its three elements of launch vehicle development, space technology and the two satellites Prospero, launched in 1971 by the Black Arrow launcher, and Miranda (1974), further strengthened the technology base and specialist advice at home to which space scientists in the UK could look for assistance and co-operation. It was the basis on which major industrial contracts were awarded to British firms, particularly by the Eurpean Space Agency (see Chapter 6). In a similar way the eventual participation of the UK in space applications projects owed much to this early investment in the relevant technologies and design techniques, which at the same time furthered the cause of UK space science.

14   Ariel 4

The next satellite in the Ariel series, generally known as UK-4 before launch, was the first to be carried through entirely under the policy direction of the SRC. Apart from this, there was much similarity with its predecessor Ariel 3. The scientific aim of the satellite was to explore the interactions between the plasma, charged particle streams and electromagnetic waves in the topside of the ionosphere. The experiments were intended to supplement and extend the investigations into the ionosphere carried out by previous satellites in the Ariel series, and being carried out by other scientific satellites. In 1969, four experiments had been proposed by British University groups, and the Radio and Space Research Station of the SRC, and agreed within UK and with NASA. In addition, and for the first time in the series, a proposal by an American group was also accepted, marking a significant step forward in Anglo–American co-operation in scientific satellites. The scientific payload comprised:

1. An improved version of the   J. Sayers
   electron density and   Birmingham University
   temperature experiments
   in Ariel 3
2. An improved version of the VLF   T.R. Kaiser
   experiment in Ariel 3   Sheffield University

3. A lightning impulse counter   T.R. Kaiser and
   experiment   F. Horner, RSRS
4. Measurement of ionospheric   F.G. Smith
   radio noise in the MHz range   Manchester University and
                                      F. Horner, RSRS
5. An experiment to measure the   L. Frank
   energy spectra of electrons and   University of Iowa, USA
   protons

   The UK-4 spacecraft was very similar to UK-3 in both its configuration
and general equipment. Amongst the differences were the provision of a
magnetic torquing coil to enable spacecraft attitude to be controlled, the
provision of redundant tape recorders to improve reliability, and the
provision of more command control channels. The contractual responsibilities for UK-4 were largely unchanged from those arranged for UK-3, as were
the general data acquisition and processing procedures. The launch by a
SCOUT rocket into a near polar orbit took place at the Western Test Range in
California on 11 December 1971. All the equipment operated satisfactorily
with the exception of the Birmingham electron temperature experiment,
which failed for reasons not positively identified. One of the two tape
recorders failed 15 days after launch, but the second recorder operated for
eight months. There was a partial loss of data from the Iowa University
experiment a few months after launch. Data was regularly received from
Ariel 4 until it was switched off in May 1973, by which time it was judged
that sufficient information had been obtained. Ariel 4 was reactivated for a
short time later in 1973 to provide additional data for use in conjunction
with sounding rocket observations. During the operational life of Ariel 4, the
Royal Radar Establishment at Malvern co-operated with the control team by
measuring the attitude of the spacecraft using an S-band radar.
   An account of some of the scientific results from Ariel 4 is given later
(Chapter 14).

15 Ariel 5
   If any particular scientific theme can be discerned in the measurements taken on the first four of the Ariel series of spacecraft, it is related to the
ionosphere and associated phenomena. This reflected both the interests of
British scientists ready to use the new technologies, and the nature of the
available techniques. However, in the 1960s, there arose a new and rapidly
expanding field of astronomy which depended upon observations carried
out through instruments mounted on vertical sounding rockets penetrating
above the earth's atmosphere. It became apparent that X-radiation from

cosmic sources promised to provide a wealth of new information hitherto quite inaccessible to earth based instruments. An historical account of these early developments is given in Chapter 15.

In December 1970 the satellite named UHURU was launched by NASA, and enabled the first comprehensive survey of the sky to be made in X-radiation. The observations were of limited precision in both angular measurements and energy spectra, and it was clear that more accurate measurements would be extremely valuable to astronomers. British space scientists had been quick to realize the potential of this new branch of astronomy and the first proposal for a cosmic X-ray scientific payload in the Ariel series was made to NASA through the SRC in July 1968. Several revisions followed, and an agreement was reached in 1970 for a satellite code named UK-5. This was to become in due course, Ariel 5, perhaps in scientific terms the most productive of the series. The importance of Ariel 5 is discussed in more detail in Chapter 15. Here we outline the sequence of arrangements and events which led to such a successful outcome.

The selected experiments, again including one from the USA, were:

1. Measurement of X-ray source positions and a sky survey in the energy range 0.3 to 30 keV — R.L.F. Boyd and A.P. Willmore[9] University College London
2. Study of the spectra of individual sources in the 2 to 30 keV range — Ditto
3. Deep sky survey in the energy range 1.5 to 20 keV — K.A. Pounds Leicester University
4. Measurement of the polarization of X-rays from 1.5 to 8 keV — Ditto
5. Study of sources of high energy X-rays up to 2.0 Mev — H. Elliot, J.J. Quenby and A.R. Engel Imperial College London
6. An all-sky monitor in the energy range 3 to 6 keV — S.S. Holt Goddard Space Flight Center USA

The original proposal envisaged that launch should be by a SCOUT rocket from Wallops Island, with a requirement that the orbit inclination should be as small as possible. The minimum inclination which seemed feasible from this launch site appeared to be about 35 degrees. As further studies of the proposed satellite were made, it became apparent that substantial advantages would accrue if a near equatorial orbit could be achieved, that is an orbit inclination near 0 degrees. This would give a gain of some 20% in observation time compared with a 35 degree inclination orbit. Also the cosmic ray background would be less obtrusive, the total weight of the satellite could be greater, and the satellite control problems would be eased. An alternative site to Wallops Island which allowed SCOUT launches into a near equatorial orbit existed in the San Marco Equatorial Mobile Range.

This facility comprised two offshore platforms, one of them a control centre, the other moored some distance away, a launch platform. The complex was established by the Italian Government to provide a SCOUT launch capability, and was at the time sited about three miles off the coast of Kenya, some 90 miles north of Mombasa. The range extended eastwards into the Indian Ocean, and allowed orbit inclinations of about three degrees, which was very suitable for the UK-5 Project.

Through the co-operation of the Centro Ricerche Aerospaziali of Rome, which controlled the San Marco Range, and of NASA, the Science Research Council was able to arrange that UK-5 should be launched from this equatorial site.

The principal responsibilities within the UK for the Project included for the first time control of the operation of the spacecraft when in orbit. A specially designed control centre at the Appleton Laboratory (previously known as the RSRS) of SRC, with a network of computers, was connected to the NASA STADAN ground stations at Quito and Ascension Island by links which would transmit commands to the satellite and telemetry signals received from it. Further processing and data links to each experimenter would give each one a close monitoring facility almost equivalent to having his cosmic X-ray telescope in his own laboratory. Although, as with the previous satellites in this series, NASA provided the rocket launch without charge to the SRC, the additional costs incurred by launching from the San Marco Range rather than from Wallops Island, were borne by the UK. The launch was successfully accomplished on 15 October 1974, and all experiments performed well, providing a wealth of new information about X-ray sources. During the third year of operation of Ariel 5, in May 1977, the supply of propane gas which was needed for the operation of the attitude control system, was exhausted. A small leakage of gas through the control valves may have caused this to happen rather sooner than expected. However, it was possible to continue observations in a satisfactory, albeit slower manner, by using the magnetic torquing system which had been installed for other reasons. Observations ceased in March 1980 when Ariel 5 re-entered the atmosphere in the sixth year of operations which had led to very significant advances in the science of X-ray astronomy, which will be discussed in Chapter 15, and had given much valuable experience to British space technologists in the overall operational control of a complex spacecraft.

As the Ariel series evolved, so did the organizational arrangements. There was a general trend to transfer increasing responsibility from the USA to the UK, and within the UK, to contract as much as possible of the technological work to industrial firms. The spacecraft was designed and constructed by Marconi Space and Defence Systems (a member of GEC – Marconi Electronics Ltd) under contract to the Ministry of Defence (Procurement Executive) which acted as agents for the SRC. The RAE acted as Research and Development Authority, carrying out technical monitoring of the work in industry, and major environmental testing. Programme management in the UK was the responsibility of the Appleton Laboratory (AL) of the SRC, and this laboratory also supported the project in other ways including co-ordination of the experimental needs, and orbital operations. The principals in the Ariel 5 programme, apart from the originators of scientific experiments themselves, who were of course the prime movers of the project, were:

In the UK

Project Scientist	A.P. Willmore
	Birmingham University
Programme Manager	J.F. Smith
	Appleton Laboratory
Control Centre Manager	A.J. Rogers
	Appleton Laboratory
Project Manager MOD(PE)	R.F. Maurice
Project Officer RAE	E.C. Semple

In the USA

Project Scientist, NASA	S.S. Holt	both of GSFC
Programme Manager, NASA	H.L. Eaker	

16   Ariel 6

We can now turn to the last spacecraft in the Ariel series, which was originated as UK-6 in 1972 mainly to meet the needs of cosmic ray scientists particularly interested in the ultra-heavy component (atomic number 30 or more) of primary cosmic rays. It was possible also to include, in addition, two X-ray experiments capable of extending the scope of the observations so successfully made by Ariel 5.

The scientific payload comprised:

1. Detection of heavy cosmic ray   P. Fowler
   primary particles   Bristol University

| 2. Observation of rapid time fluctuations and spectra in X-ray sources | K. Pounds Leicester University |
| 3. Observation of very soft X-ray emissions | R.L.F. Boyd University College London A.P. Willmore Birmingham University |

The general arrangements for the management and control of the project were similar in most respects to those described for Ariel 5. It was not necessary to specify either a near equatorial or a near polar orbit inclination, so launch from the NASA range at Wallops Island was feasible. For the first time in the series payment to NASA by the SRC for the launch of UK-6 by the SCOUT system was included in the Agreement between the two parties. The satellite was successfully launched into a most satisfactory orbit on 2 June 1979, and all the experiments transmitted good data. However, the operation of the X-ray experiments on the spacecraft has been seriously upset by spurious switching events which, it is believed, have been due to external radio frequency interference. The thermal control also proved inadequate so for most of the mission the satellite was operated outside the expected range of the high accuracy attitude sensors. Despite this it has been possible to achieve some of the planned programme by arranging for more extensive telemetry reception than originally intended, and by making full use of the flexible capabilities of the Appleton Laboratory's Control Centre. The cosmic ray experiment requiring no pointing accuracy was unaffected. The scientific results will be discussed later (Chapter 15).

The rapid advance of space science since the inception of the Ariel series in 1960 has naturally led to progressive increases in the size and complexity of payloads. Inevitably, the costs both of the experimental scientific equipment, the spacecraft itself, and the essential services of data acquisition and processing, have increased accordingly. It seems certain that the funds which the Science and Engineering Research Council can allocate to this particular method of research in the space sciences will no longer support the all-British individually designed spacecraft.

6

The European Space Research Organization

In Chapter 4 we described how British scientists were able to carry out experiments with equipment in satellites through bilateral co-operation with the United States in the Ariel programme. Although this provided opportunities which were immediately available and opened the way to further involvement in the US programme it did not fully satisfy the demand from the British space science community. The possibility of a national satellite launching programme involving the Blue Streak–Black Knight combination was still being actively canvassed in 1960 but, in addition, two other possibilities of co-operation were feasible, with the Commonwealth and with Western Europe. In a sense, Commonwealth co-operation was already involved through the use of Woomera in Australia for launching Skylark. This would also be true of a launching programme using Blue Streak because again the launching site would be at Woomera. The possibility of wider Commonwealth co-operation was actively canvassed during 1959 but although collaboration in space science did develop it was through the use of sounding rockets and did not add to satellite facilities. Its development will be described in the next chapter.

With Western Europe, on the other hand, very significant developments did take place which increased very considerably the opportunities for satellite experiments. Early in 1959 a suggestion was made that the North Atlantic Treaty Organization might organize a space research programme. There were obvious objections to this because of the military associations. Many scientists felt that it was most desirabe that the Soviet Union should continue to be openly involved in space science and a NATO programme would hardly help in this regard. Indeed, similar arguments applied to those used in the great debate in the USA about military or civilian control of space research (see Chapter 4, p. 54). The NATO Science Council at a meeting in

April 1959 decided that there should not be a 'NATO Satellite'. F. Seitz, the NATO Science Adviser, wished to re-open the matter, pointing out with justification that individual bilateral arrangements between individual European countries and the USA such as the Ariel programmes, would not be adequate to meet European requirements. He suggested the possibility of a NATO-sponsored European NASA to work with the American NASA. In default of alternative moves this might have been taken very seriously but it was overtaken by other events as we shall see.

By 1959 the success of the first large-scale scientific collaboration between the nations of Western Europe, the European Centre for Research in Nuclear Physics (European Organization for Nuclear Research, CERN), had become apparent. This was an institute devoted to research in high energy nuclear physics providing facilities too expensive to be funded nationally. The major facility was a 25 GeV proton synchrotron which came into operation in 1953 and was proving so successful that high energy physicists in Europe found themselves no longer at a disadvantage compared with their American colleagues. This was all the more remarkable because, in the early discussions about the machine, great scepticism about its practicability had been expressed by many scientists and engineers.

CERN was established primarily as an institute for carrying out research in its own right. It was not conceived as essentially providing a service for scientists in European universities although it was hoped that such scientists would nevertheless be very welcome and would be expected to make a major contribution to the whole enterprise through their research involvement. On the other hand, in Britain the National Institute for Research in Nuclear Science (NIRNS) had been established to provide expensive facilities for use primarily by university scientists. In particular a 7 GeV proton synchrotron was built by the Institute for this purpose. Massey was much involved in the work of this Institute and was a member of its Governing Board. This experience proved to be important in the negotiations which began in 1960 towards establishing a European co-operative organization for space research.

1 The commencement of European discussions

Towards the end of 1959 E. Amaldi, the leading Italian physicist who had been much concerned with the discussions which led to the establishment of CERN, after discussion with colleagues in Italy circulated a letter proposing the setting up of a similar organization for space research.

At the COSPAR meeting in Nice in January 1960 Amaldi in discussion with Massey supposed that Britain would probably not wish to take part in early discussions but might come in towards the end as with CERN. Massey, however, did not accept this and, when approached a little later in the meeting by P. Auger, the French delegate, about discussions on European collaboration, he immediately agreed to join in. As a result, some preliminary talks were held with Amaldi and delegates from other European countries. It was decided to hold an informal meeting in Auger's flat in Paris on 29 February 1960 to consider the matter in more detail after allowing opportunity for delegates to talk to their colleagues at home.

The meeting was duly set up with great secrecy. There were present P. Auger (France) E. Amaldi (Italy) J. Bartels (West Germany) M. Nicolet (Belgium) J. Veldkamp (Holland) R. Baumberg (Sweden) F.S. Houtermans (Switzerland) and H.S.W. Massey (UK). An apology was received from S. Rosseland (Norway) who was unable to attend.

The meeting began with accounts of national programmes and plans. Massey explained that in the UK no decision had yet been made as to whether the launching of satellites would be undertaken. He made it clear that British scientists were favourably disposed towards European collaboration. As all the others present felt the same way Massey suggested that, as a next step towards formalization of the discussions, he would ask the British National Committee for Space Research to consider issuing an invitation to a meeting in London, in late April, with aim of setting up a recognized Committee or working group. This was agreed and a meeting was arranged at the Royal Society on 29 April 1960.

2   Early UK reactions

Meanwhile active consideration of the pros and cons of British participation began not only within the British National Committee for Space Research (BNCSR) and its sub-committees but also within the Steering Group and in other Government departments. The situation was completely different from the US co-operation. Within Europe, the UK was relatively far more advanced in space research, including launching capabilities. In France, the research rocket Veronique was just coming into operation but the size of the space science community was still much smaller than in the UK. The proposed Blue Streak–Black Knight launching system was far in advance of any other European developments but this did involve launching at Woomera and there would be the question of the relationship of Australia to any future European organization. Then again should this organization be on lines similar to NASA in covering all aspects of space research apart from military ones or should it be concerned with space science only? In the UK at the time there was relatively little interest in space applications such as in communications and meteorology and it was the space scientists who played the leading role among the user community. The question of how to arrange for the Blue Streak launching system to be associated with a European organization concerned with space science was a complicated one particularly as the future of Blue Streak for defence purposes was being critically assessed at the time.

It was also important to realize that the analogy with CERN was far from exact, even for an organization devoted to space science only. CERN is concerned with high energy physics, a single, though very demanding, discipline. Space science on the other hand covers many disciplines including atmospheric physics, astrophysics, solar physics and planetary science. Again the central research facilities are located at CERN and all experiments are carried out there. Space scientists, on the other hand, could build up the instrumentation they proposed to fly in sounding rockets or satellites in their home laboratories while the central institution would be responsible for the provision of the satellite engineering and payload integration. It was felt very strongly by the already well-established space science community in Britain that they would not be prepared to support a central organization that specifically devoted itself to carrying out research in space science. In the Skylark and Ariel programmes they were able to operate from their home laboratories so strong university participation was possible. They were most anxious to preserve this situation.

A further very important difference was that, whereas no one regarded CERN as other than an organization for pure scientific research, with no foreseeable direct applications, there were potentially profitable applications of space research. This led to confusion between the interests of different countries in setting up a large scale collaboration. Finally, whereas it was clear that CERN should operate at a single site, this was not so obvious for a space science organization and this led, as we shall see, to an undesirable dispersion of locations.

3 Informal meeting in London, April 1960

The meeting on 29 April 1960 at the Royal Society was well attended. As Massey was visiting Australia at the time, the Physical Secretary of the Royal Society (Sir W. Hodge) took the Chair. The other UK representatives were R.L.F. Boyd, A.W. Lines, D.C. Martin, J.A. Ratcliffe, M.O. Robins and R.L. Smith Rose. Those who attended from continental Europe were P. Auger and J. Blamont (France), E. Amaldi and L. Broglio (Italy), E.A. Brunberg (Sweden), M.A. Ehmert (West Germany), H. Golay and F.G. Houtermans (Switzerland), H.C. van de Hulst, H.S. van de Mass and

A. Veldkamp (The Netherlands), L.M. Malet (Belgium), S. Rosseland (Norway) and K. Therme (Denmark).

As with the February meeting, reports were submitted, in much more detail, on present and proposed European programmes. According to Auger, the French space programme was under active consideration by the French Government, including the budget which was expected to be of order £3–4 million per annum. Scientific research in the upper atmosphere, on astrophysics and cosmic rays, from space vehicles, was being planned and satellite tracking facilities expanded. The Veronique rocket had already successfully carried payloads of 150 lb or so to altitudes of 180 km and a research programme using these rockets was being planned. A Super-Veronique rocket capable of carrying a 300 kg payload to 500 km was being developed and was expected to be operational towards the end of 1961. The main launching site for French rockets was at Colomb Bechar in Algeria though there was a smaller range available in the Ile de Levant.

The Italian plans were outlined by Amaldi. Arrangements had been made with NASA for the purchase of research rockets. There was also considerable interest in the engineering aspects of space research, particularly in the Engineering Department of the University of Rome, under L. Broglio.

These reports were followed by a lengthy discussion on satellite tracking within Europe and the possibilities of collaboration. Smith-Rose stated that 40 stations in Western Europe were already sending their observations on artificial satellites to World Data Centre C at Slough but there was still need for more stations providing accurate optical observations.

The meeting then turned its attention to the major question of collaboration in rocket and satellite experiments. The former were first considered. Boyd gave a number of reasons why a European co-operative rocket research programme would be especially desirable. These were as follows:

(a) The meteorology of the region 30–100 km above the European continent needed to be explored.

(b) Ionospheric phenomena such as sporadic E and polar blackout, which are strongly latitude dependent, need exploring in the European sector while present launching sites are mainly in the desert regions around latitudes 33°N and 33°S.

(c) Whilst the Canadian facilities at Fort Churchill gave results concerning the Arctic region, there was a very strong desire to work in the Arctic away from the geomagnetic poles and consequently launching facilities in Northern Sweden that became available would be of much benefit.

(d) In view of the great European tradition in astronomy it was necessary that work should be extended so as to survey northern skies in the ultra-violet region of the spectrum and at even shorter wavelengths.

In relation to (c) especially it was reported that a Swedish launching site near the Arctic Circle with a radius of 40–50 km was nearly ready for use and a much larger facility 70 km and 140 km long, directed towards the north, was being planned. As a further attractive feature it was pointed out that there was a Geophysical Observatory at Kiruna close to these sites.

An Italian range in Sardinia, extending for 40 km on the island but a further 250 km at sea if necessary, was also in operation.

Naturally the main question discussed in relation to collaboration in artificial satellite experiments was that of the launching system. It was stated that design studies had been carried out in the UK for the Blue Streak–Black Knight system in relation to three kinds of satellites: (a) one of 2000 lb weight in a near-earth orbit at an altitude of 200–300 miles, (b) one of 500 lb weight travelling out to two earth radii and (c) one of 100 lb in the form of a space probe travelling out 100,000–200,000 miles. In addition, the first of these types would be stabilized for astronomical studies. 1 Å resolution would be possible with the image being stabilized to a few seconds of arc and the satellite itself to a few minutes of arc. If it were decided to go ahead it should be possible to place design contracts by the end of 1961. There was unanimous agreement that advantage should be taken of the Blue Streak development. When the problem of Australian participation, if the Woomera range was used, was raised by the Chairman strong feelings were expressed that Australia, and other Commonwealth countries, should not be full members of a European organization, However, it was pointed out that, at CERN, it had always proved possible for scientists from the Commonwealth to take part.

The need for a further step towards formal recognition of the discussions was agreed and it was proposed that the next meeting should be one attended by nominated representatives of national committees. Auger was elected temporary Secretary to organize the meeting. A number of resolutions were agreed and are reproduced in Annex 7.

Following this meeting, the Steering Group invited the Minister for Science to take the initiative in convening official discussions with a view to formulating UK policy in relation to the proposed European co-operation. No further action was taken pending the outcome of the proposed meeting organized by Auger.

4 Provisional European space research group – proposed preparatory commission

The increased formality at this meeting, held in Paris, was apparent from the fact that five countries (Belgium, Denmark, Italy, Sweden and West Germany) sent government representatives. France, The Netherlands, Norway, Switzerland and the UK sent representatives of their national space organizations, Massey as well as M.O. Robins and A.F. Moore attending for the UK.[1] The main business was to work out the mechanism for establishing a Preparatory Commission which would be acceptable to the European governments involved.

The Swiss Government had already offered to issue invitations for an intergovernmental conference on space research to be held at Geneva at which the Commission would be formally established. It was realized, however, that further discussions at a technical level were necessary before the general nature of the proposed collaboration could be visualized. The meeting then constituted itself as a European Space Research Group and elected a Bureau to manage its affairs between meetings. This consisted of H. Massey (Chairman), L. Broglio, M. Golay and L. Hulthèn, (Vice-Chairman) and P. Auger (Executive Secretary). An offer from the French Government to house the Secretariat in Paris was accepted. A committee consisting of S. Campiche (Switzerland) (Convenor), J.H. Ferrier (The Netherlands), L. Malet (Belgium), A.F. Moore (UK) and P. Auger (France) was set up to prepare a draft agreement for establishing the Preparatory Commission. It was to meet in Paris on 5 July 1960. Estimates were also made of the cost of operating the Preparatory Commission, which was assumed, very optimistically as it proved, to last for one year only. Finally, it was also decided to hold the further technical discussions at a meeting in the early autumn.

It was already clear at the Paris meeting how different were the views of the various countries about the aims of the new organization. A degree of bureaucratic formality had already become apparent and indications were already present that the ultimate organization when it emerged would not follow the CERN pattern at all closely. However, these factors are more clearly realized in hindsight and there was no doubt that the whole enterprise was gathering momentum.

On 26 July the Steering Group discussed the outcome of the June meeting. The importance of the UK being involved from the outset was emphasized and it was agreed in principle to assent to the proposal for a Preparatory Commission, provided the scientific community approved of the programme which was to be made more definite at the autumn meeting of the European

Study Group. Steps were taken to study the issues raised within the Government.

Within the UK the view was still strongly held that there were two distinct aspects – research in space science on the one hand and the development of launching systems on the other. Relatively little attention was paid to space applications. As mentioned above this viewpoint was by no means held in the other countries, many of whom were concerned especially with commercial and industrial applications.

The Drafting Committee duly met in Paris on 5 July and produced a draft agreement which was circulated to members of the European Space Research Study Group for their assent so that a final version could be prepared.

5 Technical discussion meeting London 3–6 October 1960 – proposed technical organization

The technical discussion meeting of the European Study Group took place on 3–6 October 1960 in the rooms of the Royal Society in London. By that time a definite decision had been taken to cancel the development of Blue streak for military purposes. Following this the Ministry of Supply was preparing a plan for use of the Blue Streak as the main booster rocket for a European launching system. Because negotiations about this had barely begun the UK delegates to the Technical Discussion meeting were asked not to refer to the matter in any way during the meeting. This was somewhat embarrassing because some of the delegates from the continent already knew quite a lot about it. However, it meant that the discussions at the meeting were confined to consideration of the form a European organiza-tion, concerned with the provision of facilities for research in space science, might take, irrespective of how satellites would be launched. Within this remit it was a very successful discussion and in fact outlined a plan which formed the basis for the organization ultimately established.

The scientific case for European co-operation in space research was worked out in some detail as well as the nature of the co-operation. A copy of the comprehensive document prepared is given in Annex 8.

It was proposed that the core of the organization should be a Payload Engineering Centre responsible either directly or through contracts for the engineering, including the testing, of satellites and large scientific payloads, the integration of instruments within payloads and arrangements for launching. Apart from this Centre and Headquarters it was suggested that a Data Analysis Centre should also be established in addition to the tracking and data facilities which would be required. Provision for a laboratory for advanced studies was also made. These proposals were arrived at very quickly from the floor with Auger sketching them in on a blackboard chart, which finally took the form shown in Figure 6.1.

The two major bodies controlling the programme would be first a scientific committee which would examine all proposals for research, whether from Universities or other establishments or from within the organization itself. No proposal should be accepted without approval by this committee. Overall control on policy and finance would be vested in a council on which all member governments would be represented.

Suggestions were also made as to the nature of the study groups which might be set up by the Preparatory Commission. These were:

(a) to make proposals for the scientific programme related to possible payloads and satellite orbits,
(b) to make estimates of cost and technical possibilities for rockets, launching sites and facilities,
(c) to make proposals for technological research in such fields as propulsion, power sources, information storage, solid state physics, surface coatings of controlled emissivity, high vacuum technology, sensitive detecting devices and the behaviour of materials in high vacuum,
(d) to investigate possible sites of tracking and telemetry stations,
(e) to consider the general organization,
(f) to study the organization, capital facilities required and running costs of the 'Payload Engineering' establishment.

Fig. 6.1.

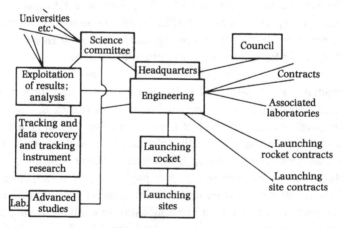

It must be remembered that only technical matters were under discussion. Administrative and legal matters still remained to be considered and at the time there was a tendency to forget the complications which these topics could raise.

6   UK decision to join preparatory commission

The Steering Group, on receiving a report of this meeting, reaffirmed their agreement in principle to joining the Preparatory Commission but asked for the views of the BNCSR on the matter. A detailed report from the BNCSR was made available on 31 October 1960. It considered the suggestions of the Study Group to be sufficiently promising to justify support for detailed investigation by experts on the lines suggested and it was believed that the UK could readily make a full contribution to the scientific and technical effort involved in the Study Group's plans. The BNCSR recommended that the UK should join the Preparatory Commission provided that the UK national space research programme was not prejudiced. It further recommended that any co-operative European programme which might emerge from the work of the Preparatory Commission should not prejudice the development and support of those parts of a UK national programme which could be justified in their own right.

The BNCSR went on to consider other possible overall UK programmes as they would be in three years' time, no important contribution from European countries being expected much before then. Taking the round figure of £10 million for the total running costs of a future European organization, the UK contribution would be £2$\frac{1}{2}$ million on the gross national product basis (at the time the contributions of UK, France and West Germany on this basis worked out as 25%, 20% and 20% respectively). The cost of a national sounding rocket programme, instrumentation for use in American and European satellites, and tracking, data acquisition, processing and analysis was roughly estimated as a further £2$\frac{1}{2}$ million. We shall see below how this analysis was modified in the light of later information and used in the consideration of how large should be the budget for the overall UK programmes.

This report was discussed in the Steering Group on 8 November 1960. The hope was expressed that the Minister for Science would give early agreement to joining the Preparatory Commission and this was soon realized. The UK therefore accepted the invitation from the Swiss Government to an inter-governmental meeting at Geneva on 28 November 1960.

7   Geneva inter-governmental meeting – preparatory commission established

The UK delegation consisted of R. Quirk, H. Massey, D.W. Haviland, Deputy Secretary, Ministry of Aviation, W.G. Downey, Ministry of Aviation, S.H. Smith from DSIR, H.C. Harnworth, P.J. Allot and D.J. Gibson from the Foreign Office, A.F. Moore from the Royal Society and J.A. Annand. Much to his surprise, as it was a governmental meeting, Massey was elected Chairman of the Conference with Auger as Secretary.

Three working groups were set up dealing with legal matters, budgeting matters and scientific and technical objectives, the respective Chairmen being S. Campiche (Switzerland) G.W. Funke (Sweden) and M. Golay (Switzerland). No special problems presented themselves in the first two groups but there was considerable discussion in the third as to whether or not the Preparatory Commission should make plans for the development of launching rockets. This was a very delicate issue for the UK in view of the negotiations which were proceeding about the use of Blue Streak as the booster for a European launching system. To many of the delegates little or nothing was known at the time about the nature and scope of these negotiations. However, after much discussion, especially in the corridors, a compromise was finally agreed and the following resolution passed.

'The Conference recommends that in furtherance of its purposes, the Preparatory Commission will:

(a) take note of the negotiations separately in progress among certain Member States of the Conference for the collaborative development of a satellite launcher.

(b) consider the closest co-operation between this organization and the contemplated European Space Research Organization (ESRO).'

Some of the Member States would have gone further and merged the two organizations but this was firmly resisted by the UK not only because they were regarded as fulfilling complementary but different functions but also on other grounds. The launcher development would not only be important for space science but also for space applications which were not within the purview of ESRO and which might involve further countries outside Europe. Finally it was considered that the launcher development organization would be largely run by government representatives concerned with development and production. ESRO on the other hand was being conceived as an organization providing resources for pure scientific research for use by independent scientific groups. It should therefore be as free as possible from detailed governmental interference in its scientific work. These views prevailed.

The final form of the Agreement having been drawn up for signature, the remaining business, apart from the signing formalities, was to choose the location of the first meeting of the Preparatory Commission. Paris was selected and Auger was nominated to organize the meeting as 'Chargé de preparer'.

The signing ceremony was much enjoyed by the Chairman as he found himself in an unusual role, calling each head of delegation in turn. Some countries signed without reservation as to ratification. Strangely enough these were all monarchies, Belgium, The Netherlands, Norway, Sweden and the UK. France, Denmark, Italy and Switzerland signed subject to ratification after a few months. West Germany was unable to sign as the delegation did not possess full powers, but signed later in Berne. Representatives from Spain attended the conference first as observers but later as full participants and signed the Agreement, a copy of which is given in Annex 9.

8 Satellite launcher development in Europe – beginnings of ELDO

A little later, on 30 January 1961, the first meeting on satellite launcher development was held at the Maison de l'Europe, Strasbourg, chaired by Sir Peter Thorneycroft, the British Minister for Supply. The organizational problem here was necessarily different from that involved in the formation of ESRO because one country had already carried out a large part of the major development. Discussions had taken place with France prior to the meeting so that the proposals made were presented jointly by the UK and France. Sir Stewart Mitchell, for the UK and General Aubinière for France gave an account of these discussions leading to the proposals which were summarized as follows.

(a) A joint European space launcher organization should be set up.
(b) A three-stage space-launcher should be developed.
(c) This should consist of the following three stages.
    (1) Blue Streak
    (2) A French Rocket
    (3) A third stage perhaps to be developed by a European consortium.
(d) The work should be done by joint teams.
(e) All the participating countries should share in the management, work, cost and results.

As an indication of a large potential application of a launcher system Captain C.F. Booth of the UK General Post Office made a statement on telecommunications. This meeting was essentially one to provide information on which further discussion between governments could take place. We shall henceforward use the acronym ELDO for this proposed organization.

9   Progress of the Preparatory Commission

Turning back now to ESRO, the first meeting of the Preparatory Commission was held in Paris on 13–14 March. A Bureau was elected to carry on business between meetings. It consisted of: President, H. Massey; Vice-Presidents L. Broglio (Italy) and H.C. van de Hulst (The Netherlands); Executive Secretary P. Auger (France). Two interim working groups were set up. One under the Chairmanship of L. Hulthèn (Sweden), concerned with Scientific and Technical matters (GTST) and the other under the Chairmanship of A. Hocker (West Germany) on Administrative and Financial matters (AWG).

It was essential that the first of these Working Groups should report as quickly as possible because the second Group could only begin seriously when the outline of the scope and nature of the scientific programme had been worked out. Fortunately the Scientific and Technical Working Group had available the proceedings of the Technical Discussion Meeting held by the European Study Group on 6 October 1960. In fact this provided the main basis for the new deliberations. Five groups were set up, on the Scientific Programme (Chairman, B. Hultquist, Sweden), the Payload Engineering Unit (A.W. Lines, UK), Data Processing and Tracking (J.C. Pecker, France) and Vehicles and Ranges (J. Vandenkirkhove, Belgium). To prepare an extensive report on which the Administrative and Financial Group could act, in time for the second meeting of the Preparatory Commission at Delft on 17–18 May 1971, the Working Group met in London on 8–9 May.

The proposed scientific programme in the steady state was divided into three sections, (1) a vertical sounding rocket programme using rockets of the Skylark and smaller types, with almost 20 firings a year, especially from the northern auroral zone (2) a light satellite (comparable with the Ariel satellites) programme for the study of the ionosphere, the solar XUV spectrum particle radiation and magnetic fields, at a rate of two to three launches per year beginning about four years after establishment of ESRO and (3) a heavy satellite and space probe programme with launches beginning two years later at a rate of 1–2 a year. The latter programme would require launchers of the Blue Streak, or comparable, type. It was specifically to include a stabilised satellite for astronomical observations and a vehicle in orbit round the moon.

In discussion of the programme complications were introduced by the different attitudes about what constituted a 'big' project. The larger countries felt that this implied a project beyond their own national means but allowance had to be made for the need to involve the smaller countries effectively. For this reason the sounding rocket and light satellite programmes were included. The scope for both 'big' and 'small' programmes in ESRO was a further factor which distinguished it from CERN, a factor which sometimes led to difficulties.

Rather similar issues arose in deciding how the scientific work would actually be carried out – in other words the familiar problem encountered in the setting up of NASA (see Chapter 4, p. 57), how much externally and how much in-house? The UK scientists, who were mainly from University research groups, were very anxious that they would be able to work with ESRO in much the same way as in the UK Skylark and Ariel programmes. They were well aware of the problems arising with NASA and they took a very firm line against the scientific work being carried out mainly by ESRO staff. For this they were accused by some delegates very much involved with CERN, in which the internal staff played a major role in research, of being anti-collaborationist. This was ill-informed criticism because each space vehicle would involve experiments from more than one country and the integration of the payloads would involve collaboration enough! Some of the smaller countries would have preferred rather more in-house science to be carried out. Eventually the UK view prevailed although it was accepted that a small scientific unit would be set up in close association with the Payload Engineering Unit. This unit would include research fellows and a small permanent staff and it was hoped that it would carry out space experiments largely in association with scientists from the smaller countries.

A closely related question was that of financial support for the instruments required for the scientific experiments. It will be remembered (see Chapter 4, p. 57) that it was arranged that NASA would cover the cost of all experiments accepted by the agency whether from outside research groups or from within the agency. To maintain the independence of the main users it was suggested that experiments by scientists of one member state, proposed and accepted by ESRO for flight in sounding rockets and small satellites, should be paid for from national sources within that member state. Only the large scale experiments carried in big satellites and space probes would be paid for by ESRO. This arrangement was agreed during the work of the Preparatory Commission.

In formulating these plans insufficient weight was given to the importance of a permanent group of scientists, within an organization largely staffed by engineers, who were to provide services for use by European scientists covering a wide range of disciplines. While still preserving the major scientific role of the outside users the size of the ESRO science laboratory should have been larger. This became especially serious when ESRO was later transferred to ESA (the European Space Agency), see Chapter 11, in which scientific research formed only a minor part of the activities.

The report of the GTST confirmed the Study Group proposals for the Payload Engineering Unit which would be responsible for design of satellite structures and all common services for the experimental equipment. For the larger projects, such as the astronomical satellite and the lunar orbiter much of the design and development would probably fall to the Unit. Based on the experience of A.W. Lines of the Royal Aircraft Establishment (RAE) it was estimated that the staff of the Unit would build up to about 300 scientists and engineers, with supporting staff and a recurring cost of about £2 million per annum. About £4 million per year would probably be spent on extra-mural expenditure in European industry.

The Study Group proposal for a large centre for data collection, collation and analysis was also confirmed. It would include responsibility for

Table 6.1. *Estimated costs of ESRO programme, years 1–8 in £ million*

	Year No							
	1	2	3	4	5	6	7	8
Payload Engineering Unit	2.9	5.0	6.7	6.4	6.3	6.3	6.3	6.3
Launch costs — Sounding rockets Northern range	0.7	0.4	0.3	0.3	0.3	0.3	0.3	0.3
Sounding rockets – other ranges	0.3	0.3	0.3	0.3	0.3	0.3	0.3	0.3
Small satellites				1.3	1.9	1.3	1.3	1.3
Space probes						1.3	2.0	2.0
Large satellites						1.4	0.7	0.7
Total launch costs	1.0	0.7	0.6	1.9	2.5	4.6	4.6	4.6
Vehicles — Sounding rockets	0.1	0.3	0.3	0.3	0.3	0.3	0.3	0.3
Light launchers				1.3	1.9	1.3	1.3	1.3
Heavy launchers						2.7	2.7	2.7
Total vehicle costs	0.1	0.3	0.3	1.6	2.2	4.3	4.3	4.3
Tracking and telemetry	0.1	0.9	1.3	0.7	0.8	0.8	0.8	0.8
Data handling	0.1	1.7	1.6	0.3	0.3	0.3	0.3	0.3
Scientific fellowships, etc.	0.1	0.1	0.1	0.1	0.1	0.1	0.1	0.1
Grand total	4.3	8.7	10.6	11.0	12.2	16.4	16.4	16.4
UK contribution	1.1	2.2	2.6	2.7	3.0	4.1	4.1	4.1

operation of some four ESRO tracking stations, designed to be complementary to the NASA network.

An additional proposal to meet requirements for the sounding rocket programme in the northern auroral zone was made by Sweden, that ESRO operate a sounding rocket range at Kiruna, and this was also recommended.

Data were obtained on the costs of launching vehicles to implement the programme, as well as range and launch costs.

The report considered all these matters in much detail and submitted cost estimates for the whole operation over a period of eight years. These are summarized in Table 6.1.

The main reason for the increase in the overall budget from the previous £10 million (see p. 117) to £16.4 million was an increase from £4 million to £8 million in estimated cost of launchers due partly to increased estimates for purchase and for launching costs and partly to taking a lower, 50%, success rate.

On this basis, assuming a percentage contribution determined by gross national product, the UK contribution would be 25% as indicated.

10   UK overall space science programme – agreed financial limits

With these estimates to hand the BNCSR were able to consider the funds required to execute a balanced programme. The discussions were based on a note submitted by the Chairman for the meeting on 3 July 1961. He considered the ESRO proposals to be realistic and judged that the launching programme could hardly be reduced without endangering the viability of the organization as a whole.

Massey then went on to say that in his view, even with the increased ESRO estimates there was still room, within a total of £5–6 million put forward (see Table 6.2), for the maintenance of the other main elements in the UK programme. These elements were complementary to, and do not duplicate, the ESRO programme. For example, in research with sounding rockets, ESRO should confine its efforts largely to auroral and other high latitude studies, leaving work in the southern hemisphere to the well-established UK/Australian Skylark programme.

Finally, he considered that, provided NASA continued to make launchers available and to launch them without charge, it would be very short-sighted not to continue with a relatively modest UK/NASA programme.

On this basis he submitted the following table of estimated UK annual costs for years 1, 3 and 6 of the ESRO programme, (Table 6.2).

Table 6.2. *Estimated annual cost of overall UK space science activities in £m*

	ESRO year		
	1	3	6
UK share of ESRO	1.1	2.7	4.1
National and bilateral experiments and their* instrumentation	0.3	0.5	0.75
Sounding rockets and firing costs	0.25	0.25	0.25
Satellite engineering for UK/NASA programme		0.15	0.20
Tracking and data handling	0.2	0.3	0.35
	0.75	1.2	1.55
Total	1.85	3.9	5.65

*This item includes all UK instruments for use or testing in sounding rockets, or for flying in NASA satellite or ESRO sounding rockets or small satellites.

The BNCSR agreed with this programme and these cost estimates and so reported to the Steering Group who in turn reported favourably to the Minister for Science. He referred the question of the size of the budget to the Advisory Council for Scientific Policy of which Lord Todd was the Chairman. On the basis of a report prepared by a small study group set up by Lord Todd to consider the matter it was agreed that, considered in isolation, the proposed programme was reasonable in scale and balance in relation to the scientific value of the research to be carried out.

Assuming a continued rate of growth of support for scientific research comparable with that in recent years a space science programme on the scale proposed would not be unreasonable. On the other hand, it would not be reasonable if the necessary growth of other fields of research had to be cut off to provide the funds for the space programme.

On this basis the estimated programme was accepted but from then on a £6 million figure for the 'steady state' was sacrosanct and it was important for the balance of the UK programme that ESRO costs be confined to the estimates given in Table 6.1.

11   Progress towards the establishment of ELDO

This stage was reached by the end of 1961 by which time substantial progress had been made towards the establishment of ELDO. At a meeting at Lancaster House in London on 29 October–2 November 1961 the main plan of the organization was agreed. It would be governed by a

Council constituted of representatives of member states, with a head-quarters in Paris under a Secretary General to supervise the operations of the organization, a Technical Director and an Administrative Director plus auxiliary staff. The first stage of the launcher system, the Blue Streak rocket, would continue to be developed by the UK, while the second and third stage rockets would be developed by France and Germany respectively. Design, development and construction of the first series of test satellites would be the responsibility of Italy. Belgium would play the leading role in the development of the radio guidance system and the ground guidance stations and The Netherlands for that of the long range telemetry link. Development launchings of Blue Streak and of the completed three stage system would take place from Woomera whose facilities would be provided by the Australian Government. The planned date for the first launching of a complete system was 1966. Provision was also made for the study of further programmes. The initial programme was estimated to cost £70 million spread over five years. Of this the UK would contribute as much as 38% even although it implicitly also contributed, apart from the initial expenditure on developing Blue Streak, test facilities at Spadeadam, Woomera and elsewhere.

As planned, the ELDO organization operated essentially as a loose federation of national activities, co-ordinated by the Headquarters staff. Because of the desire to avoid overlap with national facilities, this staff would have none under its own direct control. Again, although the ELDO convention was signed on March 1962 by Australia, Belgium, France, Germany, Italy, Holland and the UK a considerable delay was expected, and did in fact occur, before ratification by the different governments. Although it was agreed to start work as early as November 1961 the problem of co-ordination, which was very important in such a complicated system, was rendered even more difficult by the fact that, until ratification, ELDO had no legal existence.

12 ESRO – the financial protocol

We shall return again to ELDO matters on p. 129. Meanwhile we take up once more the issues which were being faced in the planning of ESRO. The UK was, in early 1962, especially concerned not only with the size of the budget proposed for ESRO, but with the selection of sites for ESRO establishments and the interim appointments of staff who would take up senior posts in ESRO when it at last began.

On the financial side there were intensive discussions, in which Quirk took a leading part, between the different member states. It seemed both to scientists and administrators in the UK that a space science programme, provided it was not on too small a scale to justify European collaboration, was open-ended. It therefore made sense to work from a given budget in planning the nature, scale and rate of execution of the scientific programme. As described on pp. 122 the UK scientists were convinced that within a ceiling of £16.4 million annually in the 'steady' state, a fully adequate programme could be pursued. Eventually a financial protocol to the draft convention worked out by the Preparatory Commission was agreed in good time before the intergovernmental conference to sign the Convention in June 1962.

According to this protocol (see Annex 10) expenditure during the first eight years of ESRO must not exceed 306 million accounting units (about £110 million) at price levels ruling at the date of signature of the protocol. Provision was made, however, for the Council of ESRO to adjust these figures on certain occasions, in the light of major scientific or technical advances, provided the adjustments were unanimously agreed by the Member States. It was also stated that states party to the Protocol would be prepared to make available to ESRO a sum not exceeding 78 million accounting units (about £28 million) during the first three years of its operation, and not exceeding 122 million accounting units (about £40.5 million) in the following three years.

This was a completely satisfactory outcome for the UK which protected the programme planned by the BNCSR as far as possible, though it was recognized that once ESRO came into being a new situation might arise.

We turn next to the selection of sites for ESRO establishments which got under way early in 1962.

13   The selection of sites for ESRO establishments

The complex of establishments to carry out the various activities as worked out by the Preparatory Commission consisted of a Headquarters, the Space Technology Centre (ESTEC), a small Scientific Laboratory (ESLAB), the Data Analysis Centre (ESDAC), a rocket range in the Arctic (ESRANGE) and at least four suitably disposed tracking stations (ESTRACK). Apart from the last two there was no a priori reason why the remaining three should not be located in a single institute rather like CERN. Indeed the Preparatory Commission had already recommended that ESLAB should be located close to ESTEC. However, unlike CERN which is concerned with a single scientific discipline there was not the same clear necessity for ESTEC, ESLAB and ESDAC to be located on the same site. From the purely scientific point of view this would certainly be a great advantage but at an early stage it became clear that the strong interests of individual countries in having an establish-ment on their soil precluded the desirable solution. There were even some countries who saw no objection in a wide dispersion of establishments.

The UK view was in favour of a moderate degree of concentration with of course the proviso that ESLAB should be close to ESTEC. Location of the latter in the UK was considered to be highly desirable by UK scientists and administrators. It was felt that the case was a good one in view of the unique experience within the UK in organizing and managing activities of a comparable nature and scale. A proposal was therefore made by the UK that ESTEC (and ESLAB) be located at Bracknell, a site 35 miles south-west of London, conveniently close to London airport.

It also had the great advantage of being close (within 20 miles) to the Royal Aircraft Establishment (RAE) at Farnborough, the foremost British laboratory working on the technological side of space research. The extensive experience of RAE could therefore be drawn upon to enable rapid progress to be made in the development of ESTEC. In addition it would be economical to share some facilities between the two laboratories.

The site was also close to the Radio Research Station at Slough with its Minitrack installation and orbital-data recording centre. Furthermore, the area was one in which there already existed small electronic, electrical and mechanical manufacturing and development facilities with experience in dealing with government contracts. Large supplies of liquid oxygen or nitrogen of importance for environmental test facilities were also available from British Oxygen at Wembley.

Proximity to a university was clearly important. The Bracknell site was within 25 miles of the University of Reading and 45 miles from Oxford while the Institutes of the University of London (University College and Imperial College), which had been engaged in research in space science since 1953, were readily accessible, London being only 35 miles away.

In addition, the Meteorological Office had a large laboratory at Bracknell. As the educational, housing, transport and recreational facilities in the area were also good it was felt that there was a strong case for establishing ESTEC at Bracknell.

No corresponding proposals were made for the Headquarters. As far as the latter was concerned, there was a general feeling that it would be found in Paris, in which city the Headquarters of ELDO was to be sited. There were misgivings on the part of British scientists about close association between ESRO and ELDO. They feared that, because of the very different aims and nature of the two organizations, science within ESRO might be dominated by engineering and technological considerations. These fears were not expressed by scientists in many of the other countries and so in those countries greater weight was given to the desirability of encouraging collaboration by locating the Headquarters of each organization in close proximity. This naturally gave great weight to the choice of Paris for the site of the ESRO Headquarters and the French Government made clear at an early stage that they would welcome this.

Firm proposals were submitted by the Belgian and Dutch Governments for location of ESTEC at Brussels or Delft respectively. These were the main competitors with the UK for this site although submissions were made by other member states – Italy near Rome; Switzerland near Geneva; France near Orly and Germany near Munich.

For ESDAC the West German Government offered a site near Darmstadt. An alternative proposal was submitted by Switzerland near Geneva and, less firmly, the UK in Bracknell.

The problem of selecting the sites was a very difficult one and it was necessary to take into account many factors apart from purely scientific ones. On 27 January 1962 the Bureau of the Preparatory Commission asked O. Dahl, the well-known Norwegian scientist, to review for them the proposals made, taking into account technical, economic, scientific, social and financial points of view. This very demanding task was carried out very thoroughly by Dahl who prepared a report, a copy of which is given in Annex 11. He visited all the member states that had made proposals and discussed with the appropriate agencies the whole question of site location.

After stressing the difficulty of arriving at a decision likely to command a consensus Dahl who, as with other Scandinavians, was impressed with the need for reasonable concentration of the establishments, recommended that on the grounds of economy and efficiency the Headquarters, together with ESTEC and ESLAB be located on the proposed site in Delft and ESDAC with ESTRACK in Darmstadt. He pointed out that the temporary quarters offered in the building of the applied physics department at the Technical University in Delft were ample and very suitable to enable ESTEC to start up quickly. In his view staff housing would not be difficult and special school facilities for the children of ESRO staff could be provided. There was a problem of shortage of skilled labour but it was not considered to be serious. The Darmstadt site for ESDAC would be close to German national facilities for computing and data processing and was centrally placed in relation to the various member states.

Dahl further remarked that he was at first in favour of the Headquarters being in Paris so that there would be continuity between the work of the Preparatory Commission and the final ESRO as well as close contact with the

ELDO Headquarters. He finally judged it to be more important to have the Headquarters and ESTEC in one location.

On 26 March the Dahl report was first considered by the Bureau at a meeting in Paris. Following this Bureau meeting, further discussion took place when the Heads of Delegations joined the Bureau. This expanded meeting was a very difficult one, conflicting interests making it hard to see how to arrive at an agreed solution taking into account the scientific as well as the political aspects of the matter. Each delegate was asked to express his views on the general principles which should govern the selection. The Scandinavian countries, especially Sweden, alone favoured concentration of establishments, though the UK favoured only moderate dispersion. Almost all were in favour of close association between ESRO and ELDO but there was no clear consensus as to whether ELDO and ESTEC should be located close together. When it came to more detailed considerations it appeared that the UK could only count on support from the Scandinavian countries for location of ESTEC at Bracknell. The UK delegate was further perturbed by the strong support which was forthcoming for ESLAB to be located in Italy despite the statement in the Convention that it should be close to ESTEC (there was no suggestion at the meeting that ESTEC also should be in Italy). No final decisions were made as some delegations did not have instructions from their governments. It was arranged for a second joint meeting of the Bureau and Heads of Delegations to be held in Paris on 4 April 1962.

The tone of the meeting had seemed so unsatisfactory in submerging scientific to political interests that, on their return, Quirk and Massey arranged for the matter to be discussed at the BNCSR and the Steering Group at a joint meeting on 30 March. As a result of this meeting it was agreed that the UK delegate at the meeting on 4 April should emphasize the UK view of the essentially scientific character of ESRO. While maintaining that Bracknell was the best site for ESTEC he should express willingness to accept a site in Belgium or Holland chosen in an area of strong local scientific activity. Finally he was to reject the possibility of separating ESLAB from ESTEC.

As is often the case, the second meeting on 4 April took place in a much calmer atmosphere. Quirk presented a strong case for Bracknell as the ESTEC site and no delegation was able to dispute this on scientific grounds. There was some feeling that such a location would give the UK too great an advantage as it was already much further advanced in space science and technology than the other member states. There was also the desirability of locating a major establishment in one of the smaller but very active countries. On these and possibly other grounds Bracknell was rejected and it became a straight choice between Brussels and Delft for ESTEC. A secret ballot was held and Delft was chosen by a majority of 6–4, Belgium and Holland abstaining.

A vote then followed on the location of ESDAC as between Darmstadt and Geneva and the former was chosen by a majority of 8–4.

There was unanimous agreement that Paris be the site of the ESRO Headquarters.

A great deal of argument, often ill informed, took place about the site of ESLAB. It was finally accepted that the agreed convention which stated that a research laboratory 'on the minimum scale deemed necessary by the Council' should be set up near ESTEC could not be disregarded but that the Council would be able to establish another laboratory for a somewhat different purpose. Accordingly it was proposed that a Research Laboratory should be established in Italy of a scope and size to be decided by the Council. This was agreed by 8 votes to 4. The UK delegate made it clear that the funds for such an establishment must come from within the financial ceilings already laid down.

These decisions, including that concerning the laboratory in Italy, were formalized at the next meeting of the Preparatory Commission in Rome on 9–11 May 1962. It was decided to set up a small technical sub-committee under J. Bartels to consider the function of the laboratory in Italy. In the view of UK scientists it could fulfil a useful purpose by studying, both experimentally and theoretically, basic processes and phenomena related to scientific investigations, being carried out or proposed, using the techniques of space research. Thus aspects of plasma physics and of atomic physics are very relevant to the interpretation of space observations. Such a laboratory need not be very expensive and could make a useful contribution. Alternative possibilities such as research in advanced propulsion would require much greater resources to be effective and in any case were more appropriate to an organization such as ELDO.

The value of a laboratory on these lines was emphasized explicitly by Massey at a meeting of the Scientific and Technical Committee in Paris on 29–30 October 1962. As he would not be able to attend the planned meeting of the Bartels sub-committee he was asked to express his views as Chairman of the Preparatory Commission. The full text of his statement is given in Annex 12. The report submitted by the Bartels Sub-committee to the Scientific and Technical Committee at its meeting on 15 November 1962 closely followed the views of the UK scientists as expressed by Massey. Thus as examples of research subjects which might be followed by the Italian laboratory (now known as ESRIN, the European Space Research Institute) the Sub-committee gave the following:

(a) physico–chemical phenomena in space such as reaction rates and surface phenomena in high vacuum,
(b) plasma and magneto–dynamic phenomena in gases,
(c) relativistic and gravitational phenomena significant in the mechanics of celestial artefacts.

From the four possible locations proposed by the Italian delegation, Milan, Florence, Rome and Naples there was a general preference expressed for Rome or Milan because of the comparative ease of communication with these centres. In particular a site near the nuclear research centre at Frascati, where research in plasma physics was carried out, was singled out as having many advantages.

While there was acceptance by the Italian delegation of the general nature of the research programme of ESRIN the scale of the laboratory was still a matter for very extensive argument. At the Preparatory Commission meeting in Paris on 3 May 1963 the Italian delegation, in the course of a lengthy discussion, expressed their dismay at the progress being made about planning for ESRIN, the leader of the delegation, Bettini, quoting Massey's statement referred to above about the value of the laboratory. With some reluctance the Commission agreed to set up a further working group reporting directly to it, under the chairmanship of van de Hulst, to complete the necessary planning.

This group reported to the meeting of the Preparatory Commission in Paris on 11–12 July 1963. At this meeting Massey took the opportunity to make his position clear in view of the references which had been made to his statement to the Scientific and Technical Committee on 30 October 1962. Reiterating his view of the value of a laboratory on the lines proposed he went on to point out the ESRIN would not be a production or development establishment and its scope and size should be related to its role as an agency engaged only in pure research. It should be comparable to university laboratories engaged in similar research work. There should be adequate theoretical and experimental provision roughly in the basis of 1:4 within a total scientific complement around 25. He strongly urged that the criterion for a successful laboratory must be quality and not cost.

In its report to the 11th meeting of the Preparatory Commission, referred to above, the van de Hulst working group recommended an increase in total staff of ESRIN, including all administrative staff, from 75 to 85 and assumed that the annual cost of equipment and personnel would be

140,000 NF ± 20%. These recommendations were accepted after some debate by the Italian delegation and formed the basis for later detailed planning following the appointment of a Director, H. Jordan (see p. 136). There remained the choice of site. Ultimately it came to be a choice between one near Rome, at Frascati, and one near Florence favoured by the Italian delegation. However, Jordan presented a strong case for Frascati because of the proximity to other scientific research institutions in that area and this was eventually accepted at a meeting of the Council on 24–25 March 1965.

14   ESRO network of tracking and telemetry stations

At its eighth meeting held in Montana, Switzerland on 29–30 June 1963 the GTST set up a sub-committee to study further the location of ESRO tracking and telemetry stations. The members were Batista (Spain) Cheynel (France) Pressey (UK) Vigneron (Belgium) and a Netherlands representative. It reported on 4 February 1964, recommending that four ESRO stations be established, in Belgium at a site (Quatre Bras) about 30 miles from Brussels[2] for tracking and telemetry reception, and in Alaska (near Fairbanks), the Falkland Islands[3] and Spitzbergen for telemetry only. In addition use could be made, by arrangement with France, of the five stations established by the French CNES namely, Pretoria and Hammaguir for tracking and telemetry, Bretigny, Brazzaville and Ougudougou for telemetry only. Reciprocal arrangements could also be worked out with NASA for the use of their stations when practicable. The proposed system would provide 80% coverage for satellites in orbits with inclination greater than 60° and altitude greater than 500 km. However, there was need for an extra telemetry station in the auroral zone and for this the sub-committee recommended that Norway be approached for use of the facilities at Tromsö. Failing this a small station should be established at Kiruna.

These proposals were accepted at the fifth meeting of the ESRO council in Paris on 27 November 1964, subject to satisfactory arrangements being made with Norway about Spitzbergen. These were concluded early in 1965.

It was planned that the ESTRACK network should enter into service at least six months before the launching of the first ESRO satellite scheduled for April 1967.

15   The organization of ESRO and its establishments – senior staff appointments

The planned organization was based on three directorates reporting directly to the Director General. These were to be headed by a Technical Director, a Scientific Director and a Director of Administration.

The Technical Director was to be responsible in particular for ESTEC of which at first he was also the Director. Figure 6.2 shows the planned organization within ESTEC. It included three Directors, of Applied Research, Space Craft and Sounding Rockets respectively who reported directly to the Director ESTEC. Below this there were 13 posts at Assistant Director level. Two of these reported directly to Director ESTEC, the remainder through the three Directors as indicated. In this scheme ESTEC assumed responsibility for both the operation of ESTRACK and of ESRANGE.

The Scientific Director was to be responsible for ESDAC, ESRIN and ESLAB, the planned organization being shown in Figure 6.3. Before agreement was reached on the plan for ESDAC there was very considerable discussion. The original proposals included in addition to the two divisions shown in Figure 6.3 on Data Processing and Data Analysis, a Mathematics Research Division. This was strongly opposed by both the UK and France as representing relatively too large a commitment to research which might be of little benefit to the main aims of ESDAC. In March 1963 the GTST set up a working group to consider the matter and on the basis of their report it was finally agreed to eliminate Mathematics Research as a separate division.

Three Assistant Directors, for Personnel and General Administration, Finance and External Relations and General Services, reported to the Director of Administration.

The interest of member states in staff appointments which until ratification of the Convention would be provisional but would then become definite, was comparable with that shown in the selection of sites. On 18 May 1962 Massey, in his capacity as President of the Preparatory Commission, wrote to the Heads of the Delegations requesting nominations for the post of Executive Secretary under the assumption that, after ratification, the incumbent would automatically become the Director-General of ESRO. In reply, Auger was nominated by several delegations, including the UK, and he was elected unanimously, without opposition.

A mechanism of selection for other posts was set up in which the Bureau acting with Heads of Delegations was constituted as an appointments commission. Next to the Director General were the three posts, of Scientific Director, Technical Director and Director of Administration. There was no difficulty in filling the post of the prospective Technical Director as it was agreed without dispute that the UK experience was so much greater than that of the other member states on the technical side. A.W. Lines from RAE was appointed without hesitation. The situation was by no means so simple for the other two posts.

All were agreed that R. Lüst from Germany would be a most suitable

134

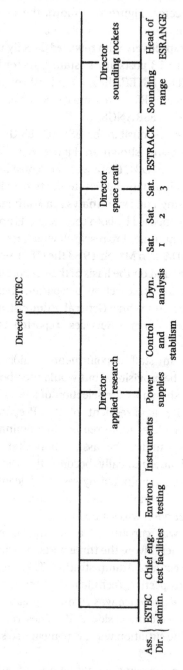

Fig. 6.2.

candidate for the Scientific Directorship but he was loath to give up his association with the Max Planck Institute. Eventually an arrangement was made whereby he was appointed on a part-time basis, devoting his time equally between the Preparatory Commission and the Institute. In fact Lüst did not wish to continue in this post after the establishment of ESRO and it remained vacant until the appointment of A. Bolin of Sweden in March 1965.

It was believed, quite wrongly, that the UK would be satisfied with one of the three top level posts. In fact there was a firm desire in the UK that a UK nominee should become Director of Administration. It was felt that it was important that the holder of this post should have had experience in dealing with administrative (including financial) problems associated with the advanced technology involved. A very well-qualified candidate was available. T.M. Crowley of the Ministry of Aviation, and his name was put forward for the post. This caused considerable embarrassment as Auger had hoped that Mussard, his assistant throughout the deliberations of the Preparatory Commission, would be appointed. Other member states did not agree immediately that the UK should have two of the senior appointments but it was pointed out firmly by Quirk, who took a strong line throughout the negotiations on this matter, that the UK while contributing the largest fraction to the budget did not have any of the establishments on its soil and should therefore be entitled to an extra post at a high level. It was arranged that in any case Crowley should be seconded part-time from 1 September 1962 to assist Auger. Meanwhile the problem of Mussard was solved by creating a new high level post, that of Secretary to the Council, to which he was appointed. There remained the problem of nominees from other countries, including particularly Italy, but with unremitting pressure brought to bear by Quirk, Crowley's appointment was agreed.

Fig. 6.3.

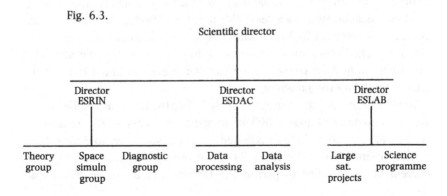

As mentioned above, at first the Technical Director also assumed the detailed responsibility of Director of ESTEC. This was natural in the build up phase of that establishment but once it was operational there was a clear need to separate the two posts. In 1966, Kesselring of West Germany was appointed Director ESTEC.

The first Director of Space Craft Projects was S.R. Shapcott seconded from RAE while the first Assistant Director of Environmental Testing was S.L. Entres also from RAE.

Of the three Directorships reporting directly to the Scientific Director H. Jordan of West Germany, D. Comet of Sweden and A. Trendelenberg of West Germany were appointed Heads of ESRIN, ESDAC and ESLAB respectively.

16   Signature of the ESRO Convention and its ratification

On 14 June 1962 at an inter-governmental meeting in Paris the ESRO Convention and Financial Protocol were signed by Austria and by all the original members of the Preparatory Commission except Norway.

The Convention provided for the operation of the organization by a Council and Director General. Each member state would nominate two delegates to the Council which should meet at least twice a year. A Chairman of Council and two Vice Chairmen would be elected annually and could be re-elected on not more than two occasions.

The Conference directed the Preparatory Commission to prepare drafts of agreements with member states on which ESRO sites were to be located as well as undertaking detailed studies concerning buildings and equipment at these sites. It was also to prepare for recruitment and training of staff and continue to plan the initial programme in consultation with universities and other appropriate institutions in the member states. Finally it was to prepare a detailed budget for the first year of operation.

Signature of the Convention was one thing but ratification was quite another. The Convention would only come into force when ratified by at least six states contributing at least 75% to the total Budget and including all countries on which ESRO sites would be located, namely – France, West Germany, The Netherlands, Sweden and Italy. It was feared that all of the necessary ratification would not be completed even in late 1963, a fear which proved all too prescient.

In view of the long time interval likely before the necessary ratifications had been made to establish ESRO on an operational basis the Secretariat of the Preparatory Commission suggested in a paper submitted to the meeting of the Commission in Paris on 21–22 February 1963 that either the agreement be modified to permit the Commission to undertake actual operations in advance of ratification or some alternative procedure be followed. At the meeting the first possibility was ruled out and the only alternative remained of encouraging member states to incur expenditure on interim operations at their own risk against reimbursement from ESRO, if and when it was established. This possibility was accepted with the proviso that any arrangement with a member state required the unanimous agreement of all the member states, especially as regards the size of the financial commitment involved.

The member states concerned in interim operations included the UK and France, who proposed to provide six Skylark and six Centaure rockets respectively, The Netherlands for site preparation and equipment at ESTEC, Sweden for buildings at the rocket range at Kiruna and possibly Italy if the Sardinia range were to be operative for launching the first rockets.

The whole situation about interim arrangements went back into the melting pot at the next meeting of the Commission in Paris on the 3 May 1963. This was a contentious meeting and many delegates emerged from it in a somewhat depressed frame of mind. It will be recalled, see p. 131, that there was a lengthy discussion about ESRIN at this meeting but the most serious matter which arose was the view expressed by the Swedish delegate that one should not try to overcome delays in ratification by the proposed interim arrangements. He considered that there was plenty for the Commission yet to do such as completing the Protocol on Privileges and Immunities, which was occupying a great deal of the time of the Administrative Working Group, as well as arranging the agreements with the different member states on whose territory an ESRO establishment was to be sited. Following this statement, several delegates expressed their disappointment and although some suggestions were made to try to break the deadlock the meeting broke up without any prospect of agreement being reached on some form of interim activity.

It was clear from this meeting that a great deal depended on speeding up the ratification procedure. Considerable discussion therefore took place in the UK as to when it should ratify to produce the best effect. Already there had been some suspicion about the UK's attitude to ESRO as it had already ratified ELDO some months earlier (March 1963). It was felt that there was a strong case for early ratification. Thus it would show that the UK was clearly in support of ESRO and it would encourage quick ratification by others including possibly Sweden whose doubts might be set at rest. The UK therefore announced on 21 May 1963 that it would ratify as soon as agreement was reached on the Protocol concerning Privileges and Immunities. Following this agreement on 26 July 1963 the UK ratified the

Convention on the 8 August 1963. Switzerland had already done so some months earlier and The Netherlands followed.

The series of ratifications had a stimulating effect but there remained the problem of Germany and Italy. The former expected to ratify in the early part of 1964 but there seemed no prospect that Italy would be able to do so before the end of 1964 at the earliest. Without some special arrangement ESRO could therefore not come into being until this late date, a most depressing prospect. In the end agreement was reached at an Extraordinary Meeting of the Preparatory Commission on 13 March 1964 so that during the period before Italian ratification Italian delegates could attend the meetings of the ESRO Council but not vote. Italian nationals would be eligible for staff appointments in ESRO and could take part in the scientific activities. Decisions about ESRIN during the period would not be made without their concurrence.

This agreement, together with some last minute ratifications, made it possible for the first meeting of the ESRO Council to take place on 23–24 March 1964. By that time all countries participating in the Preparatory Commission apart from Italy, Austria and Norway had ratified. Austria asked for observer status.

17   The first council meetings

The first Council meeting in Paris was formally opened by M. Palewski, Minister d'État Chargé de la Recherche Scientifique et des Questions Atomique et Spatiales, who, after welcoming delegates, took the chair until the first Chairman was elected. On the proposal of The Netherlands delegation Massey was unanimously elected to this post, it being understood that, in view of his long term as Chairman of the Preparatory Commission he would serve only to the end of 1964. A. Hocker (West Germany) and H.C. van de Hulst (The Netherlands) were elected as vice-chairmen. G.B. Blaker was the chief UK delegate with J.F. Hosie, R. St J. Walker and F. Bath as advisers.

Observer status was accorded to Austria and offered also to Norway.

Arrangements were made for the GTST and AWG to be replaced by permanent committees known henceforth as the Scientific and Technical Committee (STC) and Administrative Committee (AC) respectively. Further details of the establishment and early activities of the former are given on p. 140.

The subject which took up most of the attention of the delegates concerned the site of ESTEC. At the GTST meeting in Madrid on 3–4 October 1963 the results of the survey of the ground available at the proposed site near Delft were made available. They were far from satisfactory and showed that building on the site would be very expensive. At the first Council meeting the French and Spanish delegations expressed their concern at this situation but The Netherlands delegation stated that their government was confident of meeting all the technical requirements. It was even suggested by one delegation that the whole question of the site for ESTEC should be reopened, a view strongly opposed by the UK among others. After considerable, somewhat heated, discussion the matter was referred to the next meeting at which a report on the planning of the various establishments would be available.

At this meeting, held on 15–17 June 1964, the Dutch delegation offered an alternative site, again of 40 hectares, at Noordwijk 20 km north of The Hague and 30 km from Schipol. Several of the delegations expressed the view that they could not accept another site until there had been a thorough examination by experts. A report on these lines was called for by 13 July and an Extraordinary Council Meeting arranged for 28–29 July. The report was duly presented and was reasonably favourable, stating that piles would only be required if at all in a few special cases. On the other hand, it did point out that weather conditions were less favourable than in Delft so it would be necessary to prevent sand reaching equipment inside the buildings. Although the Technical Director did not consider that salt or sand would present a serious problem there was nevertheless a lengthy discussion. It was finally agreed that more detailed information be requested about the effect of proximity to the sea and about the consequences, financial and otherwise, if a transfer was made to Noordwijk. This information was available by 1 October 1964 and at an Extraordinary Session of the Council on 22 October it was agreed, informally, to accept the offer of the site at Noordwijk.

18 The initiation of the scientific programme of ESRO

During the period before ratification the GTST proceeded with the planning of the scientific programmes. This proceeded mainly through the Launching Programmes Advisory Committee (LPAC) which was set up at the GTST meeting in Rome on 9 May 1962. Initially, its terms of reference were to propose programmes of payloads for sounding rockets to submit to the GTST but this was soon extended to payloads for satellites, small and large. Unlike the main committees the LPAC was not set up as a representative body but was composed of experts in space science. For this reason the LPAC fulfilled a vital function in what was to be an organization devoted to scientific research. Its first members were Lüst (Chairman), Boyd and Dahl. To advise the LPAC a number of scientific working groups were set up as follows, the Chairman being indicated in each case – Atmospheric

Structure (M. Nicolet) Ionosphere and Aurora (B. Hultquist) Meteorology (Nyberg) Solar and General Astronomy (J.C. Pecker) Interplanetary Medium (C. de Jager) Lunar and Planetary (P. Swings) Cosmic Rays and Trapped Radiation (G.P.S. Occhialini) Geodesy, Relativity and Gravitation (J. Bartels).

The task before the LPAC and its working groups was a formidable one as sooner or later it involved selection between proposals in different disciplines. This was difficult enough in all conscience but it was also necessary in making selection to pay some attention to the principle of 'juste retour' according to which, on the average, each member state should obtain access to the ESRO facilities in proportion to its contribution. Initially this aspect was complicated by the fact that UK scientists had had longer experience in space science than those in other states so that their proposals tended at first to be the most practical.

The LPAC addressed itself not only to the sounding rocket programme but also to the programmes involving small unstabilized satellites, small stabilized satellites and the two proposed large satellite programmes.

Following ratification of the Convention, continuity in the work of the GTST and the LPAC was maintained through the establishment of the Scientific and Technical Committee (STC) to which the LPAC reported. The STC was constituted as a delegate body to which each member state nominated two delegates one of whom it was expected would be a scientist. R. Lüst of W. Germany was elected the first Chairman with B. Peters of Denmark as deputy. The permanent LPAC, consisting of four members on two-year appointments subject to maintenance of continuity, began also under the Chairmanship of Lüst, the other members being Boyd (UK), J. Blamont (France) and de Jager (The Netherlands).

We shall now describe how the various programmes which were planned in the early years developed and evolved.

19 The sounding rocket programme

While it was agreed that ESRANGE would be established at Kiruna in the auroral zone in Northern Sweden, no construction could proceed until ratification by Sweden. If sounding rockets were to be launched by ESRO in 1964 an alternative range would need to be available, from which the rockets likely to be used could be launched. These included the British Skylark and the French Centaure, some of which the respective British and French Governments were prepared to supply in advance of ratification, against later reimbursement, in order that the rocket programme could get under way as soon as possible.

For launching Centaures the French Government offered the use of their range at the Ile de Levant. It was not possible to launch Skylarks also from this range. One possibility was to launch from Woomera as with the British National Rocket Programme but this was unattractive to many potential users in Continental Europe because of the large travel costs involved. The Italian Government offered the use of their range at Salto de Quirra in Sardinia and was prepared to make the modifications necessary for launching Skylark. To assist in this work advantage was taken of the experience gained in this task at the Australian Weapons Research Establishment (WRE, see Chapter 2, p. 12). A team from the embryonic ESTEC visited WRE and two members of the team remained for some weeks to become familiar with the problems involved. In addition a consultant from WRE was seconded to assist in the early planning and preparation directed towards making the first launching possible in June 1964.

Based on this and on the availability of the Ile de Levant range as well as the necessary Skylark and Centaure rockets the GTST drew up the following planned programme for 1964 (Table 6.3).

Table 6.3. *Planned sounding rocket programme for 1964*

Rocket	Experiments	Scientific group concerned	Date of launch	Launching range
Skylark	Spectroscopy of released ammonia cloud	Univ. of Liege (Rosen)		
	Study of ion cloud produced by release of barium in sunlight	Max Planck Institute (Lüst)	June	Sardinia
Centaure	Night airglow	Univ. of Paris (Vassy)	October	Ile de Levant
	Neutral particle density	Univ. of Bonn (Priester)		
Centaure	Solar X-ray observations	UCL (Boyd) and Leicester Univ. (Stewardson)	October	Ile de Levant
Skylark	Electron temperature in ionosphere	(UCL (Boyd)	October	Sardinia
	Radio spectrometer Impedence probe	Ionospheric Inst. (Rawer)		
Centaure	Positive ion concentration	UCL (Boyd)	December	Ile de Levant
	Electron concentration Impedance probe	Ionospheric Inst. (Rawer)		
Centaure	Ion probe	UCL (Boyd)	December	Ile de Levant
	Electron temperature Phase velocity measurement in ionosphere	Ionospheric Inst. (Rawer)		

It will be seen that British scientists were much involved in this first programme. Although most of the launchings did not take place until 1965, the first ESRO rocket was launched successfully from Salto di Quirra on 6 July 1964. It was the first planned Skylark, only a month late. The barium was released at an altitude of 150 km and the ammonia at 200 km. A second Skylark containing identical equipment was also launched successfully a few days later. Thus began the operational phase of the ESRO programme.

The first Centaure was launched on 30 October 1964. Unfortunately, the scientific equipment (including airglow photometers) did not function satisfactorily and a second launching was deferred until this was investigated.

The first launching of a rocket with a UK experiment aboard was that of the SKYLARK fourth in the list in Table 6.3. This took place on 31 March 1965 but was unsuccessful. However a second similar rocket was launched on 3 April and this time everything operated satisfactorily.

Meanwhle G.V. Groves and his group from UCL prepared the ground equipment at the Sardinia range for carrying out the grenade experiment to measure atmospheric winds and temperature. Two boosted Skylarks were launched on 30 September and 2 October 1965 carrying grenades and also TMA to produce a glow which could be tracked. An instrument to release barium vapour provided by the Max Planck Institute for Extra Terrestrial Physics, Garching (Lüst and Neuss) was also included. Both flights were successful and interesting scientific results were obtained.

The first Centaure carrying UK experiments, the third in the list in Table 3, was launched on 14 December 1965. It was only partly successful but provided some information about solar X-rays in the 10–55 Å wavelength band.

The first rocket launching campaign from ESRANGE at Kiruna in Sweden took place in February 1967 (see Figure 6.4). Here again Groves organized the supply of the ground equipment for the grenade experiments, the first two of which took place on 1 and 4 February 1968 with TMA release and were successful.

A summary of all rocket flights under ESRO auspices carrying experiments from UK groups, is given in Appendix D.

20   Small spin-stabilized satellites

These satellites would be very similar to those involved in the joint UK–US Ariel programme (see Chapter 5) and it was natural to enquire whether NASA would be prepared to play a similar part in the launching of the first two ESRO satellites, using the Scout launching system.

At an early meeting the LPAC considered the payloads for these satellites so that in July 1962 when Massey was passing through the United States on his way to Australia to visit WRE he was able, as Chairman of the Preparatory Commission, to have a preliminary discussion with NASA officials, including A. Frutkin who was in charge of foreign relations, about their attitude towards a joint arrangement similar to that for the Ariel satellites. Even though Massey was somewhat embarrassed by the fact that

Fig. 6.4. Launching of a Skylark rocket from Kiruna. Note the weather protection for the launcher.

British experiments provided 75% of the provisional payloads there was an encouraging response from NASA.

It was not until nearly a year later that the selection of possible payloads had been rendered firm enough for further discussion to take place. On 15 May 1963 Auger, in a letter to Frutkin, initiated further discussions by enclosing a list of the experiments as then proposed.

The payloads were chosen so that one (to become ESRO I) satellite was primarily concerned with the study of the polar ionosphere the other (to become ESRO II) with solar astronomy and cosmic ray studies. In both cases the UK participation was very strong especially for ESRO II in which at first there were only UK experiments. However, negotiations proceeded slowly but satisfactorily with NASA and a Memorandum of Understanding was signed in July 1964.

According to the Memorandum, ESRO would provide the scientific instruments and the space craft to be launched by means of Scout rockets. These would be provided by NASA who would also carry out the launchings. This arrangement was exactly similar to that which applied to the launching of Ariel 3 (see Chapter 5, p. 96).

The solar astronomy and cosmic ray satellite (ESRO II) would be the first to be launched, probably in 1967, with ESRO I following shortly afterwards.

The final payload of ESRO II ended up by including some small input from other than UK scientists but was still substantially British. The actual experiments were as follows.

(a) Study of the intensity of solar X-rays 1Å–70Å. This involved co-operation between the Department of Physics at University College London (Boyd) the University of Leicester (E.A. Stewardson and K. Pounds) and the Observatory Sonnenborgh at the University of Utrecht (C. de Jager).

(b) Three experiments from the physics department, Imperial College London (H. Elliot). These involved measurements (i) of solar protons and inner radiation belt protons and (ii) of the ratio of the intensity of cosmic ray alpha particles to protons of the same magnetic rigidity and of cosmic ray threshold rigidities as a function of time, as well as (iii) routine monitoring of the trapped radiation.

(c) Measurement of the primary cosmic ray flux of high energy β-particles. This was proposed and would be carried out by the physics department of the University of Leeds (J.G. Wilson and P. Marsden).

(d) Measurement of flux and energy spectra of protons with energy between 35 and 1000 MeV. This experiment was the responsibility of the Centre d'Etudes Nucléaires de Saclay.

The planned orbit for ESRO II was of 350 km perigee and around 1100 km apogee.

The final experiments accepted for ESRO I were as follows.

(a) Measurement of the flux and energy spectra of electrons and protons over a wide energy range. This would be carried out in collaboration between the Ionosphere Laboratory at the Technical University of Denmark (Rybner), the Kiruna Geophysical Observatory (B. Hultquist) and the Radio Research Station, Slough (J.A. Ratcliffe).

(b) Two experiments from the department of physics at University College London (Boyd) concerned with measurements (i) of electron temperature and concentration and (ii) of positive ion composition and temperature.

(c) Auroral photometry. This would be carried out by collaboration between the University of Oslo (Omhold) and the Department of Applied Mathematics at Queen's University, Belfast (Bates).

The planned polar orbit was of 275 km perigee and 1000–1500 km apogee.

It will be seen that in these satellites the UK scientists were using their experience in ionospheric, solar X-ray and cosmic ray experiments carried out either in the National Skylark Programme or in Ariel I.

The first attempt to launch ESRO II on 20 May 1967 was unsuccessful owing to a Scout rocket failure but eventually both satellites were launched successfully, ESRO II, the space flight model, renamed Iris, in May 1968 from Cape Kennedy and ESRO I, renamed Aurora, from the Western test Range at Vandenberg Air Force Base, California on 3 October of the same year, both using Scout launchers. On the whole, the scientific equipment aboard worked well and provided interesting and important new results, some of which will be referred to in Chapters 13, 14 and 15. In fact the performance was so satisfactory that the back-up model for ESRO I, ESRO IB renamed Boreas, was launched with a similar payload by a Scout vehicle in 1969 though on this occasion the cost of the launch was covered by ESRO and not by NASA. Once again the satellite and its payload performed well. As we shall see below, the development of ESRO II was also used as a basis for a partial substitute for the proposed stabilized TD-2 satellite when that programme was cancelled.

21   The first deep-space probe – HEOS I

At an early stage the LPAC recommended that the first ESRO scientific programme should include the launching of small spin-stabilized satellites in highly eccentric orbits to function as space probes from which the properties of the Earth's outer environment, including the magnetosphere and the solar wind, could be studied. It was planned that the satellites could be launched by Delta rockets from Cape Kennedy into orbits with apogee distances as large as 225,000 km, and an inclination near 30°.

The payload for the first of these, HEOS I, was approved in March 1965. As for ESRO I and II it included a substantial contribution from UK scientists, in this case from the Cosmic Rays and Space Physics Group of Imperial College under Elliot. They provided a triaxial fluxgate magnetometer for investigation of the interplanetary magnetic field, a cosmic-ray telescope to measure the anisotropy of cosmic rays in interplanetary space and a low-energy proton telescope for the study of suprathermal particle emission from solar flares. Other instruments aboard were concerned with solar wind analysis (Universities of Rome and of Brussels), cosmic-ray observations (Centre d'Etudes Nucléaires, Saclay), primary electrons (University of Milan and CEN Saclay) and barium-cloud release at about 10 earth radii for measurement of electric fields (Max Planck Institute, Garching). The orientation of the space-craft spin axis was adjusted by radio command.

HEOS I was launched successfully on 5 December 1968 and the scientific instruments worked well. Reference will be made to some of the results obtained by the Imperial College group, in Chapter 14.

A successor to HEOS I was recommended by the LPAC on 28 September 1967 and accepted by the STC in February 1968 to be launched in an orbit of comparable geometry but at a near-polar inclination. It included again instruments for interplanetary magnetic field measurement from Imperial College and five or six experiments from scientists in other countries including one to study intermediate energy particles provided by the Space Science Department of ESTEC under D.E. Page. This was successfully launched as HEOS II in January 1972.

22   The small stabilized satellites TD-1 and TD-2

The LPAC in planning the initial scientific programme attached high priority to the development of comparatively small satellites which would be stabilized to permit steady pointing towards selected directions. As it would be necessary to use the Thor–Delta rocket system to launch these satellites they were soon referred to as the TD satellites. The availability of such vehicles would greatly increase the scope for astronomical observations as well as certain experiments in geophysics and solar physics. It was planned that the satellites would be more or less of standard form into which different payloads could be incorporated. In the early days in which financial limitations were not so obvious it was hoped to launch a series of these satellites, perhaps at the rate of one or two a year. Even as late as March 1965 the LPAC proposed an eight-year programme which included the launching of six small stabilized TD satellites and four small space probes! The first four TDs were recommended for stellar, solar, ionospheric and geophysical studies respectively but at this stage only the first was agreed by the Council. Nevertheless, there was considerable scientific emphasis on TD-2 and TD-3 in relation to observations at the next Solar Maximum.

Strong interest in the TD-1 programme was exhibited by the astronomical group under H.E. Butler at the Royal Observatory, Edinburgh, and on 30 September 1963 they presented their proposal in preliminary form (after consideration by the Solar and Stellar Astronomy Working Group) to the BNCSR. This was to build up pictures of the sky in about six months with a spatial resolution of a few arc minutes. A similar proposal had been submitted by the Institut d'Astrophysique of the University of Liège and it was proposed to merge these in a single collaboration.

The feasibility of these experiments depended on the stabilization which could be achieved in practice. A preliminary design study of a suitable vehicle was made, under contract, by RAE Farnborough and early in January 1964 the reports of this work showed that suitable stabilization could be realized, namely three-axis stabilization with one axis pointing to the sun within $\pm 5''$.

The joint Edinburgh–Liège proposal was accepted as a major experiment for inclusion in the preliminary payload of TD-1 by the STC at its meeting on 1 October 1964. Additional experiments included an infrared scan of the sky in the wavelength range 1.2–3.2 μm using the same mirror system, proposed by Liège, together with six other experiments which were as follows. (1) Ultra-violet spectrometry of the sky in the wavelength range 2000–3,000 Å (Space Research Lab., Utrecht); (2) spectrometry of primary charged particles and (3) spectrometry of celestial X-rays (Centre d'Etude Nucléaires, Saclay); (4) solar γ-rays in the energy range 80–500 MeV (University of Milan); (5) solar X-rays in the energy range 20–700 keV (Space Research Lab., Utrecht), and (6) celestial γ-rays in the energy range 70–300 MeV (Centre d'Etude Nucléaires, Saclay and Max Planck Institute, Garching). Nevertheless, the Edinburgh-Liège experiment was much the largest, including a Cassegrain telescope of 25 cm diameter.

A further technical problem had to be solved for it would be necessary in carrying out the Edinburgh–Liège programme to operate with the satellite in full sunlight. This required the design of a baffle system which would reduce the intensity of the sunlight at the detector by a factor of 10^{17}. A contract was placed by ESRO with a small company, formed by a number of members of the staff of the physics department at UCL, of which A. Boksenberg was Chairman. Thanks especially to the work of M. Esten at UCL, a Monte Carlo method of ray tracing was devised and used to work out a suitable design. As there was no means of checking this in the laboratory the entire success of the project depended upon this work which in the event proved to be completely satisfactory.

While these developments were proceeding for TD-1, comparable interest was displayed in the payload for TD-2. There was considerable agreement about the inclusion of a topside sounder to the exclusion of three experiments concerned with the observation of fast particles. At first this was accepted but later the decision was reversed and by 1968 a payload was agreed which included the following projects: (1) spectral and spatial resolution of solar X-rays in the 1.5–25 Å wavelength range (University of Leicester); (2) positive ion concentration, composition and temperature (University College London); (3) mass spectrometry of neutral gases (Bonn University); (4) the energy distribution of electrons and ions in the energy range 1–150 keV (Kiruna Geophysical Observatory); (5) the study of galactic and solar particles in the energy range 13–180 MeV (Utrecht University Observatory), 0.2–30 MeV protons and 2.5–110 MeV particles (Max Planck Institute, Garching); (6) Spectral and spatial resolution of solar ultra-violet lines in the wavelength range 150–630 Å (University College London) and (7) determination of the concentration of O_2 in the earth's atmosphere in the altitude region 100–200 km from radiative absorption measurements (University of Munich).

These two satellites and their complicated payloads were the most complex which either had been built or were in development in Europe. In TD-1 each of the three large experiments was comparable in complexity with the entire payloads of the small satellites Iris and Aurora. It is not surprising that the cost of development grew well beyond what had been anticipated. Nevertheless, the budget agreed by the STC on 8–9 November 1966 for the second third-year period of ESRO (£51 million at 1966 prices) assumed that the TD-1 and 2 projects would be given absolute priority. In spite of this, the whole situation had become critical by early 1968 when it had become clear that even the completion of both TD-1 and 2 would be too expensive.

An extensive review was carried out to decide which of the two should continue if one had to be dropped. The choice fell on TD-1 because the experiments aboard called only for stabilisation in sunlight which was considerably simpler to achieve than a system which required the solar pointing to be maintained during periods of solar eclipse as did certain of the experiments in TD-2 (numbers 1, 6 and 7 in the list given above). It was assumed that the sunlight baffle would be effective (see p. 148). The STC at its meeting on 30 May 1968 agreed that the development of the TD-1 satellite alone should proceed either as an ESRO project or as a Special Project. As some recompense to the unfortunate scientists who were working on the TD-2 payload it was recommended that, in future programmes, priority be given to providing flight opportunities for their experiments, provided they remained of scientific importance.

Following these decisions it was finally decided to continue TD-1 as a Special Project with all member states except Spain contributing in proportion to their gross national products. Careful consideration was given to the possibility of accommodating the TD-2 experiments in either an ESRO I or ESRO II satellite. It was found that all except those requiring observations in darkness could be accommodated in an ESRO II type satellite following an appropriate orbit. It was agreed that this would be done, introducing the so-called ESRO IV satellite which was planned to be launched into an orbit near 1000 km apogee and 300 km perigee at an inclination near 90°. This was successfully carried out in November 1972 using a Scout vehicle and went some way towards compensating several of the original TD-2 groups for the cancellation of the initial project which had begun with much enthusiasm.

Returning now to TD-1, a review of the situation in early 1968 revealed that the design, development and production of the main scientific payload was far behind schedule and was inadequately and inappropriately staffed for the purpose. Immediately, steps were taken by the Chairman of the Science Research Council in the UK, B. Flowers, to strengthen greatly the organization of the payload development. R. Wilson, then in charge of the Space Research Unit of the SRC at Culham, was placed in full charge of the UK side of the project with strong support provided not only from other members of his group but from the Department of Physics and Astronomy at UCL (A. Boksenberg) and the Atomic Weapons Research Establishment (AWRE) at Aldermaston (E.G. Warnke and P. Barker). This experienced team, who had been involved in the planning of the large astronomical project (see p. 155) and were engaged in other activities in ultra-violet astronomy, were able to restore the situation and even at this late stage to improve on the spectrometry. This was achieved through a suggestion by Boksenberg. He pointed out that simply by opening up the entrance slit of the spectrograph, advantage could be taken of the passage of the stellar image across the slit due to the satellite motion to cover the 1350–2550 Å wavelength band by observation in effectively 61 rather than 3–4 channels thereby greatly increasing the wavelength resolution. As a result when TD-1 was successfully launched on 9 March 1972, the joint UK–Belgium experiment, no longer exclusively Edinburgh–Liège, functioned very well indeed.

The primary telescope was an off-axis $f/3.5$ paraboloid with a primary mirror of 27.5 cm diameter. Two slits were included in the focal plane, one leading to a three-channel grating spectrometer covering the respective wavelength bands 1350–1750 Å, 1750–2150 Å and 2150–2550 Å and the other to a single photometric channel with peak response at 2740 Å. Light passing through the former, the field of view being $17' \times 12'$, was collected by a secondary mirror illuminating a plane grating so as to produce a spectrum in the focal plane of the secondary mirror/grating combination. Light passing through each of these exit slits in this plane fell on a pulse-counting photomultiplier. As the primary image moved across the entrance slit owing to the motion of the satellite the spectrum moved across the exit slit. The entrance slit to the photometer had a field of view of $17' \times 1.7'$.

TD-1 was launched into a nearly circular polar sun-synchronous orbit, the apogee being 545 km, the perigee 533 km and the inclination 97.6°. It was fully stabilized in three axes with one pointing to the sun to within $\pm 5''$. The optical axis was maintained perpendicular to the solar pointing axis and to the orbital plane. As it viewed away from the earth a great circle of sky was completely scanned in one revolution. The motion of the earth round the sun caused a complete survey of the sky in six months, successive great circles being displaced by about $4'$.

Laboratory calibrations on an absolute photometric basis were carried out on the instruments before flight to a precision of about $\pm 20\%$ both at the Royal Observatory, Edinburgh and the Institute of Astrophysics at the University of Liège.

As far as data transmission was concerned it was planned that this would be stored on onboard tape recorders to be read out as commanded on passage over ESTRACK stations. For a few weeks this was operated successfully but trouble soon began to develop on the tape recorders which finally ceased functioning on the 22 May, 10 weeks after launch. The success of the project depended on obtaining full orbital coverage of data, so frantic efforts were made to build up recording in real time to a useful fraction of the total time. Before long, up to 60% of the data were being recovered in real time and this was steadily increased to over 95% by January 1973. About 56% of the coverage was provided by ESRO through its ESTRACK network and through the enlistment of temporary receiving stations in such localities as Seymour I. in the Antarctic, Papeete, Suva and later in Greenland, Kerguelen I., Tel Aviv, Townsville, Ahmedabad and Mexico City. NASA contributed about 26% through its network, CNES about 8% and others about 10%.

Figure 6.5 shows the first ultra-violet spectrum of a star recorded by TD-1, that of γ Columba of magnitude 4.35. Reference to other results obtained with the ultra-violet equipment on TD-1 is made in Chapter 15. Despite its rather chequered history it performed remarkably well over a period of one year or more to provide two six-month sky scans during which about 30,000 stars down to the ninth visual magnitude were scanned.

23 The two large projects

The scientific programme proposed by the GTST at its meeting on 8–9 May 1961, included provision for launching of large satellites and space probes at a rate of one or two per year beginning about six years after the establishment of ESRO. Moreover, it was stated specifically that this part of the programme should include a stabilized satellite for astronomical

Fig. 6.5. Ultra-violet spectrum of γ Columba, the first star observed by TD-1.

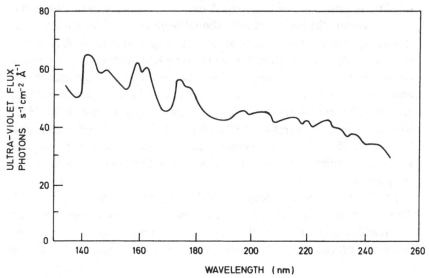

observations and a vehicle in orbit round the moon. These soon became known as the two large projects. They required the use of launchers of the Blue Streak or comparable type so there was need in working out the programmes to keep in close touch with developments in ELDO. In fact neither project was carried through to completion but nevertheless, as we shall see, thanks to the energy and ability of a UK group, a very satisfactory substitute (IUE) was eventually launched as a joint project between the US, UK and ESRO and proved highly successful. We shall here describe the course of events within ESRO which failed, despite very thorough design studies and very complex discussions, to be accepted for financial reasons. The later developments which led to IUE, the initiative for which came from the UK, we describe later, in Chapter 15.

Plans for the second large project did not proceed so far as for the first and we shall first consider the much more extensive deliberations about the large astronomical satellite (LAS).

24   The first large project (LAS)

From the outset, UK scientists showed a strong interest in the use of satellites for astronomical observations in previously inaccessible wavelength regions. The Solar and Stellar Astronomy Working Group (SASA) of the BNCSR, under the Chairmanship of H.E. Butler, was very active from its inception in 1958. At that time it was quite possible that a UK launching system using Blue Streak would be developed and SASA considered what would be the most suitable astronomical payload for a satellite launched by such a system. After discussing various alternatives it was agreed that the highest priority should be given to flying an ultra-violet spectrometer operating in the wavelength range 3300 to 1200 Å with a spectral resolution as high as 1 Å. At the time RAE was prepared to deploy resources in money and manpower to work out the technical requirements and possibilities for an astronomical satellite stabilized to a few minutes of arc. The RAE programme, worked out in close collaboration with the SASA working group, was directed by A.W. Lines who later became the Technical Director of ESRO, in which capacity this experience was to prove very valuable. Already, by 1962, the emphasis was directed towards the ESRO programme.

The first move towards definition of the LAS project was made in March 1963 when the LPAC agreed that the astronomy working groups should set up an *ad hoc* group to study the scientific requirements of the project and make recommendations on the scientific aims and instrumentation for the first LAS. Butler, referred to above, was chosen as the chairman of the group and the other members were L. Biermann (West Germany), M. Golay (Switzerland), L. Houziaux (Belgium), Y. Öhman (Sweden) and J-C. Pecker (France). It was necessary that the group should work with some knowledge of the technical possibilities and limitations. They assumed that the satellite system, involving stabilization of the axes of the satellite to a minute of arc, described by A.W. Lines at a colloquium in Paris in September 1962, would be practicable.

The main outline of the scientific package was worked out by the group at a quite early stage. In a preliminary report prepared for 1 July 1963 it proposed that the preliminary instrument should be one designed to make stellar and nebular spectrophotometric observations throughout the spectral range from about 3500 Å to about 912 Å. A second smaller primary instrument, to make broad band X-ray observations, should also be carried.

Following this preliminary report, further studies were carried out with the aim of completing a final report open to discussion at a colloquium to be held in February 1964. By October 1964 the group had largely completed its task by drawing up the following specifications. First, as regards the basic satellite vehicle, it should consist of a platform which could be pointed in any direction to within 1 minute of arc whether in sunlight or in shadow. The space available for the instrumentation should be a cylinder of about 1 m diameter and $2\frac{1}{2}$ m in length and the instrument weight permissible between 200 and 300 kg. The power requirements of the whole spacecraft would be about 300 W and environmental temperature variations kept to $\pm 10\,^{\circ}$C.

The wavelength range to be covered by the first primary instrument would be from 3500 Å to about 1100 Å, extending if possible to 912 Å. Even if this extension could only be made at low efficiency, without reducing the efficiency in the main range, it should be included in the design. Choice of the primary mirror diameter depended on the magnitude of the stars which it was hoped to observe. 50 cm was chosen, but the possibility of including parallel mirror systems was not ruled out.

Although some members of the group considered it to be too severe a requirement the majority considered that maximum spectral resolution of a few tenths of an ångstrom should be aimed at, though it should be possible to degrade the resolution to a limit of 100 Å when required.

Whereas when individual stars are being observed a spectral resolution corresponding to a stability of a few arc seconds should be obtainable by lock-on techniques, it was realized that for spectra of extended objects the stability would be of the order of 1 arc minute.

On the basis of rather slender information obtained from US rocket flights it was assumed that stellar observations could be made both in sunlight and in darkness.

No further details were given about the proposed second primary instrument for X-ray observations.

While these intensive studies were proceeding, very animated discussions had been taking place about the organization of the whole programme, the first one in which ESRO would be responsible for the cost of the scientific as well as the technical equipment. One possibility might be to carry out the entire programme within the organization by recruiting a strong scientific team to operate in-house, albeit with input from external advice, much as with NASA. However many scientists were aware of the problems which this had raised, as most were in touch with American colleagues and, from the outset, ESRO had been planned to avoid these difficulties as far as possible. An alternative which was suggested at an early stage was that of contracting the design and development of the scientific package to a national group with adequate resources for the task. It was considered at the time (1963) that the LAS programme would be an ongoing one in which successive scientific packages, differing more in detail than in their main purpose, would be launched in satellites of basically the same form. There would then be opportunities for several national groups to participate in designing the different packages. Responsibility for the co-ordination of the scientific package with the satellite would rest within ESRO, by establishing a small team of scientists in ESLAB especially for this purpose. The proximity of ESLAB to ESTEC made this very practicable. The ESLAB team could be expected to design and develop one of the later payloads.

These ideas were well received by the larger member states but with considerable reserve by the smaller ones who looked somewhat askance at the idea of national groups bearing primary responsibility for the scientific packages. This fear was not fully allayed even when it was pointed out that it would be expected that nationals of other member states could join the national groups selected, so retaining some of the advantages of an international organization. It was also made clear that groups of different countries might join to form an international group which would operate singly just like a national one.

At its meeting on the 10–11 September 1964 the STC issued a statement concerning the organization on the lines sketched out above which was quickly approved by the Council. The statement included the following recommendations.

1. Several complete astronomical payloads should be developed as primary experiments for launching in the LAS. These payloads would be launched successively without back-up launchings.
2. For this purpose ESTEC should develop a basic astronomical satellite capable of accepting different primary experiments.
3. The responsibility for development of each primary astronomical experiment should be exercised by a single group.
4. The co-ordination of the scientific payloads will be exercised through a team of scientists in ESRO under the control of the Scientific Director.
5. When it has the necessary capacity this scientific group in ESRO will be expected to undertake the task of providing a primary experiment for the LAS.
6. To ensure an early start ESRO will place contracts for this work with scientific groups in member states, which should be supplemented by scientists from other member states.
7. In the light of the work and recommendations already made by the LAS *ad hoc* working group, member states are to be invited to submit proposals for carrying out primary experiments on the LAS from groups suitable to undertake this responsibility, along the lines set out above. An advisory group to be established by the STC and LPAC will examine and report on the scientific and technical merits and on the feasibility of the proposals.

In relation to 7, it was recommended that proposals should be submitted by 15 December 1964.

There was naturally very strong interest in the UK throughout the deliberations, both of the *ad hoc* working group and of the LPAC and STC, about organization. As mentioned earlier, UK scientists had been concerned with the design of ultraviolet spectrometers and detecting systems from the beginning. Both the University College group under Boyd and the Culham group under Wilson were well equipped on the basis of their experience with rocket experiments and both were anxious to join in working out a practicable proposal. As soon as it became clear that such proposals would be requested, Massey called a meeting in his room in the Department of Physics at UCL to initiate concerted action on the matter. The meeting was not only attended by Boyd and Wilson but also by J.B. Adams, the Director of the Culham Laboratory and I. Maddock from the Atomic Weapons Research Establishment (AWRE), which also wished to collaborate. It was agreed that a proposal should be worked out for submission to ESRO. While Wilson and his group would concentrate on the design of the instrumentation for ultraviolet astronomy, Boyd and Boksenberg at UCL would provide the design for a detector for the main payload as well as the instruments for the subsidiary X-ray experiment. Wilson, being responsible for the major part of the payload, was chosen to be chairman of the Scientific Group.

The UK consortium duly presented its proposal to ESRO before 15 December 1964, incorporating a design meeting the scientific requirements laid down by the LAS working group. A second proposal was submitted by a German–Dutch group which also met these requirements. A French–Belgium–Swiss proposal was also submitted which was directed more to the observation of faint objects using a spectral resolution (10–300 Å) considerably poorer than the 1 Å proposed by the LAS group, and included provision for wide field viewing.

The three proposals were discussed at the STC meeting on the 20–21 January 1965. Although there was some discussion about the departure of the French–Belgium–Swiss proposal from the specification it was recommended to Council that contracts should be placed with all three groups to produce detailed designs within six months. This was agreed by the Council in February 1965 but there was some delay in placing the contracts because of uncertainty about the launch vehicle.

Contact between ESRO and ELDO had improved considerably following a meeting between the ELDO Bureau and that of the ESRO Preparatory Commission in Paris in October 1963 and it had been assumed that the proposed ELDO A (later Europa I) launching system would be available in time to launch the LAS. This was a three-stage system with a total lift-off weight of 112 tonnes and the planned launching site was Woomera. By early 1965 it became clear that this would not be adequate, so the technical specification to be assumed by the contractors designing the scientific package could not be spelled out. However, by June 1965 the likely possibility of a French equatorial launching site at Kourou in French Guiana becoming available for ELDO by 1970 had arisen. From such a site an equatorial orbit at 700 km would be possible for an easterly launching with a payload of 800 kg and a payload compartment of 2 m diameter, $2\frac{1}{2}$ m long. These specifications were adopted and the contracts with the three groups finally placed in August 1965.

It was agreed that an assessment of the three designs would be made by a Board, working under the control of ESLAB, which need not be constituted entirely from scientists working in the member states. The members finally chosen were K. Bochasten, Uppsala University, Sweden (spectroscopist); H. Butler, Royal Observatory, Edinburgh, Scotland (astronomer); P. Connes, CNRS Bellevue, France (spectroscopist); H. Elsasser, Landessternwarte auf den Königshuhl, Heidelberg, W. Germany (astronomer); O. Petersen, Space Institute Laboratory, Lyngy, Denmark (astronomer); J. Rogerson, Executive Director, Princeton OAO[4] Package, Princeton University Observatory, Princeton, USA and L. Husain, Large Satellite Division ESTEC. In addition, in framing the recommendation based on the assessment, the Director General of ESRO would be assisted by a panel of three astronomers, H.C. van de Hulst, Professor of Theoretical Astronomy, University of Leiden, Holland; L. Gratton, Director, Laboratory of Astrophysics, Frascati, Italy and L. Goldberg, Director, Harvard College Observatory, USA.

The three groups submitted their final design reports by 31 January 1966 and the recommendations from the Assessment Board were circulated in a document dated 15 April 1966 that was discussed at an STC meeting on the 9 May 1966 at which H. Massey and M.O. Robins were the UK delegates. It was recommended that the UK design be adopted for the first LAS instrumental package with one back-up unit. The project should have a strongly international character with the management and part of the execution being carried out within ESLAB.

The meeting was understandably a very tense one and some agreement was only reached after more than five hours' discussion. It was accepted that the LAS should be basically of high spectral resolution but should aim further at wide field, broadband (10 Å or more) spectrophotometry.

The way in which the project should be managed presented much more difficult problems. Thus, on the one hand, it was clearly desirable to keep together the UK national team responsible for the accepted design while on the other hand giving the project the greatest possible international character through assigning responsibility for its development to ESRO, presumably through ESLAB. The only obvious solution, that of seconding the UK team to work within ESRO, was not practicable. A strong plea was made to avoid taking the development of the scientific package completely out of the hands of ESRO. On the assumption that the structure of ESLAB would be suitably modified it was finally agreed, against the opposition of France and the UK, that the direction of the project would be a responsibility of ESRO, conducted in ESLAB, and that part of the work on the scientific package would be done within ESRO.

It was further agreed to recommend the appointment of a Science Project Manager and set up a Programme Group to work out a scientific programme

158   *European Space Research Organization*

for the LAS. The Science Director of ESRO would negotiate the division of responsibilities with the UK team leader, Wilson, with the aim of submission to the next Council meeting towards the end of June.

The management problem remained. It was clearly out of the question to appoint a Project Manager from outside the UK group at this stage who would have any effective control of the main work to be carried out within that group. Equally well, the only way to use the experience and special expertise of the group would be for them to continue the development of the project. It is not surprising that the special meeting of the STC on the 21 June, although very animated, was quite inconclusive. Nevertheless it was agreed by most delegations that the UK group would work under contract from ESRO, responsibility for which would rest with a Project Manager in ESLAB. Although the UK delegation attempted to make clear how unrealistic it was to talk of scientific control by ESLAB at this stage – at this time it was quite clear that the leader of the UK team, Wilson, would not be prepared to transfer to ESLAB – most delegations agreed that the Secretariat should continue to seek a Scientific Project Manager to work at ESLAB.

At its meeting on 20 July 1966 the ESRO Council passed a resolution, based on the STC discussions. As far as management was concerned it accepted that a contract would be negotiated with the Culham Laboratory, at which the UK group was based, on the following lines. The Culham Laboratory would be responsible for the design, development and manufacture of one flight unit of the scientific package, and a back-up unit. It would also provide a team at the disposal of ESRO to assist during the integration phase with the spacecraft and the launching. Responsibility for the evaluation and approval of the final design and its compatibility with the scientific objectives remained with ESRO. The resolution also required that a post as overall LAS systems manager should be created and filled as quickly as possible. Finally, the principle of establishing a ceiling of expenditure for both the scientific package and spacecraft was accepted. Proposals were called for to be worked out for submission to STC and Council meetings in November.

By the time the STC met on 8–9 November 1966 the stringency of the financial situation facing the organization had become apparent. The budget agreed for the second three-year period was 719 mF (£51m) at 1966 prices, that for 1967 being 230 mF (£16.5m) as against 260 mF requested. It was apparent that, with the TD-1 and TD-2 projects being given absolute priority (see p. 148), little remained for work on the scientific package for LAS during 1967. A proposal was made by the UK whereby the United Kingdom Atomic Anergy Authority (UKAEA) which operated the Culham Laboratory would finance the further work on the scientific package at

Culham during the whole of 1967 (about 250,000 F) without expecting reimbursement from ESRO until January–March 1968. Included among the work so financed would be development of half-scale models of the scientific package which would be launched by stabilized Skylarks (see Chapter 9).

This proposal was not accepted. Finally, the Chairman of the STC (Lüst) proposed that, while work on the scientific package be left in abeyance for the time being, tender action could be initiated for the design of spacecraft incorporating the Culham design. He further proposed that a recommendation should be made to the Council to hold a Special Conference in late 1967 to examine the conclusions drawn from the tender action and decide on the future of the LAS. This was accepted by seven votes to three, the UK being among the latter.

To all intents and purposes this marked the end of the LAS project. Nevertheless the Council accepted the recommendations of the STC. A Conference of Ministers from member states would be held. To provide reliable cost figures for the Conference, tenders would go out for the LAS spacecraft. Meanwhile the contract with Culham would be extended by at most two months after the Conference on the basis of a yearly expenditure of 1 million F.

To further this preparatory work W.G. Stroud from NASA was appointed as a consultant to the Director General on the LAS project, taking up his post at Noordwijk in March 1967. A full team was assembled under him which included all aspects of the project, the UK group being its scientific component.

In June 1967 Stroud presented to the Council his estimate for the cost of two flight models as 455–560 million F (£31.5–40 million) with an additional 200 million F (£14 million) for ground support.

A detailed document detailing the Project Development Plan was prepared by the LAS Project staff under Stroud as Project Leader, for submission to the Ministerial Conference on the 11–13 July. Before describing the final stages of the LAS saga it is of interest to summarize a number of aspects of the plan.

The scientific specification followed by the Culham design was essentially the same as that proposed by the original LAS working group. It involved an 80 cm Cassegrain telescope feeding a Paschen-Runge spectrometer operating in the spectral range 900–3000 Å. Fine guidance was provided either by movement of the secondary mirror of the telescope or by control of the spacecraft by error signals from the telescope. The best spectral resolution would be between 0.1 and 0.2 Å, capable of degredation to 0.4, 1.6, 6.4 Å and one higher value. The statistical photometric accuracy would be between 0.01 and 0.15 and the absolute accuracy about 0.1. Stars with magnitude <9 would be observable at lowest resolution. The payload dimensions would be $3\,m \times 1\,m$ weighing $210\,kg$. The fine pointing accuracy would be $\sim 0.1''$. 40 watt of power would be required on the average to operate the system, which would have a design lifetime of about 1 year.

The detector system consisted of an array of photomultipliers which would be replaced by an image storage system under active development at the time at UCL, giving at least a 100-fold improvement in data collection, if this could be proven in time.

A subsidiary X-ray telescope array using detectors at the prime focus of paraboloidal mirrors was also included to operate in the spectral range 2–$100\,\text{Å}$.

The cost of the full scientific package was estimated as 36 mF (£2.6m) for the primary experiment plus 2.3 mF (£165,000) for the X-ray experiment. Within the UK, the project would be set up within the UKAEA. The project team would report to a project management board, under the Chairmanship of J.B. Adams, Member for Research UKAEA. The other members were R.L.F. Boyd (UCL), W.J. Challens, Asst. Director UKAEA Aldermaston, V.H.E. Cole UKAEA London Office, J.F. Hosie, Science Research Council, R.S. Pease, Director UKAEA Culham Laboratory, R. Wilson, the Project Leader and Head of the Astrophysics Research Unit, Culham, and V.C. Birchall, UKAEA Culham Laboratory (Secretary). This Board and the Project Team at Culham were to play a vital role in revitalizing the concept of an ultra-violet astronomical satellite after the demise of the LAS as we shall see in Chapter 15.

The Project Development Plan included preliminary recommendations about the basic spacecraft. Apart from meeting the requirements of the scientific package as regards power, weight and space, it should contain a steerable platform which could be pointed in any direction with an accuracy of the order of $1''$ whether in sunlight or shadow.

The Ministerial Conference set up an Advisory Committee on Programmes under the Chairmanship of J.P. Causse to consider the status of such programmes as the LAS within the organization. At the level of operation which they recommended the LAS project would have required about 40% of the available resources for the four or more years beyond 1968. On this basis it was not possible to proceed with the project. The way in which a project of comparable effectiveness was finally completed with great success, thanks to the never-say-die attitude of the UK group which had worked out the original proposal, is described in Chapter 15. The further deliberations of the Ministerial Conference are referred to on p. 162 of this chapter and on p. 229 of Chapter 11.

25 The second large project

In 1962, an *ad hoc* working group was established under the Chairmanship of P. Swings (Belgium) to consider proposals for this project. The other members were J.E. Blamont (France), L. Biermann (W. Germany), Z. Kopal (UK), R. Minnaert (Netherlands), G.P.S. Occhialini (Italy), Y. Öhman (Sweden) and J-C. Pecker (France).

At its first meeting (December 1962) the group widened the scope of its studies to include not only lunar satellite missions but the possibility of establishing a lunar ground station as well as of a planetary, including cometary, mission. After carrying out these studies the group recommended to the GTST that a flyby mission to a comet should be considered as the second large project.

This proposal came somewhat as a surprise and to many, including the UK delegates, seemed to be rather ambitious, particularly in view of the short time of observations which would be available in relation to the great expense of the mission. On receiving the report from the working group the GTST at its meeting in February 1964 agreed that further study of the feasibility of a cometary mission be carried out by a study group chaired by Biermann. The UK delegation suggested, however, that other proposals should be considered, as for example a large satellite for geophysical studies, in case the feasibility study showed that the comet project could not be carried out for a long time. This suggestion, supported by some other delegations was implemented at the first meeting of the Scientific Committee of ESRO and alternative proposals were called for. However, the worsening financial position of the organization caused the concept of the Second Large Project to be shelved, and it was not until 1980 that a cometary flyby, to Halley's Comet in 1986, was agreed.

The working group under Swings also included a suggestion that ESRO should take the initiative in setting up a programme of international co-operation for a flyby mission to Jupiter. The GTST agreed that ESRO should invite COSPAR to initiate preliminary enquiries as to the practical possibilities in this matter but nothing significant was achieved.

26 The Bannier Committee and the re-organization of ESRO

We have already referred to the Ministerial Conference, which through the Advisory Committee on Programmes had a profound influence on the plans for the ESRO programme beyond 1968, including the LAS. This conference met first in 1966 and set up a study of the organization of ESRO under Bannier of The Netherlands.

The Bannier Committee, reporting in 1968, recommended a substantial re-organization of ESRO. It has been described on p. 132 how the existing organization was based on three directorates, Science, Technical and Administrative, reporting directly to the Director General. While agreeing that this clear separation between science and technology was completely justified in the initial phase of building up large technical resources and facilities with the associated infrastructure, the Bannier Committee considered that with ESRO involved now in mainly operational activities, the time was ripe for setting up what they regarded as a more flexible organization.

They proposed that four directorates, reporting directly to the Director General should be set up – two, the Directorate of Programmes and Planning and of Administration located in Paris and one, that of Space Research and Technology at ESTEC in Noordwijk and one, that of Space Operations, at ESDAC in Darmstadt. Responsibility for ESTRACK and ESRANGE would be transferred from ESTEC to ESDAC.

These proposals were accepted and marked the end of the first phase of ESRO development, the build-up from scratch as it were. Auger resigned as Director General in 1967, Bolin as Scientific Director in the same year and Lines in March 1968. H. Bondi (UK) followed Auger while in the new organization J. Dinkelspiler, W. Kleen, U. Montalenti and M. Depasse became the respective Directors of Programmes and Planning, Space Research and Technology, Space Operations and Administration respectively. The emphasis of the further deliberations of the Ministerial Conference was on the integration of all civil space activities, including applications in one single European organization.

This is a convenient point at which to terminate the present account, returning to describe the transformation of ESRO into ESA in a later chapter (11). In this chapter we have only strayed into later years than 1968 to follow through to the ultimate stages the major scientific projects initiated much earlier.

7

Commonwealth co-operation in space research

Reference has already been made in Chapter 4 to the statement made by the Prime Minister to the House of Commons on 12 May 1959 about space research in Britain. In this he referred, among the different possibilities for the provision of satellite launching systems, to that involving co-operation with some Commonwealth countries. Following this statement, enquiries were made through the UK High Commission in various Commonwealth countries about possibilities of collaboration, not only as regards launching systems but in other aspects of space research such as tracking, data analysis, provision of launching sites for rockets and satellites, and the preparation of scientific experiments to operate in space. Canada, Australia, New Zealand and India expressed interest in different aspects of the matter. Thus Canada, while favourable to collaboration in general, drew attention especially to the existence of their topside ionospheric sounder programme that offered many opportunities for UK participation. Australia was interested in the use of Woomera for launching satellites and a proposal for a space experiment concerning cosmic rays was made by H.G. Messel of Sydney University. India welcomed co-operation in the first instance in tracking and the preparation of scientific equipment. New Zealand welcomed the suggestions for collaboration and looked forward to receiving further details. No country indicated any interest in participation in the development of a satellite launching system as such.

The first opportunity for discussion between scientists from Commonwealth countries occurred at the same COSPAR meeting in Nice in January 1960 as that at which the first discussion on European collaboration took place. At the informal discussion the UK was represented by R.L.F. Boyd, A.W. Lines, A.C.B. Lovell, H.S.W. Massey, M.O. Robins and A.P. Willmore; Canada by J.H. Chapman and D.C. Rose; Australia by H.J. Higgs and L.G.H. Huxley; India by A.P. Mitra and S. Africa by D. Hogg. By and large the opinions of the scientists from Canada, Australia and India were consistent with the information gathered by the UK High Commission. It was agreed that a further meeting of interested parties should be held after opportunity had been afforded for information discussions within the different countries. A likely time was suggested to be in the third week of August in 1960 in London.

The British National Committee for Space Research approved this suggestion and the Council of the Royal Society invited the principal scientific academy, or equivalent institution, in different member countries of the Commonwealth to send representative space scientists to a meeting to discuss collaboration in space science, to be held at Burlington House on 25 and 26 August 1960.

1   Royal Society meeting on Commonwealth co-operation in space science

This meeting was well attended. The full list of those present is as follows. In such case the delegate is named first, any other having observer status.

Australia: L.G.H. Huxley, D.F. Martyn and F.W. Wood; *Canada:* D.C. Rose and A.F.B. Stannard; *Ghana:* R.W.H. Wright; *New Zealand:* Sir E. Marsden, R.S. Unwin, J.R. Gregory and M. Gadsden; *Nigeria:* N.J. Skinner and G.S. Kent; *Rhodesia* and *Nyasaland:* O.G. Reitz; *Union of South Africa:* J.A. King and J.A. Gledhill; *United Kingdom:* H.S.W. Massey, R.L.F. Boyd, E.B. Dorling, H.G. Hopkins, D.C. Martin, A.F. Moore, B.G. Pressey, J.A. Ratcliffe, M.O. Robins, R.L. Smith-Rose and A.P. Willmore. J.G. Mallock, representing the British Commonwealth Scientific Committee, was also present. Unfortunately, no delegates were able to attend from India or Pakistan and apologies were also received from Ceylon and the British West Indies. Massey was elected Chairman of the meeting which soon adopted an agenda taking into account informal discussions which had occurred between the scientists at University College London (UCL) on 22 August prior to visits to the laboratories in the Department of Physics there on 24 August.

1.1   Co-operation in satellite tracking and telemetry reception

The first item to be discussed concerned co-operation in satellite tracking and telemetry reception. The importance of obtaining accurate information on satellite orbits by optical tracking was stressed by Ratcliffe and there was considerable discussion of ways and means for improving the quantity and quality of data. Wood drew attention to the important arrangements at Woomera where a Baker–Nunn camera and Minitrack apparatus had been set up side by side. It was agreed that a representative from each country should prepare a statement on the present position and possibilities for optical tracking in his country, listing possible sites and existing facilities as well as climatic conditions and the availability of operating personnel. These statements, which would in no way commit the author, would be forwarded in the first instance to the Royal Society.

Attention was also drawn to the importance of radio as well as optical tracking for Commonwealth scientists engaged in ionospheric research and it was agreed that the statement concerning optical tracking should also include information from the latter scientists about what information they would require to further their work through use of satellite data.

Discussion then turned to telemetry reception. This was opened by Robins who described the arrangements made in this respect for the first and second joint UK–US satellites S-51 and S-52 (see Chapter 5, p. 79). Of the continuous telemetry facility to be included, about 10% would be received by ground stations. All such signals recorded at Commonwealth stations would be of great value. Again Wood referred to the telemetry command facility at Woomera as well as in Western Australia. Possible means for achieving maximum effectiveness of such a scheme in terms of the provision of simple equipment were considered.

The Chairman assured Commonwealth scientists that the codes used in the S-51 and S-52 telemetry systems would be made available to all Commonwealth stations which requested them.

1.2 Canadian topside sounder satellite

A very important item for ionosphere physicists, involving telemetry reception, was then introduced by Rose who gave an account of the Canadian topside ionosphere sounder[1] satellite which would be launched by NASA in 1961 and tracked by the NASA system and three additional Canadian stations. With the high orbital inclination of 80° it was hoped to obtain important new data about the auroral zone and polar cap phenomena. No recording equipment would be carried by the satellite. UK plans to command the telemetry facility at RRS, Slough and at the overseas stations in Singapore and possibly the Falkland Islands were outlined by Ratcliffe and Smith-Rose. This was the beginning of a very fruitful collaboration as the topside sounding satellites proved to be very effective, providing a huge amount of interesting data (see Chapter 13, p. 312). In fact the first of these, Alouette I, was the first international scientific satellite launched by the USA to be designed and constructed by another country.

Ariel I was the first international satellite containing instruments designed and constructed by one country and launched by the USA but in this case the satellite itself was designed and manufactured in the USA.

So much interest was expressed in the topside sounding satellite that it was even suggested that, if the Blue Streak – Black Knight launching system became available, it might be desirable to consider whether a topside sounding satellite should be planned on a Commonwealth basis.

1.3   Co-operation in research using rockets

The meeting then turned to the discussion of Commonwealth co-operation in research using sounding rockets, taking account of the existing British Skylark programme and of the Canadian rocket launching range at Fort Churchill. The possibility of extending the British programme to make it possible for scientists in other Commonwealth countries to participate was raised. Huxley welcomed this possibility on behalf of Australian scientists and Marsden hoped to secure official approval in New Zealand for taking part in such a co-operative programme. Massey reported on a visit to India which he had just made on his way home from Australia in April 1960. He had discussed the possibilities with Sir K.S. Krishnan who considered it of sufficient importance to arrange for Massey to raise the matter directly with Pandit Nehru.

While Canadian scientists would naturally concentrate their activities at the Fort Churchill range Rose pointed out the complementarity of this range to Woomera. Moreover he could see little difficulty in launching rockets containing scientific payloads prepared by scientists in other countries. It might also be possible to adapt launching equipment so that Skylark could be launched from Fort Churchill. Alternatively, an instrumented Skylark nose cone might be attached to Canadian rocket motors for launching.

Attention was drawn to the development of small rockets proceeding in Britain and it was thought that standardization of a small rocket for Commonwealth use might be of value.

It was agreed that the case for consideration of a Commonwealth rocket programme would have to be based on information about the degree of interest in each country and the following were invited to prepare reports on this matter within their individual countries and forward them to the Royal Society not later than 31 December 1960: Australia, F.W. Wood; Canada, D.C. Rose; Ghana, E.A. Boaling; India, A.P. Mitra; New Zealand, W.H. Ward; Nigeria, N.J. Skinner; Rhodesia and Nyasaland, O.G. Reitz, and the Union of South Africa, F.J. Hewitt.

1.4 Co-operation in research using earth satellites

Next came the question of co-operation in research using earth satellites. Massey outlined the potential of a Blue Streak–Black Knight launching system (see Chapter 4, p. 66) and enquired what scientific interest there would be in Commonwealth countries in using satellites launched in this way. Most interest was expressed by Huxley on behalf of Australian astronomers who would welcome the opportunity to observe stars in the southern hemisphere in the ultra-violet and X-ray wavelength regions. Martyn remarked that it should be understood that scientists from those countries contributing to the tracking of satellites should have a claim on part of them for their experiments.

2 The Commonwealth Consultative Space Research Committee

These discussions showed the need for the establishment of some continuing link between space scientists in the different countries and of a body of informed opinion from which proposals for presentation to the various governments could be prepared. Mallock described the organization of the British Commonwealth Scientific Committee whose main committee members responsible for distinct fields of scientific activity carried out their general business in London through deputies. Massey therefore suggested that a Commonwealth Consultative Space Research Committee should be set up with its secretariat in London. Members of the committee would be nominated by the appropriate academies or their institutions in the different countries. Working groups could be set up as required in which deputies could serve if necessary. This was agreed – Massey was elected interim Chairman of the Committee and requests were sent to the appropriate bodies of the different countries to nominate members. Three working groups were set up on Tracking and Data Recovery, Vertical Sounding Rockets and Satellite Ionospheric Sounding with Ratcliffe, Robins and Beynon as the respective convenors.

While the meeting had taken place in a most co-operative atmosphere, progress was handicapped by the fact that there was no obvious source of funds within the different Commonwealth countries to match their interest in participating in co-operative rocket programmes. It is a remarkable fact that the two countries which later did participate in such programmes with the UK – India and Pakistan – were not represented at this first conference.

3 The use of small rockets

These same two countries were however, able to send representatives to an informal meeting held by the BNCSR in London on 24 and 25

January 1961 (see Chapter 8) to consider the use of small vertical sounding rockets. At this meeting representatives attended not only from Australia, Canada, New Zealand, Rhodesia and Nyasaland and South Africa, but from several European countries (Denmark, France, Italy, The Netherlands, Norway, Sweden and West Germany) as well as the UK. The extensive discussion which took place at the meeting is outlined in Chapter 8. During the meeting, information was made available about the 5″ and 7.5″ rockets under consideration by Bristol Aerojet Ltd. The history of the further development of these rockets is given in Chapter 8. They are mentioned here because they were examples of rockets which, while cheap, they were priced at £300 and £720 each respective in quantity production, would nevertheless be very useful for research on the middle and lower atmosphere.

4 Developments in India and Pakistan

The second meeting of the Consultative Committee was held in London on 29 September 1961. This time both India and Pakistan were represented, the former by S. Chandrasekharan and the latter by C. Kemal Reheem and A. Salam. It was reported that a space research organization would soon be set up in India and that discussions were taking place with NASA about the loan of small rockets and ancillary ground equipment. Indian scientists had fully realized the importance of small rocket launchings near the equator. In Pakistan a national space research committee had just been formed and here again there was considerable interest in the possibility of using small sounding rockets. These were encouraging developments which later bore fruit in active collaboration.

Early in 1962, V. Sarabhai, the Chairman of the new Indian Committee on space research, INCOSPAR, visited Massey in London and brought news of a rapidly expanding interest in India, an interest backed by resources in money and manpower. At about the same time, A. Salam, who, although a Professor of Physics at Imperial College London, was closely involved in planning of science and technology in Pakistan, explained why there was an interest in the use of sounding rockets by that country. He considered it to be important to maintain some advanced research activities in order that the best local scientists should not all be tempted to seek posts abroad. Following up these discussions, Massey visited both India and Pakistan in September 1962 on his way back from Australia.

In India he attended a meeting of INCOSPAR and visited the Research Institute at Ahmedabad of which Sarabhai was the Director. He also was shown the plans for an equatorial rocket launching site at Thumba near Cape Comorin on the Southern tip of India. Despite the fact that it would be located in the province with the densest population no serious problem was expected as far as safety was concerned.

By this time a small range suitable for launching sounding rockets was being set up at Sonmiani, close to Karachi in Pakistan. On his arrival in Pakistan, Massey was taken to see the site and had discussion with Pakistan scientists and technologists. Here again there seemed to be a real chance for small rocket experiments.

5   Possibility of a Commonwealth synoptic rocket programme

By the time the next meeting of the Consultative Committee took place in London on 14 June 1963, both Ariel I and the Canadian Topside Sounder satellite Alouette had been successfully launched, the latter on 29 September 1962. The Committee congratulated the Canadians on this achievement which they proposed to follow up by launching further satellites for ionospheric investigations.

Both India and Pakistan were strongly represented at this meeting by V. Sarabhai and M. Rahmatullah respectively. A considerable amount of time was devoted to the possibility of establishing a Commonwealth synoptic rocket sounding programme which, because of the developments on the subcontinent, was now no longer purely academic. Firings of small rockets from ranges in the UK, Australia, Canada, India and Pakistan could provide world-wide coverage, especially if co-ordinated. In this way the wind and temperature structure of the neutral atmosphere at heights up to 80 km or so could be investigated. The D region of the ionosphere could also be explored and the ranges in India and Pakistan would be especially valuable for studying the equatorial electrojet.

It was agreed that each country would be invited to nominate individuals to co-ordinate the scientific aspects of a proposed programme as well as the facilities which would need to be made available. Names proposed at the meeting were, for Australia, D.F. Martyn and E.C. Montgomery, India, E-V. Chitnes and H. Marthi, Pakistan, T. Mustafa and for the UK, G.V. Groves and A.C. Ladd. In each case, apart from Pakistan, the first-named was the Scientific Correspondent and the second the Technical. Pakistan nominated only a Technical Correspondent. Finally a resolution formulated as follows, was agreed: 'The Commonwealth Consultative Space Research Committee, noting the scientific need for co-ordinated global observations for the study of the structure and composition of the atmosphere, including the ionosphere at altitudes which cannot be studied from satellites and *noting* the international programmes in this regard set up for the IQSY[2] and for COSPAR, and *taking account* of the geographical distribution of present and potential launching facilities in the Commonwealth, and recognizing the need for low cost rocket systems to meet the requirements of such a programme, *recommends* that (a) a co-ordinated co-operative programme of suitably instrumented small rocket launchings be initiated in Commonwealth countries as soon as possible and (b) every encouragement be given to the production and provision of means for ready procurement of such rockets.'

The co-operative rocket launching programme was further discussed at an informal meeting held during the COSPAR conference in Florence on 20 May 1964, at which Massey, Rose, Sarabhai and R.S. Rettie (Canada) were present, together with A.W. Frutkin from NASA who attended by invitation.

It was agreed that grenade techniques were the most suitable to obtain the required accuracy in wind and temperature determination as had been fully established by the experiments of Groves and his colleagues using the Skylark Rocket at Woomera. In fact, discussion had started between the British National Committee for Space Research (BNCSR), the Pakistan Space and Upper Atmosphere Research Committee (SUPARCO) and the National Aeronautics and Space Administration (NASA) for carrying out as a joint project, grenade experiments using American Nike–Cajun rockets launched from Sonmiani. It was proposed that the UK would be responsible for the provision of the grenade bay and the grenades. The design of a suitable grenade bay that might be interchangeable with the British $7\frac{1}{2}''$ rocket was in an advanced stage. These grenade bays could be used in preliminary experiments at the different launching sites which would include South Uist in Scotland, Sardinia, Pakistan, Southern India, Darwin and Woomera in Australia. It would be necessary to design smaller grenade bays and grenades if such vehicles as the American-boosted Arcas rocket, capable of taking a payload of 10 lb to an altitude close to 100 km, were used. These requirements could be considered in the UK.

The tripartite discussions between NASA, Pakistan and the BNCSR were successfully concluded and an agreement signed. The full text is reproduced in Annex 13. By and large, SUPARCO was responsible for launching the vehicles, NASA provided six Nike–Cajun and Nike–Apache vehicles while the BNCSR, through Groves, would provide the grenade payloads and the special ground equipment needed for the grenades and train SUPARCO personnel in the acquisition of the data.

The first two experiments under this agreement were carried out on 29–30 April 1965 using grenade payloads manufactured in the USA but the remaining four firings used grenade bays manufactured in the UK to Groves' design.

In the first experiment 9 out of 12 grenades were ejected and flashes were recorded at four stations. The second was only partly successful, only two flashes being recorded optically. In the remaining flights carried out under the agreement a further payload provided by Groves' group at University College London was included. This was an ejector for trimethyl aluminium (TMA) (see Chapter 13, p. 278) which produced a glow in the atmosphere from observation of which information about atmospheric winds and temperature could be obtained.

Flights on 24 and 27 March 1966 were successful, those on 26 April 1966 and 29 November partially so.

Analysis of the data[3] obtained from these flights provided interesting information about the near equatorial atmosphere. Thus, the altitude profile of atmospheric temperature showed two maxima in spring and was observed to change markedly during a 72-hour period. The meridional component of wind above 55 km showed marked diurnal changes confirming earlier wind observations made with Judi-Dart rockets. The principal maximum in wind speed was found to occur at 105 ± 5 km altitude at which altitude in winter it is a little less than $100 \, \mathrm{m \, s}^{-1}$ towards NW and in summer near $118 \, \mathrm{m \, s}^{-1}$ towards E or NE.

This campaign showed that the combined grenade–TMA technique could be used very effectively in a co-operative programme.

As described in Chapter 10 the responsibility for the UK space science programme was transferred from the BNCSR and the Steering Group to the newly constituted Science Research Council (SRC) in 1965. Nevertheless the BNCSR continued to take a close interest in Commonwealth co-operation and maintained close contact with the Chairman (P. Sheppard) of the Space Policy Committee which was the body within the SRC concerned with the space science programme. A small sub-committee of the BNCSR was set up, consisting of Massey (Chairman), Groves, J.T. Houghton, Robins and G.D. Robinson to ensure that this good liaison persisted.

In 1966 Massey paid a further visit to India and Pakistan on his way out to Australia. In India he visited the range at Thumba and had many discussions with Indian space scientists, including Sarabhai. On his second visit[4] to the range at Sonmiani in Pakistan he was able to see at first hand the developments which had taken place since his first visit four years earlier.

6 Proposal for a conference of Commonwealth space scientists

The following year a well-attended meeting of the Consultative Committee took place on 25 July 1967 on the occasion of the COSPAR meeting in London. There were present H. Massey, W.J.G. Beynon, R. Frith, D.C. Martin, B.G. Pressey and A.P. Willmore (UK), J.H. Carver and V.D. Hopper (Australia), R.S. Rettie (Canada), K.R. Ramanathan (India), S.A. Jafri and M. Rahmatullah (Pakistan), S. Gnanalingam (Ceylon) and A.N. Hunter (Kenya). After describing the new situation arising from transfer of responsibility to the Science Research Council, Massey referred to the difficulty of initiating co-operative rocket programmes because of the large distances separating those involved and considered that what was now required was a meeting of those who would be directly concerned. Following expressions of support for this idea, especially from India and Pakistan, Massey suggested that the meeting might be held in Colombo, Ceylon, in January 1966. This site had the special advantage of being in a Common-wealth country in equatorial latitudes to which scientists from both India and Pakistan would have no difficulty in travelling. The representative from Ceylon welcomed this proposal confirming that it would be practicable and agreeing to indicate as soon as possible what dates in January would be most suitable.

It was agreed to proceed with the organization of a meeting on these lines involving four or five representatives from each country and lasting three to four days. The conditions for success of such a meeting were improved very much by the co-operative attitude taken up by Sheppard on behalf of the SRC, recognizing the need for expenditure on a co-operative rocket programme if clearly worthwhile and of value to the UK. A sum of the order of £10,000 was included in the SRC's space budget for this purpose in the financial year 1967–8 and sums of this order were likely to be available in subsequent years. Under these circumstances it became possible to consider the supply of launchers for Skua or Petrel rockets to one or two countries.

7   The Colombo Conference

The conference was opened in Colombo on 7 February 1968 by the Governor-General of Ceylon and was attended by the following scientists from the Commonwealth; B. Rofe (Australia), B. Wilson (Canada), S. Gnanalingam, L.A.D.I. Ehanayake and P.C.B. Fernando (Ceylon), K.R. Ramanathan, P.D. Bhavsar, S. Pisharoty and T.S. Sastry (India), H. Rahmatullah (Pakistan), A.P. Willmore, G.V. Groves, K. Burrows, F.N. Byrne and R. Frith (UK). The Chairman of the Consultative Committee was unable to attend and in his absence Willmore acted as Secretary and organizer of the meeting which lasted through to 10 February. It was very successful and apart from agreeing a number of recommendations, also discussed in some detail a number of projects involving co-operation by Commonwealth scientists.

Among the recommendations were that launching facilities for Skua and Petrel rockets be installed at Thumba and Sonmiani and that the UK examine the problems involved with the countries concerned. It was also recommended that a telemetry receiving station be established at Sonmiani. To facilitate the conduct of co-operative experiments, it was recommended that communication links be established between South Uist. Sonmiani, Thumba, Gan and Woomera launching sites. The study of the equatorial electrojet was of obvious importance and emphasized the need for the establishment of ground magnetic observatories at Colombo and the island of Gan in the Maldives where there was a Meteorological Office station. A substantial extended programme of grenade and TMA launchings spread over several years at Sonmiani was regarded as valuable. It was also recommended that for studying winds and temperature and other atmospheric parameters over the Indian Ocean area a series of meteorological rockets (such as Skua I) launchings be carried out from Thumba, Gan, Sonmiani, Carnarvon in West Australia and possibly also from the Italian San Marco launching platform off the coast of Kenya and Soviet ship stations in the Indian Ocean.

In more detail, the conference discussed a number of projects proposed for the initial phase of the programmes. These included studies of the equatorial electrojet, the D region of the ionosphere, and atmospheric temperatures and winds. Leading scientific participants from each country were identified in each programme, those from the UK being Burrows (Equatorial electrojet by ground-based and rocket-based measurements), Frith (Magnetic observatories and meteorology), Groves (Temperature and wind measurements) and Willmore (the D region). Interest was also expressed by Indian and Canadian scientists in the study of cosmic X-ray sources.

Having identified the initial projects the conference proceeded to consider what was involved logistically in carrying them out. For the work on the equatorial electrojet, Centaure or Petrel rockets would be required with a special head release mechanism for the latter, costing up to £300 per payload. With the strong interest in participation expressed by the Radio and Space Research Station (RSRS) it was expected that assistance would be forthcoming, both in this matter and in the establishment of the magnetic observatory at Colombo. That at Gan would be the responsibility of the Meteorological Office.

No difficulty was anticipated with Australian and Indian co-operation on D region projects apart from the possible need to provide some Skua-2 motors for India but there would need to be further development of facilities in Pakistan.

As far as atmospheric wind and temperature measurements were concerned, India already manufactured TMA dispensers and Indian scientists were carrying out this type of experiment while Pakistan was anxious to continue the co-operative arrangements carried out with the BNCSR (through Groves) and NASA.

8   Contribution to conference recommendations from the UK

The outcome of the conference was considered on 26 February at a meeting held in London which was attended by Massey (Chairman), Burrows, Byrne, G.W. Eastwood, Frith, Groves, J.A. Saxton, Sheppard and Willmore. They paid particular attention to the costs which would be incurred if the recommendations of the conference were met. These were estimated to be close to £28,000 for the first two years to provide the necessary facilities made up as follows: £11,000 for India including £2000 for launching tubes, £4000 for two Petrel rockets and £4000 for four Skua-2 rockets, £17,200 for Pakistan including £2000 for launching tubes, £1000 for telemetry, £2000 for two Skua-2 rockets and £6200 for other facilities.

The question of the administrative machinery required to deal with these matters was also considered and it was suggested that the SRC provide the necessary secretariat and that the Space Policy and Grants Committee (SPGC) of that body appoint an *ad hoc* working group to implement UK participation in a synoptic programme of rocket-borne experiments, including direct negotiations with the appropriate overseas authorities with whom joint projects were arranged. At the same time the Consultative Committee would appoint an *ad hoc* co-ordinating sub-committee consisting of two persons for each participating country.

9   The Commonwealth co-operative programme in operation

The first joint experiments at Thumba were carried out by the group under A.P. Willmore at the Mullard Space Science Laboratory of UCL, in collaboration with Indian scientists. These consisted in measurements of electron concentration and ion composition in the D region, using instruments prepared by the UCL group and carried in Skua-2 rockets (see Chapter 8, p. 185). It was planned to launch exactly similar instrumented Skua-2 rockets from South Uist in the Outer Hebrides, the aim being to compare the behaviour of the D region parameters at high and low magnetic latitudes and at different solar zenith angles.

The six Skua rockets were launched successfully from Thumba between 14–17 April 1970, and apart from the loss of telemetry in one rocket the instruments functioned well.

In 1968 the Meteorological Office established facilities for Skua rocket launchings on the island of Gan, which is within 100 km of the equator. Over the following four years they undertook studies of temperatures and winds over the height range 20–65 km. In March 1970 a short series of co-ordinated firings of Skua rockets was made at Gan and Thumba, the latter in collaboration with the Atomic Energy Department of India.

The first Petrel rockets to be launched from Thumba carried barium and strontium vapour dispensers and the experiments were carried out jointly by the group from the University of Sussex (A. Martelli and P. Rothwell) and scientists from the Physical Research Laboratory at Ahmedabad. The magnitude and direction of motion of the neutral and ionized vapour clouds produced were measured from photographic observations at three sites. These flights carried out on 4, 6 and 7 January 1972 were successful and the results obtained are referred to in Chapter 13, p. 276.

Shortly after these flights the first Petrels carrying experiments to investigate electrical and magnetic states of the lower and middle atmosphere over Thumba were successfully launched (on 2, 5 and 13 February 1972). An account of the magnetic measurements made and their application to studies of the equatorial electrojet is given in Chapter 14, p. 319.

The programme at Thumba culminated in a launching campaign in February 1975 in which it was planned to launch 12 rockets in a single day between dawn and dusk to investigate atmospheric and ionospheric processes including neutral winds and the equatorial electrojet. The experiments were performed in a collaboration between scientists at the Indian Space Research Organization, UCL and the University of Birmingham. In the event, five Petrels were launched on 9 February four of these with complete success and the remaining one with partial success. The experiments carried made measurements of neutral winds and temperature at dawn using the TMA trail method and in daytime using a lithium vapour trail, of electron density and temperature and of electric fields. These were followed on 19 February with two further wholly successful Petrel flights measuring wind distribution at dusk and daytime. Two Centaure rockets were also flown, one completely successful, to measure electron density and temperatures and electron currents using a scalar magnetometer and the other, also mainly successful, to measure electron currents with a vector magnetometer. A preliminary account of the results from these flights has been given by Rees *et al.*[5]

Following the successful development of these programmes much of the responsibility was gradually taken over by the Indian space scientists so that by 1976 all hardware operations, including the design and construction of experiments, were carried out locally. Nevertheless, the foundation was laid for a much wider Anglo–Indian collaboration in space research which was initiated in 1981.

We have already referred to the tripartite experiments involving co-operation between NASA, SUPARCO and the BNCSR which were carried out in 1966. Further grenade and TMA release experiments were carried out using Petrel rockets and the grenade bay designed by Groves and his colleagues (see p. 170). The first such launchings were successfully carried out on 7 and 25 November 1971 at Sonmiani but the further programme was curtailed through increased preoccupation of the UK scientists involved with high latitude studies.

8

Smaller rockets for scientific purposes – Skua and Petrel

We have described in Chapter 3 how the Skylark rocket was developed and became the mainstay of the research programme requiring comparatively large and complex scientific payloads. In this chapter, we review the initiation and development of a series of smaller rockets of more modest performance. These met a need for a cheaper means of carrying out experiments in the neutral atmosphere and in the lower and mid-ionosphere. Simpler launchers allowed them to be more easily used to give wider geographic coverage, and programmes of synoptic observations were feasible in ways not possible with the more expensive Skylark. The basic rockets in question, with some nominal figures to give an approximate indication of the relative sizes and capabilities. are listed in Table 8.1.

1 The scientific case

The case, in general terms, for rockets small enough, and cheap enough, to be used widely and launched frequently was well recognized in the late 1950s. Massey had introduced the subject to the Government

Table 8.1.

Rocket	Diameter	Payload weight	Peak altitude
Skua 2	13 cm	6 kg	100 km
Petrel 1	19 cm	18 kg	143 km
Fulmar	25 cm	50 kg	250 km
For comparison			
Skylark[a]	43 in	150 kg	700 kg

[a] The payload altitude characteristics depend on the particular combination of boost and sustainer motors in use. The figures quoted are for the Skylark 12, a three-stage rocket system.

177

Steering Group on Space Research in November 1960. In a paper dealing primarily with the topic of European co-operation in space research, he referred to the need to develop for scientific purposes a rocket smaller than Skylark, costing no more than £1000 per rocket, to be suitable for world wide use. The British Meteorological Office was also interested in the use of small rockets as a tool in their research programme. Design studies by the firm of Bristol Aerojet Ltd and an active interest by Ministry of Aviation research establishments provided a sound technological background against which scientists could prepare their case.

The anticipated world-wide use of the rockets required that the specification should, as far as practicable, be widely discussed and agreed by the potential users. Massey, as Chairman of the British National Committee on Space Research (BNCSR) took the lead in proposing an informal discussion meeting of interested scientists from countries in the Commonwealth and Western Europe. A two-day meeting was held in London under the auspices of the Royal Society on 24 and 25 January 1961 to consider the use of small vertical sounding rockets. In addition to United Kingdom scientists, the meeting was attended by scientists from Australia, Canada, New Zealand, Rhodesia and Nyasaland, and South Africa. From Western Europe, there were scientists from Denmark, France, Italy, The Netherlands, Norway, Sweden and West Germany. The Chairman also had prior indications of the possible interest of India and Pakistan. One important aspect of the meeting was that concerned with the organization of co-operation in small rocket programmes between the various interested participants. This is described in more detail in Chapters 6 and 7; here we consider only the relevant scientific and technical matters.

It was natural that the UK members of the meeting should take the lead in outlining the scientific case; Massey and Boyd in particular gave a summary of the situation as they saw it. For the purposes of the discussion it was agreed that the term 'small rocket' should imply a basic vehicle significantly cheaper than Skylark, costing no more than £1000 in quantity production. It was described how the atmospheric properties of interest varied greatly both in time and in location. Instruments carried in balloons could record phenomena up to altitudes of 30 km, whilst satellites would enable regular surveillance of the regions above 150 km to be carried out. The altitude band between 30 km and 150 km would best be explored by rocket-borne instruments. The higher frequency of launching and wider geographical coverage of the launching sites possible with the small rockets would enable an overall picture of transitory phenomena to be obtained, particularly if regular observations could be made. The relatively short observation time and limited payload capacity could to some extent be offset by the greater number of launchings.

A breakdown by Boyd of the fields of interest showed five main groups of experiments into which a programme could be divided.

Physical properties of the atmosphere
Hitherto, most rocket launching sites had been in the desert areas of the world, around latitudes of 33 degrees. This had resulted in most of the existing data being relevant to these latitudes, with the exception of some measurements in the Canadian auroral zone. Measurements were now needed on a world-wide scale, at latitudes between the desert regions and the polar caps. World-wide measurements were required of atmospheric characteristics, including winds, tides and temperatures. The distribution of atmospheric gases such as ozone, molecular and atomic oxygen and nitrogen and carbon dioxide, needed investigation, together with information about minor constituents such as hydroxyl radicals and the rare gases. The altitude band between 90 and 130 km would probably include most of the features of interest from the point of view of atmospheric composition. R. Frith of the Meteorological Office explained that much of meteorological interest lay in the altitude band up to 60 km and for this purpose a rocket of even more modest performance would be adequate.

Electrical properties of the atmosphere
There was much to be explored in the D region of the ionosphere, up to about 100 km, whilst the sporadic E phenomena in the altitude band 100 km to 150 km posed many problems. The magnetic effects of electric currents in the E region required investigation. Whilst ideally rocket measurements up to 150 km were required, there was no doubt that it would be extremely useful if observations could be made up to 130 km should the higher apogee prove beyond the reach of the type of cheap rocket under consideration.

Photo phenomena in the atmosphere
The origins of some of the more interesting spectral components of the airglow radiation were not known with any certainty, and the relations between air glow variations and temperature required study. These effects were believed to occur in the 70 to 90 km region, and a rocket apogee of 100 km should be sufficient.

High latitude phenomena

Phenomena due to incoming particles, from the sun or perhaps from the radiation belts (the van Allen belts), were generally sporadic, and many rocket launchings would be needed to examine the nature of the ionizing particles and the auroral and magnetic storm effects which resulted. Much of the interest was believed to lie in the altitude band 100 to 120 km, but some effects may occur at altitudes as low as 40 km. A rocket apogee of 130 km was thought to be adequate.

Astronomical observations

It was recognized that only limited use could be made of small rockets for astronomical observations, but topics such as the relations of solar ultra-violet and X-radiation to atmospheric variations could be studied, particularly if simultaneous measurements could be made from multiple launchings. Peak rocket altitudes of at least 100 km would be required.

The consensus view coming from this discussion was that a ceiling altitude of at least 130 km was highly desirable, but if this led to unit costs above the £1000 mark, then a ceiling of 100 km would still enable much useful work to be done.

2   The payload specification and other design features

Peak altitude and a target unit cost were only two of the constraints on the rocket designers; a third of equal importance was the payload weight to be carried. In this context, the term 'payload' is taken to mean the scientific instruments, the telemetry transmitter and the power supplies, but the weight of the rocket nose cone is excluded. Experience suggested that about half the payload so defined could be allocated to the scientific instruments, and that a total payload weight of about 30 lb would be a suitable figure on which to base further considerations. The rocket diameter should be as small as possible to achieve an efficient design, and it was thought that 7.5 in would be a reasonable compromise in this respect. As regards variations in the performance parameters, RAE had noted that as a guide it seemed that the product of maximum height and payload was roughly proportional to cost for the type of rocket under discussion. This implied that for about the same cost, a rocket could take either 30 lb to 100 km, or 23 lb to 130 km, a height band which straddled the regions of main interest. A feature of rocket performance which was bound to be of importance in circumstances in which launching from a number of sites was contemplated, perhaps not all of them regular firing ranges, was that of dispersion. In other words, the variation of the impact point from that predicted. This would crucially affect the safety area required for rocket launchings to be acceptable and could seriously limit the geographical sites which might be desirable from a scientific point of view. Thus acceptable dispersion also emerged as a necessary feature of a successful design.

3 The technological background

We now turn from the scientific considerations leading to the requirements for small rockets, to the state of affairs in relevant rocket design and availability. The first moves in Britain had been made by the Meteorological Office, which in 1959 had obtained approval for the establishment of a High Altitude Research Unit. This Unit aimed to use vertical sounding rockets to extend measurements in the atmosphere up to an altitude of 80 km, well above the 30 km altitude limit attainable by balloons. The requirements for this, and other more advanced needs, had resulted, by early 1961, in design studies by the British firm Bristol Aerojet Ltd for two rockets.

The smaller of the two, provisionally designated as 'BAJ Synoptic' was 5 inches in diameter and intended to carry a payload of 10 lb to 80 km for a unit cost of some £400. The larger rocket, 'BAJ Research' was 7.5 inches in diameter and intended to carry 40 lb to 100 km, for a unit cost of £800 or so. It was, of course, no coincidence that in general terms the specification to which the space scientists had been led was quite close to that of the larger of the two rockets of interest to the Meteorological Office. There was naturally close scientific and technical liaison between the scientists represented by the Royal Society's British National Committee on Space Research, and the scientists of the Meteorological Office, but the two parties were separately funded and administered. This should be borne in mind as we continue to describe the closely linked progress of the two rocket programmes which in many respects appear to coalesce.

Following the discussions between Western European and Common-wealth scientists, and further considerations within UK, the Ministry of Aviation, at the request of the British National Committee, reported towards the end of 1961 on ways and means of meeting the specified needs. The Ministry report brought out clearly some of the problems of logistics and safety which understandably did not fully emerge from the scientific discussions. Below are summarized some of the points which were brought to the attention of the British National Committee in December 1961.

3.1 The rocket vehicle

The use of a solid propellant rather than liquid propellants for the rocket motor would be likely to result in a cheaper and more easily handled vehicle in the size range of interest. Safety aspects of the launch operations would require that the rockets should have a low dispersion. This implied that the rocket should leave the launcher at high speed; the larger the rocket, the larger and more expensive the launcher.

3.2   Ranges and range safety
For the type of rocket under discussion, to be launched near to the vertical, the scale of the safety area likely to be necessary would be a circle of radius some 20 miles. This would probably ensure that the chance of a rocket falling outside the area would be about one in a million. In practice it would be necessary to check that the area was clear before each rocket launching, and this would require some system of surveillance; in the case of a sea range, a suitable surveillance radar would be the most likely method. As an aid in aligning the launcher immediately prior to the launch fairly accurate wind measurements both at the surface and at higher levels would be necessary.

3.3   Instrumentation
Two types of instrumentation would probably be needed; telemetry from the rocket to a ground receiver for recording experimental measurements, and a method by which rocket trajectory could be determined. Trajectory data may be needed to establish an altitude/time relationship for the scientific payload, and also as a check on the safety aspects of the operation. No basic problems could be foreseen in providing each type of instrumentation, but the optimum economic methods would depend upon a number of factors such as the nature of the scientific experiments and the particular features of the launching range.

3.4   Buildings
In addition to the general facilities such as accommodation for staff, special purpose buildings would be needed for the safe storage of explosive stores, the safe assembly and testing of complete rockets and payloads, and a firing point at the range head for the firing team.

3.5   Costs
It was estimated that the capital cost of setting up a sea range along the lines summarized above, suitable for launching small rockets of the size under consideration, might be some £300,000 if there were no existing facilities on which to build.

4 Alternative ways ahead

By early 1962 a clearer picture of the possibilities had begun to emerge from the many discussions and studies which had taken place in the previous twelve months. The scientific needs had crystallized into two specifications. The smaller of the two rockets, for measuring temperatures and winds up to a height of about 60 km and mainly of interest to the Meteorological Office was the first to be approved. Arrangements were made for its development and production by Bristol Aerojet Ltd, using a rocket motor developed by the Rocket Propulsion Establishment of the Ministry of Aviation. Plans were made for this rocket, known as Skua 1, to be launched from an Army firing range at South Uist in the Outer Hebrides. Although not of direct interest to the space scientists working through the BNCSR, the progress of Skua 1 was to be important because it led the way in the development of the South Uist range for scientific purposes, and a later version of the rocket, Skua 2, was eventually to be used in the BNCSR space science programme itself.

The provision of a rocket to meet the second and more advanced specification posed in the first instance a problem not uncommon in this field. Should an all-British development be undertaken, or would it be preferable to purchase rockets already developed abroad? In offering advice on this matter in 1961 the Ministry of Aviation listed 28 rockets of various types, sizes and stages of consideration, development and use. Of these, 21 were of United States origin, three British, two French and one each from Canada and Australia. Only eight promised to meet the performance specification for altitude and payload, and all had limitations or disadvantages of one sort or another. It was particularly difficult to assess the safety factors in relation to the relatively unsophisticated ranges from which it was hoped that the rockets could be launched. Most of the American rockets used propellants which were either complicated (such as nitric acid and alcohol) or were security classified, which restricted the conditions under which they could be operated.

The outcome of such considerations led to the conclusion by the BNCSR that the best course of action would be to support development of a British rocket specifically designed for the performance and conditions of operation envisaged. In June 1962 the Steering Group on Space Research recommended in principle that the development of a rocket to meet the specified needs of the space scientists and meteorologists should be undertaken, and the Ministry of Aviation was invited to make the necessary arrangements. The cost of the development was tentatively estimated by the Ministry as about £80,000.

5   The development of Skua

At this point we digress somewhat to outline some of the factors influencing the particular design of rocket, bearing in mind the special needs for simplicity, cheapness and safety. We have already referred to the choice of solid propellant motors for simplicity. The type and combination of motors has considerable bearing on the dispersion characteristics of the rocket. The aim always is to ensure that the rocket can be launched in the prevailing wind conditions at such an angle that not only is the desired altitude achieved, but also that the rocket strikes the ground (or sea) within the defined safety area and that any boost motors separated during flight also land safely. There are three combinations relevant to this situation:

(a) Unboosted – single-stage rockets.
(b) Boost-coast – two-stage systems, the first being a short burning booster rocket giving a high initial acceleration; the second stage, a coasting dart having a low drag/weight ratio.
(c) Boost-sustainer – two-stage systems, comprising a short burning booster rocket, the case of which separates from the main vehicle at the end of burning. The latter contains a slow burning sustainer motor which then ignites.

The unboosted rocket is the simplest type, but is inefficient and heavy compared with two-stage rockets, and a particular disadvantage in the present context is that a long (and expensive) launcher may well be required to minimize dispersion. The unboosted Skylark with its large launch structure is an example. The boost-coast system gives relatively low dispersion, but is only suitable for small diameter payloads, not acceptable for the proposed programme. The boosted two-stage system has a high launch velocity from a short launcher, giving low wind dispersion, whilst a good altitude/payload performance results from the lower continuous acceleration given by the sustainer motor in the lower atmosphere. Since the rocket is launched nearly vertically, the impact of the boost motor case can cause problems, but there are methods of avoiding this hazard.

We have already mentioned the development of Skua 1 by Bristol Aerojet Ltd for the Meteorological Office. A small rocket, of 5 inches diameter, Skua 1 was a boosted two-stage system, launched from a tube. The first stage boost motor, code named 'Chick', burned for 0.2 seconds and separated from the second stage about 20 metres above the ground, falling close to the launcher. The second stage motor, code named 'Bantam' burned for a little over half a minute, and resulted in the second stage, carrying a payload of some 6 kg, reaching an altitude of about 65 km. It is not intended to describe here the programme for which Skua 1 was used by the Meteorological Office, and it must suffice to record that its use began in January 1964 with launchings from South Uist in the Outer Hebrides. A radio sonde with a fine wire temperature sensor was released at apogee and descended on a partially metallized parachute which could be tracked by radar. Winds at altitude could be computed from the observed drift of the parachute.[1] The programme developed in nature and scale; in the year ending July 1967, for instance, 80 Skuas were launched. Skua 1, as used by the Meteorological Office, did not meet the needs of experimenters requiring telemetry from equipment in the rocket, but its economical performance and simple launch arrangements were attractive in situations where its small diameter and limited altitude were adequate.

Accordingly, a modified version named Skua 2 was developed. This was basically similar to Skua 1, with an additional short rocket body section carrying a 465 MHz telemetry transmitter, made by EMI Ltd. The antenna consisted of a resonant slot cavity in the skin, so that the effect on rocket performance was minimal. An improved rocket motor allowed altitudes up to 100 km to be attained with a payload of 6 kg. The Science Research Council ordered nine of the Skua 2 rockets in 1967. The first launching took place in March 1968, and by the end of that year 11 launchings for scientific purposes had taken place at the South Uist range. Instruments to measure positive ion densities had been installed in six of the rockets by the Mullard Space Science Laboratory (MSSL) whilst the University College of Wales had instrumented the remaining five rockets to measure electron density profiles in the ionosphere. From 1968 onwards, Skua 2 was widely used both in the UK space science programme and in co-operative programmes with Commonwealth countries as described in Chapter 7. Some 40 launchings of Skua 2 took place before the termination of the programme in 1972. The use of Skua 1 by the Meteorological Office continued until November 1980, when the last rocket of the series was launched from the South Uist range. Since the inception of the programme, more than 500 such rockets had been launched. Figure 8.1(a) shows in outline the Skua meteorological rocket.

6   The development of Petrel

We now return to the development of the research rocket for space science purposes which was to become known as Petrel. The first substantive moves took place after the agreement of the Steering Group on Space Research in June 1962, and the processes of design, costing, development and testing could begin. This was to be a long drawn-out operation and the proving flights were not to be completed until January 1968. The lapse of nearly six years was due in part to the development time required for the new motor, and in part to the continual battle against rising costs at a time when the budget for UK space science was stretched by the competing claims of the Skylark rocket programme, the satellite programme, contributions to ESRO and grants to university research teams.

The vehicle design studies were undertaken by Bristol Aerojet Ltd with the Rocket Propulsion Establishment responsible for the motor design. Both possible types of configuration for the 7.5 inch diameter motor were considered; the end-burning (cigarette-type) and the radial-burning design in which burning takes place along an axially symmetrical internal surface. The former has the advantage of a constant propellant burning area, whereas the latter generates a much smaller shift in centre of gravity of the motor as the propellant burns. The basic problem facing the motor designers was to produce a long burning, high performance rocket of low frontal area. About mid-1963, the Ministry reported that the problems seemed to have been mastered, and gave a tentative estimate of £750 for the unit cost of the basic rocket vehicle in quantity production. Although the technical outlook appeared bright, the financial forecasts given by the Ministry in March 1964 gave reason for much concern. The development costs estimated in 1962, of £80,000 had risen to £220,000 by 1964. The Ministry of Aviation believed that £190,000 of this total should properly be attributed to the space research budget. The remaining £30,000 would be attributable to the Ministry's budget to cover its own interests in the rocket.

This unwelcomed increase in costs caused the various Committees of the Royal Society to pause and reconsider the case for supporting the development of an all-British rocket. In a paper by the Space Research Management Unit (SRMU), the present (1964) scientific demand for this type of research rocket was reviewed, and the cost and feasibility of using an alternative rocket were explored. The paper confirmed the scientific interest in the rocket, particularly in the fields of meteorology and ionospheric physics. In reviewing possible foreign alternatives to the proposed British rocket, the paper emphasized that not only had the costs to be considered, but also the limited facilities and the safety requirements of the South Uist launching range. It seemed that the only feasible alternatives might be the American Nike–Cajun, the Japanese Kappa 6S, or one of a new series of American sounding rockets, of which the Archer appeared the most suitable. However, such cost estimates as were available suggested that any suitable alternative rocket with a proven record of launchings was likely to cost about £6000, several times the estimated unit cost of the British rocket.

The reasons for the relative cheapness of the British rocket lay in the choice of an end burning sustainer motor, recoverable boost motors, simple telemetry and a simple launching system. The SRMU paper concluded by pointing out that the high cost of alternative rockets meant that if the funds available for the British rocket were used to purchase an alternative, only a very limited programme of launchings would be possible. Even this assumed that all other safety and logistic matters presented no serious problems.

After considering the situation and the evidence available, the BNCSR decided to continue to recommend development of the British rocket. The main grounds were that a firm scientific requirement was foreseen for at least ten years, that no suitable alternative was available at an acceptable price, and that the additional development costs could be found by savings on the budget item for satellite computing costs. There were some, such as J. Sayers, who had misgivings about the decision, but they were in a minority and in July 1964 the Steering Group on Space Research approved the development of the rocket. The way was then clear for the Ministry of Aviation to take the necessary contract action. The Rocket Propulsion Establishment was to develop the solid propellant sustainer motor, to become known as the Lapwing, and the boost motors. Bristol Aerojet Ltd became the main contractor and design authority for the rocket system and motor hardware. The Co-ordinating Research and Development Authority was to be the Atomic Weapons Research Establishment (AWRE) which also was to collaborate with EMI Ltd in the adaptation of the latter's 24-channel telemetry system.

The development was expected to take nearly three years, and it was planned to launch eight proving rockets from the South Uist range to clear the design before use for scientific purposes. Work proceeded in 1965 in a promising fashion, and the Science Research Council made arrangements with AWRE for the latter to provide a complete user service of assembly, launching, data recording and analysis. In 1966 SRC obtained agreement to place a production order for 28 rockets, eight to be delivered in 1967/68, and 20 in 1968/69.

The early proving flights of Petrel were delayed by bad weather, but the first two launchings were successfully accomplished in June 1967. The last of the eight development flights took place in March 1968 the final four of these including scientific payloads. The operational phase of the scientific programme began in May 1968, and by the end of that year, scientific payloads had been carried in a total of eleven Petrel rockets launched from South Uist. The experiments, most of which were concerned with electron and ion measurements in the ionosphere, originated in the Universities of Birmingham, Southampton and Sheffield, University College of Wales, and the Radio and Space Research Station. No rocket failures occurred, and of the 22 experiments carried, complete or partial success was achieved in 13 cases.

The basic Petrel rocket is shown in outline in Figure 8.1(b). Figure 8.2 shows the Petrel and Skua rockets and Figure 8.3 a typical Petrel launching.

Fig. 8.1. Outline of (*a*) a Petrel rocket and (*b*) a Skua meteorological rocket.

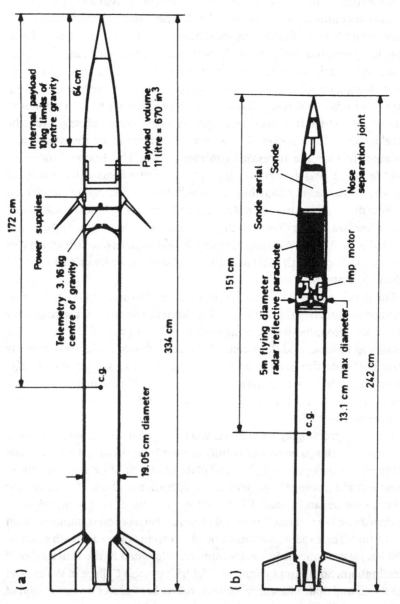

Fig. 8.2. Photograph of a Skua and a Petrel rocket with booster motors, parachute and sabot.

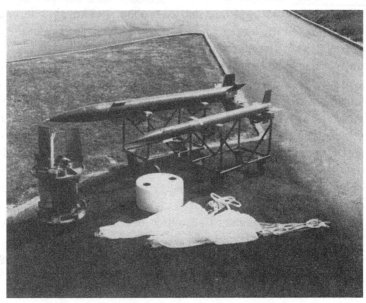

Fig. 8.3. A Petrel rocket leaving the tubular launcher.

7   The main characteristics of Petrel 1

A payload of 18 kg is carried to about 143 km altitude, or 25 kg to 115 km, when the launch angle is effectively 5 degrees from the vertical. The main sustainer motor, the Lapwing I, burns for about 31 seconds and the rocket is stabilized by six fins at the rear of the motor. These can be set to give a nominal spin rate of nine revolutions per second at the end of motor burning. The Petrel 1 launch is normally boosted by a cluster of three Chick motors, which burn for about 0.2 seconds before separating from the main vehicle. The boost assembly descends by parachute and can be recharged and reused. The launcher is a 10-metre long tube, of a diameter sufficient to take the boost carriage which is fitted with wheels for location in the tube. The payload section of Petrel normally remains attached to the motor for the duration of the flight.

Originally the telemetry transmitter, operating at a nominal 465 MHz was of a standard EMI type and used a multiplexer to sample 22 channels at 80 Hz and 23 sub-commutated channels at 3.3 Hz. Later in the programme a pulse code modulated telemetry system with a data rate of 125 kilobits/second was introduced. Since the development of the basic Petrel system, a number of variants have been produced to give a wider range of performance. For instance, Petrel 2 uses four Chick motors (instead of three) for the boost phase with a lengthened Lapwing motor and can take a payload of 18.5 kg to an altitude of 175 km. It was first used in the scientific programme in 1977. Thus the Petrel rocket was successfully introduced as a valuable facility available to space scientists in the UK and eventually through co-operative arrangements, to scientists in the Commonwealth and in Europe.

Since 1968, some 234 Petrels have been launched from sites in South Uist, Kiruna (Sweden) Andøya (Norway) Thumba (India) Sonmiani (Pakistan) and Greenland. About two-thirds of the launches have taken place from South Uist. The programme of Petrel launchings terminated in August 1982. Further details will be found in Appendix B.

8   The Fulmar rocket

One further rocket in this series needs to be mentioned briefly although the numbers launched have been insignificant compared with the numbers of Petrel and Skua launches.

In the mid-1970s it became apparent that the programme of high latitude launches to obtain observations from rockets, being planned as part of the geophysics research programme, would greatly benefit from a rocket intermediate in performance between Petrel and Skylark. The need was met by the development of the Fulmar rocket, 10 inches diameter, and intended to carry payloads of some 50 kg to heights in excess of 250 km. Fulmar, a product of British Aerojet and based on the Flamenco rocket developed by that firm for Spain, used a modified mobile Skylark launcher. For the purposes of the high latitude campaign, this was sited at Andøya.

The first two Fulmars were launched in late 1976, with only partial success, in each case the apogee being well below 250 km. Modifications were made, and in the autumn of 1977 three more were launched from Andøya with much greater success. The scientific payloads were provided by University College London and Appleton Laboratory (neutral winds and temperatures and electron characteristics); by Sussex University and Appleton Laboratory (electron and positive ion characteristics); and by Sheffield University (very low frequency radio waves). Only a very limited future use was foreseen for Fulmar, the proposed scale of activity being about three launches every three years. The sixth Fulmar, launched in 1979, was a failure, so it must be said that the project did not achieve the high standard of performance set by its companion rockets, Skua, Petrel and Skylark, and the Fulmar series was terminated after the six launchings to which we have referred.

9

Attitude controlled Skylark rockets

1 The scientific case

We have described in Chapter 3 the genesis of the Skylark programme, and the development of the rocket into a reliable and versatile vehicle for space science experiments requiring a duration of a few minutes at altitudes up to a few hundred kilometres. Here we follow in more detail one important technological development, that of attitude stabilization of the rocket head. This transformed the vehicle from a mere carrier of equipment to high altitudes, to a sophisticated facility enabling astronomical instruments such as spectrometers in the ultra-violet, and X-ray detectors, to be aligned accurately on the sun and on selected stars. Attitude controlled Skylark became the forerunner of the astronomical satellites Ariel 5, Ariel 6, and The International Ultra-violet Explorer (IUE).

On 21 April 1955, when it was clear that a scientific programme using the high altitude rocket being developed by RAE Farnborough would indeed become a reality, a group of interested scientists met informally in Massey's room at University College London (UCL). In addition to Massey and R.L.F. Boyd of UCL, the group included J. Sayers of Birmingham University, D.R. Bates of Queen's University, Belfast, and W.J.G. Beynon of The University College of Wales, Aberystwyth. Possible experiments were discussed in relation to the expected performance of the rocket. It was decided that RAE should be asked to provide, if possible, either telemetry of the angular aspect of the rocket axis during flight, or preferably stabilization of the aspect 'by means of a large gyro having its axis along that of the vehicle'. This particular suggestion for the method may not have been appropriate (amongst other reasons, the roll aspect of the rocket would have been indeterminate) but the objective of attitude stabilization was clearly seen as desirable even at this very early stage.

In June 1955, the same group of scientists met F. Hazell and T.L. Smith of RAE, and explained how the scientific value of the rocket experiments would be increased by a knowledge of the aspect or preferably by control of the rocket attitude. The accuracies being discussed were very crude at this stage. For instance, the photometric experiments proposed by Queen's University, Belfast required a knowledge of aspect to within 5° or control to within 10°. Ion sampling experiments would also benefit from similar characteristics. Later that same year, in December, RAE reported that the possibility of obtaining 'sunseeker' equipment from the USA was being explored and UK research groups were invited to send proposals for possible experiments using such a facility.

There seems to have been a lull in the proceedings, at least from the scientific user point of view, for the next two years or so. Then in April 1958, H.E. Butler at the Royal Observatory Edinburgh (ROE), explained to the group of scientists meeting at UCL, that he was interested in using high altitude rockets to carry cameras above the atmosphere. These would photograph the sky in light of much shorter wavelength than that which could penetrate to ground level (about 3000 Å being the lower limit of wavelength). In particular he would hope to obtain photographs at the Lyman α wavelength. Whilst there was as yet no direct evidence of the brightness of stars at such wavelengths, there was reason to expect considerable radiation from at least some sources. If these observations were successful, the next step would be to photograph the sources through a small angle prism to give low dispersion spectra. Clearly the camera axis would need to be stabilized during the exposure time of at least a few seconds. A stability of the order of one minute of arc would probably be necessary. Such accuracy was greater than any which had been discussed so far, but Hazell and E.B. Dorling of RAE inferred that a stabilizing device might be available, capable of the accuracy needed, although it would not be possible to point the rocket in a predetermined direction.

The next year or so saw big developments in the establishment of a space research programme in Britain, including the start of the joint Anglo–American satellite series, and on the organizational side, the inception by the Royal Society of the British National Committee for Space Research (BNCSR) and its various sub-committees. It was at a meeting of the Working Group on Astronomy, in October 1959, that further scientific problems involving the need for stabilized high altitude rockets were discussed. J. Ring proposed solar observations, using a coronagraph carried in a rocket stabilized to 1 minute of arc. Boyd had joined with Butler in planning to operate a camera suitable for photographing in the ultra-violet, from 1550 to 2700 Å, which would require stabilization to a few minutes of arc; it would also be necessary to recover the film. There was also at this same time, a hope that X-ray measurements from astronomical sources might be made from the Skylark rocket even though no such objects had yet been discovered. Thus by the end of 1959, there was no lack of scientific proposals which needed an attitude stabilized rocket, and the question was how and when this need could be met.

2   The engineering approach

It was known that H. Friedman of the Naval Research Laboratory in the USA was, amongst others, very active in this field. At this time (1959) experimenters in the USA were responsible themselves for any attitude control or pointing equipment needed in the rockets being used. The University of Colorado had developed a 'sunseeker' and Ball Bros. of Boulder manufactured stabilization equipment for vertical sounding rockets. None of this was directly applicable to Skylark without substantial adaptation. The design engineers at RAE approached the problem *ab initio*, mainly in pursuit of their own research and development aims, partly and informally (since there was no formal commitment at this stage) to meet the expressed needs of the scientists so far as practicable.

The first International Space Science Symposium, sponsored by COSPAR at Nice in January 1960, was an opportunity for RAE to show how the work was progressing. The most relevant RAE exhibit was of an auxiliary jet attitude control set using compressed air. This was designed to control the head of a Skylark rocket about three axes, after separation from the spent rocket motor at a height of about 61 km. The head would be expected to coast upwards to some 150 km before falling back to earth. Several minutes would be spent effectively above the atmosphere during which time the attitude would be stabilized by the three-axis air jet control to a datum direction established by two free gyroscopes. This method was not to be the main line of the ultimate successful developments, but it was a valuable first step.

3   Uncontrolled sky scanning

Whilst the detailed engineering assessments and experiments aimed at attitude stabilization were proceeding, some scientists sought ways of making limited observations using only knowledge of the uncontrolled rocket attitude, as distinct from controlling to a specific direction. In this way, a notable first record of ultra-violet radiation from stars in the Southern sky was made by Heddle and his collaborators at UCL.[1] A battery of five 'telescopes' each consisting of 4-inch lengths of aluminium honeycomb, were fixed at different angles in the rocket head, as illustrated in Figure 9.1. Ultra-violet sensitive photocells, with sharply peaking sensitivity at about 1900 Å, detected radiation collimated by the telescopes as the rocket rolled and changed attitude during flight. This attitude was to be measured by a combination of a moon detector, a magnetometer, and a camera photographing the star background. Of these three, only the moon detector records were received satisfactorily during the flight on 1 May 1961. Two of the telescopes detected ultra-violet radiation from 22 stars, and painstaking reduction of the data enabled the stars to be identified and the ratio of fluxes at 1900 Å to the corresponding fluxes at 5390 Å to be determined. Further

Fig. 9.1. Diagram of the ultra-violet radiation detectors as used by Heddle and his collaborators in a Skylark rocket.

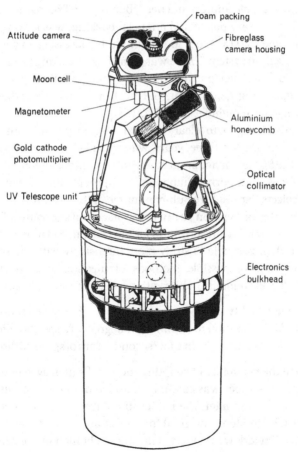

scientific details of this work will be found in Chapter 15. Although greatly encouraging, reliance on the uncontrolled scanning of the sky by instruments carried in the rocket was clearly no substitute for accurate control of attitude.

4 Further engineering developments

In June 1960, both the Astronomy Working Group of the BNCSR chaired by Butler, and the Rockets Working Group chaired by Massey, discussed experiments needing attitude stabilized rockets and explored with A.W. Lines of RAE Farnborough how these needs might be met. The main scientific interests continued to centre on experiments to photograph the sun in X radiation and in ultra-violet light, and to detect ultra-violet emissions from stellar sources. Boyd of UCL and Butler of ROE were taking the lead in proposing and preparing equipment. There was by now no question of the scientific interest and need, but as yet a firm requirement on RAE, with the necessary policy and financial backing, was lacking. It was thought that some £20,000 to £30,000 might be necessary to develop an accurate stabilization system. There was some talk of seeking industrial co-operation in such a development, together with specialist advice from a university engineering department, but no firm plans along these lines emerged during the remainder of 1960.

The way ahead began to clear when, in January 1961, an item of £10,000 per annum for three years was included in the 1961–1962 estimates for the Space Science programme to cover development of attitude stabilization for Skylark-borne equipment. In approving these estimates, the University Projects Expenditure Sub-committee of the BNCSR recommended that the Ministry of Aviation be invited to place and supervise a Contract for the work. With formal authorization for the allocation of money and staff to the project, RAE was able to study more intensively with the scientists the pros and cons of possible methods of meeting their needs. The requirements for astronomical observations were of three kinds:

  (a) attitude stability of the rocket about an arbitrary direction,
  (b) attitude stabilization to within a few degrees of a specified direction,
  (c) attitude control to within a few seconds of arc of specified directions.

At this stage the possibility of adapting the stabilization system designed for the American Aerobee was carefully considered. In essence this used a three-axis jet control to align the rocket to a datum direction defined by gyroscopes. Unfortunately, the combination of aerodynamic and inertial forces to which Skylark was subjected during the launch phase could, and frequently did, cause large conical angular motions of the rocket sufficient to topple the gyroscopes and lose the datum direction. Whilst similar problems existed for Aerobee, they were less severe. The proposal simply to purchase Aerobee and use it for experiments requiring stabilization was examined and rejected on the grounds of expense and the unproven Aerobee stabilization system. The first problem to be tackled by RAE therefore would concern the choice of possible options; to attempt to control the large angular motions of the Skylark rocket, or to devise a direction datum system which would survive these motions. The use of the earth's magnetic field and a stellar reference direction offered a possible solution along the lines of the second option.

5   The UK Atomic Energy Authority interest in stabilized rockets

The first moves in the development of a stabilization system for Skylark were accompanied by a new scientific interest in the matter by the UKAEA. A paper by R. Wilson, D.B. Shenton and R.S. Pease, of the Controlled Thermonuclear Research Division (CTR) at Harwell, entitled 'Ultra-violet Spectroscopy of the Solar Corona and the Stars; Proposals for Experiments using Rocket-borne Equipment' was submitted to the BNCSR of the Royal Society in March 1961. The researches of the CTR Division of UKAEA were aimed at the detailed understanding and eventual control of thermonuclear fusion. A particular aspect of that work was the study of the physics of plasmas, formed in laboratory conditions, by the use of quartz and vacuum ultra-violet spectroscopic techniques. It was explained in the paper by Wilson *et al.* how it was desired to extend the observations to naturally occurring plasmas, particularly those in the solar corona and eventually those in certain bright stars. This would involve carrying the spectroscopic equipment to heights above the atmospheric absorption zone, above 100 km, alignment of the equipment on the sun, or a bright star, recording the data photographically and recovering it by parachute.

Observation of the sun, in Phase 1 of the proposed programme, would require the rocket to carry a coarsely stabilized platform aligned within two degrees of the required direction. This platform would carry a normal incidence grating spectrograph. The image of the solar corona would be reflected on to the slit of the spectrograph via a fine optical alignment control with an accuracy of better than 30 minutes of arc. Light from the photosphere would be excluded from the spectrograph by light traps. It was planned to observe in the spectral regions from 100 to 3000 Å, changing the grating and focusing arrangements as appropriate. Phase 2 of the proposed programme, the observation of stellar spectra in the wavelengths below

3000 Å, would be in a completely new field. The main technical problem appeared to lie in the design of a fine alignment system to keep the star image aligned within about 10 seconds of arc.

The UKAEA requested the agreement of the BNCSR to the inclusion of the experiments in the British Space Science programme and to the necessary allocation of Skylark rockets together with the coarsely stabilized platform needed for the experiments. A detailed plan for the allocation of work and financial responsibilities accompanied the proposals, which were welcomed and accepted by the Astronomy Working Group and the BNCSR. As will be described later, they not only succeeded very effectively in their own right, but laid the foundations for the much more advanced ultra-violet astronomical observations from the IUE satellite.

6   The RAE proposals for a staged programme of development

A paper by J.J. Gait of RAE to the BNCSR in July 1961 outlined the views to which the RAE team had been led by the study of options to which we referred earlier. The proposal, which at first sight seemed attractive, to combine the RAE air jet control system with the gyroscopic reference system being developed for Aerobee, suffered from the difficulty already mentioned concerning the large angular motions of the rocket likely to topple the gyroscopes. Further study failed to reveal a ready solution to this problem, added to which the special gyroscopes had only been made in very small numbers and were likely to be expensive.

The RAE team recommended an alternative course of action. This comprised the step-by-step development of an attitude control system for Skylark, starting with the simplest form of stabilization and proceeding to refine this in stages. Each proposed stage would be capable of meeting some scientific needs. It was envisaged that the nose cone and forward payload compartment would be separated from the rocket motor after the latter had ceased to burn. The nose cone and forward compartment alone would be controlled and stabilized. Five possible stages were outlined:

Stage 1
To achieve pointing at the sun with errors of less than two degrees the existing RAE air jet control system would be used, with directional information from sun detectors and either roll control gyroscopes or magnetometers detecting the earth's magnetic field. This system could meet the needs of the UKAEA group in their solar UV spectroscopy programme, and of the UCL group in their X-ray spectroheliograph observations.

Stage 2

This would be similar to Stage 1, although more sensitive detectors would allow pointing at the moon rather than the sun.

Stage 3

This would aim at increasing the accuracies of the sun and moon pointing capabilities, so that errors would be reduced to minutes of arc rather than degrees. Such a performance would allow ultra-violet photography of the stars as proposed by Butler of ROE.

Stage 4

The use of horizon scanners as direction detectors would allow alignment of the Skylark head to the local vertical. Observations of the earth and atmosphere below would thus be facilitated.

Stage 5

The culmination of the development would, by a combination of techniques from the previous stages allow experimental equipment to be pointed at, and 'locked on' to preselected stars.

It was thought that such a staged programme might be completed in four or five years; more detailed studies would be necessary to confirm the feasibility of the various stages and assess the likely costs. It was agreed by the BNCSR that RAE should assess in more detail the suggested five-stage programme, particularly with regard to technical feasibility, time scale and costs. Towards the end of 1961 a Contract was placed with Messrs Elliott Bros. Ltd. (now Marconi Space and Defence Systems Ltd) for a design study with estimated costs, to be completed by March 1962. This marked the beginning of a co-operative effort on the stabilized Skylark project between RAE, Messrs Elliott Bros Ltd, and the various groups of scientists interested in using the facilities for astronomical observations. It will be evident that the successes which were achieved were only made possible through the closest co-operation by the interested parties. Whilst it is convenient to describe events in terms of rocket engineering and the scientific payloads, this is in many ways an artificial concept which obscures the essential unity of the system.

After considering the results of the design study, the BNCSR decided in May 1962 that the development of Stages 1 and 3 of the proposals should be supported. The Steering Group on Space Research accepted this recommendation from the BNCSR and the necessary contractual arrangements were made with Messrs Elliott Bros Ltd, the research and development aspects of the work being monitored by RAE. It will be recalled that Stage 1 involved sun pointing to the relatively coarse accuracy of 1 or 2 degrees, whilst Stage 3 would aim to reduce the errors in sun or moon pointing to a few minutes of arc. The development of Stage 2, which involved pointing at the moon with low accuracy was not supported by the BNCSR and no further action was taken upon it. It was judged that the potential scientific value did not justify the expected cost. Action on Stages 4 and 5 was held in abeyance pending progress on Stages 1 and 3.

7   Technical features of stage 1

We do no more here than describe in brief outline the technical features of the Stage 1 Attitude Control Unit (acu) as it was developed; a more detailed description of the system is given by W.R. Thomas.[2] The general concept involved the stabilization of the nose cone and payload section of the Skylark rocket after it had been separated from the main body of the motor casing at the end of motor burning. This separation occurred about 77 seconds after launch, at an altitude of about 52 miles. The control system then came into operation, and the attitude was stabilized, pointing at the sun, at about 111 seconds from launch. This attitude was held for about 270 seconds during which time the altitude of the head varied from about 76 miles to a peak of about 120 miles and back to 45 miles. The deviation of the longitudinal axis of the nose cone from alignment with the sun was detected by silicon solar cells mounted on the body. Signals from these detectors, together with stabilizing signals from rate measuring gyroscopes, controlled on/off poppet valves which released high pressure nitrogen through pairs of jets so as to align the axis of the nose cone with the sun. The necessary stability about a third axis, roll about the longitudinal axis, was again controlled by gas jets responding to signals from magnetometers detecting the earth's magnetic field. Since many of the experiments proposed for installation in the stabilized Skylark rocket required the examination of photographic records made in flight, parachute recovery of the appropriate section of the payload was included in the overall design.

8   The first three launchings of stage 1

Substantive development of the Stage 1 equipment began about mid-1962, and the first three launchings of proving models, together with selected scientific experiments, took place from Woomera in August and December 1964, and in April 1965. The Skylarks were designated SL 301,

SL 302, and SL 303 respectively. All were successful in so far as the operation of the stabilization equipment and the scientific instruments were concerned. Only the failure of the parachute recovery system for SL 301 marred an otherwise perfect trio of launchings. Even this did not prevent recovery of the photographic cassettes which survived the impact, and allowed some recorded observations to be recovered. All three Skylarks carried experiments from the Culham Laboratory of UKAEA and from the University of Leicester. The Culham equipment included the optical fine alignment system described earlier in the Chapter.

Culham Laboratory obtained from the first two flights spectra of the solar limb and of the complete solar disc in the extreme ultra-violet wavelengths.[3] Perhaps the most significant scientific results came from the third Skylark, SL 303, in which the chromospheric/coronal spectrum was recorded over the full wavelength range, with over 200 emission lines and a continuous spectrum. Some of these new data were given by Wilson to the meeting of COSPAR at Mar del Plata in Argentina in May 1965.[4] Soft X-ray photographs of the sun were also obtained from the second and third flights by both the Culham group and a group from Leicester University led by K. Pounds. Figure 9.2 shows the solar limb spectrum.

8.1 Further developments in the Stage 1 programme

From 1965 onwards, the Stage 1 Stabilized Skylark played an important role in the total British space science programme and over the years from 1964 to 1974 on average four Stage 1 rockets were launched annually. During this period there were significant scientific, technical and

Fig. 9.2. Solar limb spectrum (1500–2300 Å) obtained by the Culham Laboratory group during a flight of a sun-stabilized Skylark rocket (SL 303) launched on 9 April 1965.

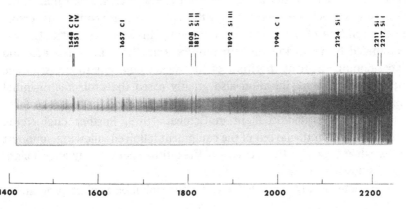

administrative developments. The Culham Laboratory team and the X-ray astronomy groups at Leicester University and the Mullard Space Science Laboratory (MSSL) of UCL were particularly active in exploiting the capabilities of Stage 1 to obtain new information about solar phenomena in the ultra-violet and X-ray wavelengths. As an example of the scientific work being carried out in 1966, it is interesting to note the variety of solar instruments installed in two Stage 1 Skylarks (SL 304 and 307):

> Normal incidence spectrograph
> Plane crystal X-ray spectrometer
> Proportional counter X-ray spectrometer with reflection optics
> X-ray pinhole camera
> Grazing incidence spectrograph
> Extreme ultra-violet spectroheliograph
> Extreme ultra-violet pinhole camera with reflection filter

The records from the sun pointing Skylark SL 304 gave a number of high resolution spectra of the solar emission over the wavelength range 11 to 25 Å. The stabilization performance of Skylark SL 307 was particularly good, and image resolution down to 1 arc minute was achieved throughout the 300 seconds of controlled flight. The instruments provided a range of sensitivities and covered a range of X-ray wavelengths.

Design improvements to Stage 1 continued, and benefited from the work in progress at RAE Farnborough and at Elliott Bros on the development of stages 3 and 5. As a result the performance of Stage 1 was refined to the benefit of both the scientific users and the assembly teams which were responsible for the complex alignment and prelaunch test procedures. For instance the first version of the Stage 1 attitude control unit (acu) involved mounting the fine sun sensors on the skin of the acu some distance to the rear of the scientific sensors at the front of the rocket. Distortion of the structure between the sensors during the stresses of flight could cause errors in the pointing of the scientific equipment. This source of difficulty was eliminated by removing the sun sensors from the skin of the acu and repositioning them at the front of the rocket adjacent to the scientific equipment. This modification also greatly eased the problems of initial angular alignment. The first of these 'fine eyes forward' Stage 1 assemblies successfully operated in flight in October 1969. Further engineering improvements in the layout of the equipment followed, allowing, amongst other advantages, for the recovery of the attitude control bay after a flight and its possible re-use.

The administrative arrangements for the stabilized Skylark programme allowed for a progressive shift of responsibilities for the assembly, testing and launching operations, as they became more routine, from the research and development staff to regular assembly teams. For instance, in 1965, whilst British Aircraft Corporation (BAC) had a contract for the assembly of unstabilized Skylarks, and were soon to have this extended to cover supply and management, the combined teams of RAE and Elliott Bros Ltd were carrying the major load of assembling, testing and pre-launch preparation work for the stabilized Skylarks. Towards the end of 1967 however, the preparation of the Stage 1 acu was in the process of transfer to BAC and in July 1968 the first Stage 1 prepared by BAC was launched and successfully operated. In future, all Stage 1 Skylarks were to be prepared by BAC at Bristol. In early 1969, a contract was placed with BAC and Elliott Bros Ltd to carry out Skylark launchings from the Woomera range on a campaign basis. This became effective in July 1969, when responsibility for the conduct of the Skylark Trials passed smoothly and successfully from the Australian Weapons Research Establishment to the British firms.

9   The Stage 3 stabilization system

The refinement of Stage 1 to transform it into the much more accurate sun and moon pointing system known as Stage 3 depended essentially on the development of suitable error sensors. Much of this development was carried out by RAE with the co-operation of Messrs Elliott Bros Ltd. Several different techniques were employed, including the use of detectors based on silicon photocells, cadmium selenide, and photo-multipliers. As an example, a successful moon sensor (used also subsequently in the Stage 5 design) comprised a single lens which focused an image of the moon on to a quad of silicon photocells etched from single chip. The cell outputs, suitably amplified and differenced, enabled the optical centre of the moon to be detected with errors below 1 minute of arc. There were of course limitations inherent in such a system; for instance it was necessary that the moon should be within seven days of full to give adequate signal to noise ratio in the detectors. Development of Stage 3 proceeded comparatively slowly, and although it started in early 1963, and the sun and moon sensor development was completed by mid-1965, the first launch of a sun pointing stage 3 Skylark (SL 403) did not take place until July 1968. This was a successful operation, and the University of Leicester group obtained a wealth of new data concerning X-radiation from cosmic sources, including a high resolution scan across the galaxy M 87.[5]

The launch of a second sun pointing Skylark (SL 404) in May 1969 resulted in a vehicle failure and a disappointment for the Culham Laboratory and Leicester University experimenters. The first moon pointing Stage 3 Skylark (SL 401) was, however, successfully launched in March 1970 and enabled the Royal Observatory Edinburgh to photograph two adjacent areas of the sky in ultra-violet light. The rocket carried an ultra-violet camera of the objective prism Schmidt type and an associated scanning photometer.

Fig. 9.3. Objective prism ultra-violet spectra obtained with equipment on the stabilized Skylark rocket SL 401 launched from Woomera at 01.18 hrs on 20 March 1970.

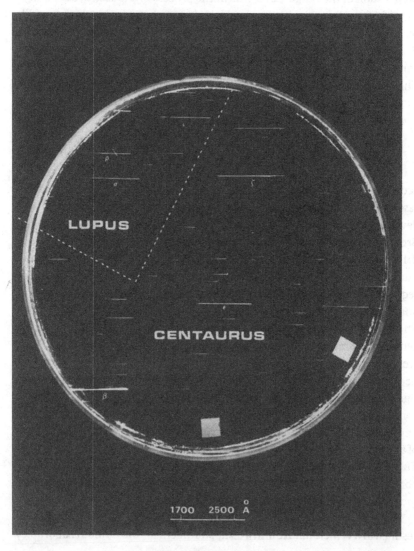

Spectra of some 70 stars down to about 1800 Å were obtained in the region of Lupus/Centaurus, as shown in Figure 9.3. The second moon pointing Stage 3 Skylark (SL 402) was successfully launched in October 1972, and the Meteorological Office was able to measure the lunar albedo in ultra-violet light, and the night time concentration of ozone above 65 km. This was the fourth and last of the Stage 3 launches, it having been decided in 1969 to concentrate resources on the star pointing Stage 5 rather than continue an extensive Stage 3 programme. This decision was influenced by the length and complexity of the Stage 3 development programme, the success of Stage 1 with comparable accuracy for sun pointing, and the pressure from scientists for star pointing to the Stage 5 specification. Nevertheless Stage 3 had enabled successful observations to be made in the ultra-violet and X-ray regions of the spectrum, on stars and X-ray sources, much earlier than might otherwise have been the case.

10   The development of Stage 5, the star pointing system

It was an important feature of the RAE proposals for a staged programme that each stage should contribute to the succeeding stages, and the designers of Stage 5 were able to take advantage of the successes and lessons of Stages 1 and 3. Discussions between the potential scientific users of Stage 5 and the design teams at Elliott Bros Ltd and RAE proceeded in 1966, concentrating on the desirable performance specifications and cost effective techniques which could be used to achieve these aims. It was agreed that in the first instance, the system should be capable of finding and 'locking on' to two specified stars during each stabilized flight. The first star should be of magnitude 2.5 or brighter; following this, the system should find and stabilize on a second star of magnitude 5 or brighter. An observation time of at least 170 seconds per star would be required. The permissible angular errors were also specified:

	Lateral axes	*Roll axis*
Bias	2 arc min	1 degree
Drift	10 arc s/min	1 arc min/min
Noise	10 arc s RMS	1 arc min RMS

In March 1967 Elliott Bros Ltd proposed what became known as the 'Starling' system to meet the specification. This used the Stage 3 attitude control unit, with an additional star pointing adaptor containing gyroscopes for inertial reference angles with moon and star sensors at the front of the payload. A contract was placed with the firm in mid-1967 for two flight-worthy prototypes to prove the basic design. This particular design was identified as Phase 1, because a modified improved version, known as Phase 2, was conceived and became the subject of a further contract let in November 1969 for two further flight prototypes. The essential difference between Phase 1 and Phase 2 was that the latter would not use the optical coarse moon sensors of the Stage 3 design, but rather would acquire the moon by coning around the local magnetic vector at an appropriate angle.

A detailed technical description of the operation of the Phase 2 Stage 5 system will be found in the references by J.K. Abbott[6] and C.T. Farr.[7] However, a brief outline here of the sequence of events will indicate in general terms how the system searched for and found the preselected target stars. As with the Stage 1 and Stage 3 systems, the payload is separated from the rocket motor case after the end of motor burning, and high pressure nitrogen jets respond to signals from the control system to align the axis of the astronomical instruments in the payload with the desired star. The control depends on acquisition firstly of the moon, which must have a brightness greater than one half of full moon. This is a significant constraint on the number of nights suitable for a launch. The unit contains a three-axis magnetometer which, after payload/motor separation, signals the control system so as to constrain the longitudinal (twist) axis to describe a coning motion about the earth's magnetic vector. Thus at some point in the coning motion the moon is 'seen' by forward looking sensors, and the control system then 'locks on' to the moon. This establishes the payload attitude in known space axes, and it can then be repositioned so as to point near to the chosen target star. This repositioning uses information from an inertial reference system in the payload based on accurate integrating rate gyros. When pointing near (within about 3 degrees) to the chosen star, the unit performs a scan of a small area (6° by 1.5°) of sky, finally locking on to the brightest star in that area. If all has gone well, the brightest star will be the chosen target star. After sufficient observation of this first star, the attitude can be changed so as to seek and find the second star, again using the inertial reference system to measure angular change. Figure 9.4 outlines a typical configuration for a Stage 5 star pointing Skylark.

10.1   The test flights of Stage 5

The first two test flights of the Phase 1 system, which used the coarse moon-sensors rather than the magnetometer system of Phase 2, took place in July 1970 and June 1971. The first Skylark (SL 811) developed a fault in the roll control loop, and the wrong target star was selected. However, the inertial and star-sensing modes functioned correctly and established confidence in those new features in the design. The second

Fig. 9.4. Typical star pointing vehicle configuration for Skylark.

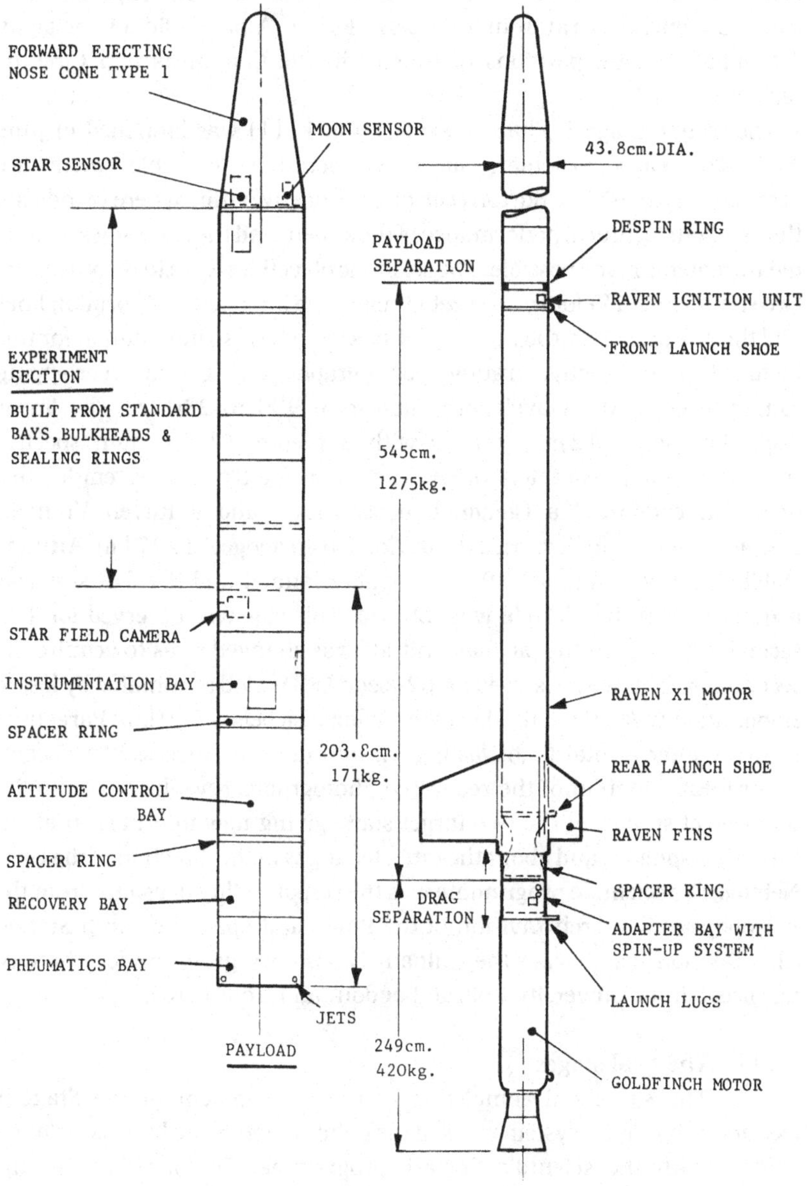

FORWARD EJECTING
NOSE CONE TYPE 1

MOON SENSOR

STAR SENSOR

43.8cm.DIA.

PAYLOAD
SEPARATION

DESPIN RING

RAVEN IGNITION UNIT

FRONT LAUNCH SHOE

EXPERIMENT
SECTION

BUILT FROM STANDARD
BAYS,BULKHEADS &
SEALING RINGS

545cm.
1275kg.

STAR FIELD CAMERA

INSTRUMENTATION BAY

SPACER RING

RAVEN X1 MOTOR

203.8cm.
171kg.

ATTITUDE CONTROL
BAY

REAR LAUNCH SHOE

RAVEN FINS

SPACER RING

SPACER RING

RECOVERY BAY

DRAG
SEPARATION

ADAPTER BAY WITH
SPIN-UP SYSTEM

PHEUMATICS BAY

LAUNCH LUGS

JETS

PAYLOAD

249cm.
420kg.

GOLDFINCH MOTOR

Skylark (SL 812) carrying Leicester University X-ray equipment was a failure, in that moon acquisition was not achieved. The problems with these two flights confirmed the wisdom of changing to the Phase 2 design for moon acquisition, and the first of this type, Skylark SL 1011, was successfully launched and operated in April 1973. This again carried Leicester University X-ray detectors, and was the first UK experiment to study low-energy X-ray sources (below 1 keV). A parabolic mirror array determined precise positions of sources in the Centaurus and Circinus regions.

The second Stage 5 Phase 2 Skylark (SL 1111) was launched in June 1973. This was a particularly successful operation. The achievements are described in Notes[8,9] by Burton *et al.* and are briefly outlined here to indicate the excellent technical performance of the system and the resulting scientific measurements made possible. The scientific objectives were to record for the two stars Gamma Velorum and Zeta Puppis, the interstellar absorption lines and the line profiles in the ultra-violet spectra. The instrumentation for this included three objective grating spectrographs, giving three overlapping ranges to cover the wavelength band from 900 to 2300 Å. The linear dispersion was 0.12 mm per Å, with a theoretical optimum spectral resolution of approximately 0.2 Å. The two-stage Skylark system for this operation comprised a Goldfinch boost rocket and a Raven VI main sustainer motor which carried the payload to an apogee of 227 km. Attitude stabilization was initiated 89 seconds after launch and the first star was acquired when the altitude was 220 km. This star was observed for 103 seconds after which the payload attitude was changed so as to acquire the second star. This was observed for 62 seconds, a period terminated by loss of stabilization when the altitude of the payload fell below 120 km. Parachute recovery after a total flight lasting 12 minutes was successfully accomplished. Examination of the recovered photographs revealed high quality ultra-violet spectra of the two target stars, giving new information about their atmospheres, and about the interstellar gas in the direction of the Gum Nebula. The scientific programme was the responsibility of groups from the Astrophysics Research Division of the Radio and Space Research Station (this Division was in fact the Culham Laboratory group under changed auspices), from University College London, and from ESRO.

11   The final stages

The successful completion of the development of the Stage 5 Skylark stabilization system marked the culmination of the long association of RAE with the scientific Skylark programme. Responsibility for the procurement of Skylark rockets for Science Research Council purposes was formally transferred from the Ministry of Defence, Procurement Executive, to the Appleton Laboratory (previously the Radio and Space Research Station) of SRC in April 1974. Space Department of RAE provided an advisory service until April 1976. Stabilized Skylarks continued to play a most important part in the UK space science programme, as will be evident from the record of launchings in Appendix A2. The X-ray groups at Leicester under Pounds and at the Mullard Space Science Laboratory of UCL under Boyd, and the UV astronomy group at UCL under Wilson were particularly active. Successful though the Stage 5 system was, it was recognized that the limitations imposed by the phasing of the moon and the rather complex pre-launch alignment procedures and the weight of the system were disadvantages. As early as 1971 it was realized that the great strides made in the development of gyroscopically stabilized inertial platforms for other purposes such as aircraft navigation systems, might lead to a simpler and cheaper system for Skylark. In response to a request from the SRC in 1972, a feasibility study contract was placed by MOD with Messrs Ferranti to assess the potential use of an aircraft inertial reference unit for the attitude control of Skylark. The results of the study were favourable and suggested that a performance comparable to that of the Stage 5 system could be achieved at lower cost and without some of the disadvantages mentioned above. Development of a flight unit was put in hand for installation in Skylark SL1611. The scientific payload, made jointly by Leicester University and the Max Planck Institute at Garching, was intended to observe the first (predicted) dust halo around an X-ray star. Unfortunately, the design of the software for the inertial stabilization unit proved more difficult than expected and this led to delays and the eventual abandonment of the launch. This development was in effect overtaken by events, in particular the rundown of the scientific Skylark programme as part of the overall policy under financial pressure of the Science Research Council, and the transfer of the scientific interest in X-ray and ultra-violet astronomy to the even more powerful and comprehensive facilities for observations offered by scientific satellites such as Ariel 5 and the International Ultra Violet Explorer (IUE). So ended a long and successful chapter for UK space scientists.

10

The Trend Committee and the Science Research Council

The organization which had been set up to deal with the management and financial administration of the British space research programme had already become quite complex by the beginning of 1961 (see Chapter 4, p. 72). The impact of the negotiations for the establishment of the European Space Research Organization (ESRO) threw a greatly increased burden on the system and it was clear that, before long, some substantial modifications would be necessary. This matter was a subject of frequent discussion between R. Quirk, H.S.W. Massey and M.O. Robins, the establishment of the Space Research Management Unit (SRMU) (see Chapter 5, p. 93) being one consequence. However, in 1962 the government set up a committee to examine the whole question of the organization of the support for civil science in Britain. Pending the issue of a report by this committee it was not worthwhile to consider any major changes specifically concerned with space science.

1 The Trend Committee

The Committee was set up under the Chairmanship of Sir Burke Trend, Second Secretary at the Treasury, the other members being Sir Keith Murray, Chairman of the University Grants Committee; Sir Thomas Padmore, Second Secretary at the Treasury; Lord Todd, Chairman of the Advisory Council on Scientific Policy; F.F. Turnbull, Secretary, Office of the Minister for Science; C.H. Waddington, Professor of Animal Genetics, University of Edinburgh and E.G. Woodroofe, Director, Unilever Ltd. The terms of reference were to consider (a) whether any changes were desirable in the existing functions of the various agencies for which the Minister for Science was responsible concerned with the formulation of civil scientific policy and the conduct of civil scientific research; and whether any new agencies should be created for these purposes, (b) what arrangements should be made for determining, with appropriate scientific advice, the relative importance in the national interest of the claims on the Exchequer for the promotion of civil scientific research in the various fields concerned and (c) whether any changes were needed in the existing procedure whereby the agencies concerned are financed and required to account for their expenditure.

2 The problem of 'big' science

The need for a thorough re-examination of the support of civil science was perhaps most clearly obvious in relation to what is now known as 'big science', the growth of which had not been foreseen in setting up the existing research councils and the Department of Scientific and Industrial Research (DSIR).

Already, before the Second World War, the demands of research in nuclear physics were calling for the development of means for accelerating particles to energies beyond those attainable through the application of direct voltages as in the pioneering experiments of J.D. Cockcroft and E.T.S. Walton in 1932. The way to achieve these high energies had already been indicated by the invention of the cyclotron. An immense filip to large-scale scientific research was provided during the war by the Manhattan Project – the scale of the research laboratory at Berkeley concerned with the electromagnetic separation of the uranium isotopes was breathtaking to those who were used to laboratory research as carried out for example in the Cavandish Laboratory in the 1930s. After the war the trend continued and indeed the rate of expansion of indirect accelerating equipment speeded up. The exploration of the nucleus proved so much more difficult and led continually to such unexpected results that there was a steady demand for bigger and larger accelerators. While ingenious methods were introduced at some stages to cut the cost substantially, by and large, research in nuclear physics grew steadily more and more expensive. This was especially true of the most fundamental part of the subject, now termed particle physics, concerned with the study of the relationships between the many new particles discovered using high energy accelerators, and their role in the structure of matter.

In 1951, a number of European physicists proposed that the countries of Western Europe should pool their resources to establish a laboratory at Geneva equipped with an accelerator of such a high energy as to match the facilities available in the USA and the Soviet Union. It was proposed that this accelerator should provide protons with an energy of 30,000 million electron volts – a big rise from the few megaelectron volts available just before the war! The UK did not take part in the formal preparatory discussions, but kept closely in touch and joined very soon after the organization, CERN, had been formally established in 1954. The initial subscription, in proportion to gross national product, was £0.8 million for the UK rising to £1.2 million by 1961.

While an international collaboration within Europe was required to build and operate an accelerator on the 30,000 MeV scale there remained an important role for machines of somewhat smaller size, within the reach of a single country such as the UK but nevertheless much too large for any single university to operate. The value of such a facility was well recognized by the nuclear and particle physics community in Britain. In 1958 the National Institute for Research in Nuclear Science (NIRNS) was established by Royal Charter to provide facilities for research in high energy physics by staff and students at Universities. The principal item of equipment was to be a 7000 MeV proton accelerator, known later as NIMROD, but provision was also made for auxiliary equipment to enable the users from universities to take full advantage of the accelerator.

NIRNS which operated a large laboratory, the Rutherford Laboratory, at Chilton near Abingdon, was administered by a Governing Board with Lord Bridges, lately Head of the Treasury, as Chairman, including strong representation from the universities. The Institute was financed from the Atomic Energy vote to the tune of nearly £7 million but direct financial responsibility for its operation was taken directly by the Minister for Science. On the other hand the responsibility for providing grants to university research workers to use the NIRNS facilities lay with DSIR, the body which was responsible for the UK contribution to CERN and for certain other national facilities in nuclear physics.

The problem of how best to handle 'big science' within the system was compounded by the rise of space research which, through the need for large rocket boosters, required financial and other resources comparable with those involved in the production of large accelerators. Here again the subject at first grew outside any recognized government establishment but an elaborate organization had been evolved largely outside DSIR, as with NIRNS, in which the Royal Society through the BNCSR, and the Ministry for Science through the Steering Group, played the key roles. It was especially anomalous that, while accounting responsibility was asigned to DSIR that body did not have the final say in the selection of grant awards.

A further subject which had not yet become a branch of 'big science' but was already close was ground-based astronomy, both on the optical and radio sides. In connection with the former the position of the Royal Greenwich Observatory within the Admiralty, which was natural in the days when methods for accurate navigation were being developed, had also become quite anomalous. Many of the resources of the Observatory were concerned with research in solar and galactic astronomy and it would therefore be much more reasonable for it to be associated more directly with an organization concerned with pure research.

We need not concern ourselves here with discussing problems associated with industrial research though they were many. It is necessary, however, to draw attention to another important matter which concerned all scientific research. Which body within government was ultimately responsible for deciding the priorities between the requests made by different government scientific organizations, such as the Research Councils, for financial support? At the same time this task was performed by the Treasury, without direct scientific advice. Arrangements for securing an adequate input from the scientific community in these matters were clearly called for.

3   Royal Society discussions and proposals

Soon after the Trend Committee was set up, the Royal Society under the energetic leadership of its President, Lord (then Sir Howard) Florey initiated a number of discussions between its Fellows with the aim of presenting a written statement to the Committee as well as presenting oral evidence. There was a strong and growing ground swell of opinion among physical scientists, including particularly those involved in 'big science', that the industrial and scientific responsibilities of DSIR should be separated and the latter expanded to include NIRNS and space science. Thus, Sir John Cockcroft, who had been asked by the Minister for Science to chair a committee to report on the future level of support for research in nuclear physics, submitted a further memorandum on behalf of a number of members of his committee to the Trend Committee. This document, signed by J.B. Adams, C.C. Butler, J.M. Cassels, H.S.W. Massey, T.G. Pickavance and D.H. Wilkinson, in addition to Cockcroft, all of whom had been much concerned with NIRNS, submitted on 10 July 1962, recommended the establishment of a Council for Fundamental Research on the lines referred to above. It was agreed that the organization of government support for applied research must be closely involved with foreseeable financial benefits and commercial policy which do not arise in fundamental research. It would therefore be logical to deal with these different aspects of research through different bodies.

The Royal Society referred the issues to a number of *ad hoc* committees representing different subject disciplines. Those for astronomy and geophysics physics, mathematics and chemistry were strongly in favour of the formation of a Fundamental Sciences Research Council. There was almost unanimous agreement that there was a need for greater scientific participation in high level decision-making on science policy, including financial issues. Because the biologists were much less concerned with the idea of the new research council a sub-committee was set up to draft the written evidence for the Royal Society to submit to the Trend Committee consisting of three physical scientists, two physicists, P.M.S. Blackett and H.S.W. Massey, and one electrical engineer, F.C. Williams. Their final draft which was generally accepted by the Royal Society Council, envisaged an organization under a Minister for Universities and Science, advised by a Civil Science Board, of four Research Councils, the existing Medical and Agricultural Research Council together with the new Science Research Council (SRC) and one constituted from the remainder of DSIR. The Civil Science Board would be composed of four or five academic and industrial scientists and engineers together with a few civil servants attached on some appropriate basis. It would be responsible for recommending to the Minister the financial resources required for the prosecution of their programmes by the Research Councils and their subdivision between these Councils. In addition it would be in close contact with bodies such as NEDC, NRDC etc. The proposed Science Research Council was exactly on the lines discussed earlier.

The University Grants Committee according to this draft would also be responsible to the Minister, who would now become Minister for Universities and Science.

Members of the space science community, who would certainly be affected by any proposed new organization, were closely in touch with developments in which Massey and others participated fully. There was general agreement with the Royal Society's proposals. On 25 October, E.C. Bullard, Massey and R. Quirk gave oral evidence to the Trend Committee, especially about the place of space research in any new organization. There was no doubt that there was full consultation on all sides.

4   The Trend Committee Report

The report of the Trend Committee was published in October 1963. It recommended, just as proposed by the Royal Society and other bodies, that a Science Research Council (SRC) should be set up taking over responsibility for CERN, NIRNS, the space programme, as well as for postgraduate awards in science and technology. In addition the SRC would also take over direct control of the Royal Observatories of Greenwich and Edinburgh from the Admiralty and the Scottish Authorities respectively. On the other hand, it was considered best that the Air Ministry should retain responsibility for the Meteorological Office.

A more unexpected recommendation was the establishment of a new Natural Resources Research Council. This would take over the functions exercised by the Nature Conservancy and be responsible for research into hydrology, fisheries and related aspects of aquatic biology, forestry, oceanography as well as for the Geological Survey and the Soil Survey. As it later developed this recommendation did impinge somewhat on space research, leading to demarcation problems as we shall see.

In place of the rump of DSIR it was recommended that a new Industrial Research and Development Authority should be created. This would not only include the present activities of the industrial and applied research side of DSIR but also take over the work of NRDC.

As in the proposals of the Royal Society it was recommended that the Minister for Science should have additional responsibilities for the allocation of funds for civil research and development. In rather similar vein to the Civil Science Board proposed by the Royal Society it was recommended that a new advisory body to the Minister for Science should be set up with enlarged responsibilities as compared with the existing Advisory Council. It would advise on scientific manpower, on allocation of resources between agencies, on national scientific needs, on international scientific policy and on organizational matters. With the proposed structure this advisory body would be responsible for both pure and applied research impinging on industrial and commercial matters. It would be made up of independent members of whom half should be scientists.

Government reaction to this report was complicated by the fact that elections were held in 1964 at which Sir Alec Home's government was defeated. Before that the only action had been to welcome the report and to fuse the responsibilities of the Ministries for Science and Education under Lord Hailsham, the incumbent Minister for Science.

The new government accepted the recommendations as far as they concerned pure scientific research but rejected the proposal for establishing an Industrial Research and Development Authority within the Ministry for Science and Education. Instead they established a Ministry of Technology which would take over responsibilities for the DSIR applied research stations and other industrial activities of DSIR. This left the top-level advisory body, which was called the Committee for Science Policy (CSP), with little direct responsibility for any applied science apart from that involved in the Agriculture and Natural Resources Councils. While the Chairman of the CSP was an independent scientist the corresponding body within the Ministry of Technology was chaired by the Minister with an independent adviser as Deputy.

These questions, though very important, did not affect the prosecution of space science which depended largely on the new Science Research Council (SRC) whose terms of reference were established by Royal Charter as described below. As mentioned earlier the terms of reference of the new Natural Environment Research Council (NERC), as it was finally called, did raise the spectre of demarcation problems because that Council assumed responsibility for supporting research in geophysics, including meteorology, but not middle or upper atmospheric physics.

The new Research Councils were formally established in April 1965 and the Science and Technology Act 1965 specified the responsibilities and power. The transfer of responsibilities for space science from the BNCSR and the Steering Group to the SRC occurred smoothly, without delay. As Massey was appointed the first Chairman of the Council for Scientific Policy he relinquished most of his responsibilities for the management of the space science programme. He did, however, continue to be Chairman of the BNCSR and acted as assessor for it on the SRC committee most closely concerned with space science, the Space Policy and Grants Committee, and from 1966–68 represented the UK on the scientific committee (STC) of ESRO.

5   The Science Research Council and space science

We now proceed to describe the new organization and how space science developed through it. One important new aspect of the situation was that all financial support for pure scientific research would come from the Science Budget administered through the Department of Education and Science. Assessment of the merits of scientific proposals within its sphere of interest would be the responsibility of the SRC, rather than being dispersed as had been the case in the past. Thus space research, nuclear and particle physics and astronomy would compete directly for the resources made available. This was especially complicated when international commitments were involved as with CERN and ESRO because the contributions to such bodies were far from being under national control. Nevertheless, it was considered that there would be great advantage in requiring all scientific developments, international or otherwise, to be financed from one vote. It could be hoped that their priorities would then be judged on scientific rather than political grounds.

In the words of the Science and Technology Act 1965: 'The Science Research Council shall be a body established wholly or mainly for objects consisting of or comprised in the following, namely, the carrying out of scientific research, the facilitating, encouragement and support of scientific research by other bodies or persons or any description of bodies or persons and of instructions in the sciences and technology, and the dissemination of knowledge in the sciences and technology.' Later in the Act 'scientific research' is stated to mean 'research and development in any of the sciences (including the social sciences) or in technology'.

Such wide terms of reference might appear to give the SRC almost unlimited freedom of action, but as with all government funded organizations, ultimate control rested with parliament. 'The Secretary of State may, out of monies provided by Parliament, pay to any of the Research Councils such sums in respect of the expenses of the Council as he may with the consent of the Treasury determine, and so far as relates to the use and expenditure of sums so paid the Council shall act in accordance with such directions as may from time to time be given to it by the Secretary of State.'

Thus in the last resort, the Treasury and the Secretary of State for Education and Science could control the activities of the Research Councils. From the point of view of scientists supported by the SRC and of the staff administering the funds, very important aspects of the arrangements would be the total sums of money made available annually, and the degrees of delegation of authority to commit and spend money at the various levels of the Council structure. Responsibility for advice to the Secretary of State for Education and Science on the size of the science budget and its subdivision among the different research councils rested with the Council for Scientific Policy (CSP) which was a body composed of independent scientists. At a later stage in 1972 the CSP was replaced by a body known as the Advisory Board for the Reseach Councils (ABRC) which included representatives from certain government departments as well as those from industry and universities.

The Royal Charter granted to the SRC defined the objects of the Council rather more precisely, including the charge 'to provide and operate equipment or other facilities for common use in research and development in science and technology by Universities, Technical Colleges or other institutions or persons engaged in research' and 'to make grants for postgraduate instruction in science and technology'.

More specifically in the field of space science, the consideration of space research proposals and the award of grants to university groups, together with the responsibility for general administration relating to the British space science programme was transferred to the SRC from the Royal Society and the other organizations involved. The SRC also assumed responsibility for UK participation in ESRO and for UK co-operation with NASA in the joint US/UK co-operative programmes in space research. The Radio Research Station (RRS), to be renamed the Radio and Space Research Station (RSRS), was transferred to SRC and continued to provide data handling facilities and other services of a general nature.

The Steering Group on Space Research, no longer having a part to play, was abolished. Government interest in and influence on space research was to be exercised by the Department of Education and Science, advised by the CSP, and the SRC itself. The Royal Society's British National Committee on Space Research (BNCSR) continued in being to foster scientific interest in the general fields of space science, particularly in relation to the activities of the Committee on Space Research (COSPAR) of the International Council of Scientific Unions.

The first Council met under the Chairmanship of Sir Harry Melville, who had been Secretary of DSIR, and who now assumed the full-time duties of head of the SRC staff and Chairman of the Council. The membership of the first Council was:
Sir Harry Melville, Chairman, A. Caress, D.G. Christopherson, S.C. Curran, G.C. Drew, M.R. Gavin, Rt Hon. the Earl of Halsbury, Sir Ewart Jones, Sir Bernard Lovell, K. Mather, C.F. Powell and J.E. Smith, with W.L. Francis as Secretary.

The Council member who was very familiar with the needs and aspirations of space scientists, Sir Bernard Lovell, was appointed Chairman of the Astronomy Space and Radio Board. This was one of the three Boards operating under the Council, the other two being concerned with nuclear physics and university science and technology.

The responsibilities of the ASR Board included the formulation of policy, the assessment of grant applications, and advice and recommendations in its field of interest to Council. The Board brought together the two major scientific disciplines of astronomy and geophysics, which spanned most of the activities of space scientists. At this early stage in the evolution of the organization, the dichotomy between space science and astronomy was maintained by the establishment of two committees to report to the ASR Board and Council.

The Astronomy Policy and Grants Committee (APGC) was chaired by H. Bondi, and the Space Policy and Grants Committee (SPGC) by P.A. Sheppard. The responsibilities of the SPGC resembled in many respects those of the Royal Society's BNCSR prior to 1965. It was able to call on the experience of a number of space scientists, in particular the Chairman,

Sheppard, W.G. Beynon, R.L.F. Boyd, H. Elliot, J.A. Ratcliffe and J. Sayers, who had been involved in the British programme since its inception. Close links with the Royal Society were maintained by the appointment of Massey as Royal Society Assessor.

The wide scientific field of interest to the SPGC is indicated by the list of working groups set up to consider and advise on specific grant applications and detailed policy issues. These Groups are included in Table 10.1.

By contrast, the Astronomy Policy and Grants Committee had no formally established working groups. One common member with the SPGC (H. Elliot) and one with the Solar and Stellar Astronomy Working Group of the SPGC (J. Ring) linked the two Committees together rather loosely. As we shall describe later, the growing use of space technology for astronomical measurements rapidly broke down any imaginary barriers between the space science and astronomical communities.

Membership of the Council, Boards and Committees was composed almost entirely of practising scientists and engineers from universities, industry and various research establishments. The tasks, among others, of administering the policies and putting into effect the decisions of the Council and its subordinate bodies fell to the permanent staff of the Council. In the space science field these came within the responsibility of the ASR Division, and at the formation of the SRC in April 1965, J.F. Hosie was appointed Director of this Division and of the Nuclear Physics Division. Hosie had for several years previously played a major role in the space research section dealing with administrative and financial matters in the Ministry for Science and had been much concerned with the negotiations for the establishment of ESRO. He now moved to State House, High Holborn, to be joined by Robins and the staff of the SRMU from premises at Chester Gate, Regent's Park. Robins became Head of the ASR Division and the SRMU was assimilated into the Division.

Table 10.1.

Subject	Chairman
Terrestrial and planetary atmospheres	G.D. Robinson
Ionosphere and radio propagation	J. Sayers
Energetic particles and magnetic fields	H. Elliot
Lunar, planetary and interplanetary matter	S.K. Runcorn
Solar and stellar astronomy	R. Wilson
Determination and study of orbits	J.A. Ratcliffe
Technical facilities	M.O. Robins

Within the major reorganization therefore, the detailed space science affairs continued to be handled by the same group of people already experienced in the Skylark and early Ariel satellite series and in the co-operative activities with NASA and ESRO; little or no interruption occurred in the smooth flow of business.

6   Beyond the SRC

We have described at some length the organizational changes immediately following the formation of the SRC, and these were of course very important, affecting as they did the channels through which public funds flowed to the various space science activities. However, the bulk of these activities and much of the technological support continued as before.

In 1966 there were active space science groups in about 17 universities and in about nine SRC and government establishments. The latter were funded directly by SRC or government. University funding was based on the dual support system, whereby the University Grants Committee financed the basic university needs for staff, buildings, and basic equipment whereas SRC supported specific research by making direct grants for essential equipment and services which could not be financed from university funds, and provided central facilities on a national basis or through international co-operation. In 1966, the total value of SRC research grants in space science current in universities was about £1.4 million. For comparison, the corresponding figures for astronomy and nuclear physics were about £1 million and £3 million respectively; the total for all subjects was about £17 million.

Public funding was not the only source of finance which supported space science in the 1960s, although it was by far the largest. Several industrial firms had, through favourable contract arrangements, effectively subsidized some of the advanced technology required for space experiments devised by university groups. The Mullard Company made a significant and lasting contribution when in 1965 it donated a sufficient sum to enable University College London to purchase a country mansion at Holmbury St Mary near Dorking, Surrey. This was converted to a laboratory to house the scientific space research group of the Department of Physics, which was at that time the largest in Britain, and hard pressed for adequate accommodation in the main College in Gower Street. The resulting Mullard Space Science Laboratory (MSSL) of the Department of Physics, University College London (UCL), opened in 1967 with R.L.F. Boyd as Director, has remained amongst the leading laboratories of its kind in the world.

Space science outside the university system in the 1960s flourished in a number of Government research laboratories, particularly in the Culham Laboratory of UKAEA, Space Department of RAE and in the Meteorological Office Research Group. These were unaffected at the time by the emergence of the SRC. So was the provision of essential services by RAE and UKAEA.

7 Further changes in the SRC organization

The new arrangements instituted by SRC settled down and functioned well for the first few years following the formation of the Council. In 1968, an additional working group, on Commonwealth Collaborative Programmes, chaired by A.P. Willmore was added to the existing seven working groups of the SPGC (see Chapter 7, p. 174).

Events outside the SRC led in 1969 to a significant addition to the scientific strength of the Council's staff. Financial pressures on the UKAEA had forced a critical review of the research programme of the Culham Laboratory, which was devoted to a search for a solution to the controlled nuclear fusion problem. It had been reluctantly decided that the work on ultra-violet solar spectroscopy, to which we have referred in Chapter 9 in relation to the Stabilized Skylark programme, could no longer be supported. In view of the high quality of the work and its relevance to other astronomical and space science programmes, the SRC decided, with Government agreement, to accept formal transfer of the staff and pro-gramme of the Astrophysics Research Unit (ARU) as it was now known. This took effect on 1 April 1969, and the staff under R. Wilson remained at Culham but were henceforth to be financed by the Council as part of the 'in-house' research effort.

The somewhat unwieldy nature of the SPGC and its eight working groups, and the growing coherence of the space science programme as a whole eventually led the Council to change the structure. In October 1969, the eight working groups were abolished and replaced by three:

Astrophysics Working Group	Chairman R.L.F. Boyd
	(later J. Ring)
Geophysics Working Group	Chairman F.G. Smith
Facilities Working Group	Chairman H. Elliot
	(later F. Horner)

These reflected the two main divisions of science served by the facilities of space technology. The separation of ground-based astronomy, the subject matter of the APGC, and astronomical observations from space vehicles, the subject matter of the SPGC and its Astrophysics Working Group, remained. Indeed by this time, 1969, members of Council chaired both the ASR Board (Lovell) the APGC (F. Hoyle) and the SPGC (Sheppard).

8   The Radio and Space Research Station

In 1970 the question as to how the RSRS could most effectively play its part in supporting the space science programme was causing concern. The ASR Board appointed a Review Panel under Sir Eric Eastwood to examine the situation and report. The RRS had been set up by the Radio Research Board, itself established by the DSIR in 1920, to try to understand the ionosphere and advise on radio propagation, which at that time was very much affected by ionospheric phenomena. When experimental rockets and satellites became available, the Council of DSIR decided in 1959 that half the effort of the Station should be devoted to space research, including the provision of services to the space science community. The change of name to the Radio and Space Research Station (RSRS) in 1965 described more accurately its activities.

The Review Panel reported in May 1971, and the Council approved its recommendations that the RSRS should play a greater role in the provision of central support for the space research programme. Two consequent actions implemented in early 1972 were the transfer of the SRMU from the ASR Division in State House, to be assimilated into the RSRS at Ditton Park, and the formal incorporation of the Astrophysics Research Unit, still at Culham, to become the Astrophysics Research Division (ARD) of RSRS.

These changes were not made without some questioning, but an event of a different kind which was unanimously welcomed was the renaming of the RSRS as the Appleton Laboratory, in honour of Sir Edward Appleton the great pioneer of ionospheric studies. The dedication by the Secretary of State for Education and Science, the Rt Hon, Margaret Thatcher, took place in November 1973.

9   Space science and astronomy

Rapid progress was being made in X-ray and ultra-violet astronomy using rockets and satellites, and there was promise of infra-red and gamma ray observations from satellites. Together with the advances in radio astronomy and results from the increasing power and sensitivity of optical telescopes, they served to emphasize the unity of astronomy irrespective of the techniques of observation. The Council and ASR Board sensed that the APGC and SPGC structure, which had served well whilst the new techniques of space science were establishing their place alongside the centuries old traditional optical astronomy and the newer radio astronomy, needed a close examination and possible revision.

After much consultation with the community of users in both ground-based astronomy and space science, a new committee structure was devised to operate under the ASR Board. A fair division of responsibilities and as much logic as the subject matter allowed led to the establishment of three committees which came into operation in October 1974. They were:

Astronomy 1   X-rays, gamma rays, cosmic rays, radio
               astronomy
               P.L. Marsden, Chairman
Astronomy 2   optical, infra-red, ultra-violet astronomy
               D.W.N. Stibbs, Chairman
Solar system   solar physics, lunar and planetary physics, solar/
               terrestrial relations, radio wave propagation
               A.P. Willmore, Chairman

At the same time the role of the SRC establishments concerned with the ASR Board programme was reviewed. A three-fold task was foreseen for them: to support university research, particularly in the planning, construction and operation of new facilities; to carry out research on a reasonable scale, preferably in collaboration with universities and certainly taking account of university programmes; to carry out work in the national interest for instance the time service at the Royal Greenwich Observatory. In addition it was expected that the Appleton Laboratory would have a prime responsibility for supporting the space science programme by operating facilities and managing projects. This committee structure and the nature of Establishment participation in the space science programme has served the community well, with little or no change up to the present time (1982).

During the 1970s, two trends brought additional SRC establishment resources to bear on the programme. In the first place, the Rutherford Laboratory, originally devoted only to high energy physics, diversified its activities in a way which allowed its extensive advanced engineering experience to contribute to other programmes, including that of space science. Then the increasing need for engineering support for space science and the pressures for economy in the use of manpower led to a merging of the Rutherford and Appleton Laboratories. Both are now sited at Chilton, the Ditton Park site of the Appleton Laboratory having been vacated. This concentration of resources was perhaps not to the liking of every one concerned, but it has meant that space scientists could benefit directly from one of the most powerful groups of research engineers in the country, with ready access to extensive computer and other facilities.

Some may regret the passing of the heady days of the late 1950s, when organization was fairly rudimentary, and the excitement and novelty of the new enterprise carried everyone forward in the face of many difficulties. In return for today's tightly controlled system, the British space scientist has a claim on massive resources and the entrée to a system of international collaboration without which much of value could not even be attempted.

11

The transformation of ESRO into ESA

In Chapter 6 we recounted the steps which led to the establishment of ESRO as an organization for the support of European space scientists by the co-operative provision of most, but not all, of the technological facilities, project management services and contractual services needed in the 1960s. In so far as the vertical sounding rocket and scientific satellite programmes were successful, bearing comparison with any contemporary similar activities elsewhere, ESRO was very successful. But on a broader basis, there was by 1968 much dissatisfaction both among the space scientists making use of ESRO facilities, and amongst the member States providing the funds. Many scientists were critical of what they believed to be poor value for money, although these criticisms were not always easy to substantiate because of uncertain subsidies supporting alternative ways of proceeding. Govern-ments tended to look critically at the build-up by ESRO of expensive technical and engineering facilities which clearly could be of great value to co-operative European programmes in applied space technology, whereas their restriction to the ESRO programmes of space science alone could, it was believed, lead to under-utilization and expensive duplication. In the background the faltering progress of ELDO, the desire of many Europeans to see a comprehensive space capability in Europe, including European satellite launching rockets, and the ever-present industrial and commercial interests in exploiting a new and potentially powerful technology, all contributed to a very confused scene in the late 1960s.

In this chapter we try to unravel the strands of importance to space scientists, and to the British space science programme in general, as ESRO passed through several critical phases, finally to be subsumed in the European Space Agency, ESA. It will be necessary to refer to the problems of ELDO and the influence of NASA programmes and opportunities for co-operation but we shall not dwell on these more than is necessary for an understanding of our general theme.

1   The European Launcher Development Organization

By far the most important organization with which the fortunes of ESRO were inevitably linked was ELDO, and an outline of some of the events in its chequered history and eventual collapse helps in understanding how ESA came into being.[1]

Most satellite launching rockets of the period were based on military rocket designs. The British medium-range ballistic rocket intended as a military weapon was started, with technical assistance from the USA, in 1956–57 under the code name of Blue Streak. A smaller rocket code named Black Knight was also developed for purposes of back-up research. The development programme for Blue Streak for military purposes was cancelled on 13 April 1960. In the intervening years, studies were carried out both at RAE Farnborough, by King-Hele and his associates, and also in industry, on ways and means of adapting Blue Streak as the first stage of a satellite launching system, with the second and third stages based on Black Knight technology. We have referred to related studies in Chapters 2 and 4. In view of the eventual success of the Blue Streak first stage in the ELDO programme it is more than likely that an all-British launching system based on Blue Streak and Black Knight technology would have been very successful and relatively economical, but this was not to be (see Chapter 4). The political tide of the time was flowing strongly in favour of international co-operation, particularly with Western Europe.

As described in Chapter 6, p. 115, in the autumn of 1960 discussions were initiated by the UK with other Western European countries to explore the possibilities of co-operative development of a satellite-launching system based on the proposed Blue Streak, modified to form the first stage. It is interesting to note that the discussions between European scientists, which were to lead to ESRO, were well under way by this time. There seems to have been rather little direct contact initially between the groups discussing scientific co-operation in space and those discussing co-operation in launcher development, at least in Britain. One reason may have been the heavy political and military overtones associated with rocket developments compared with the traditional free exchange of scientific views.

There was undoubtedly a reluctance on the part of the potential users of satellites, particularly the scientists and the telecommunication and meteorological authorities, to become financially involved in the development of launching rockets, which in their terms would be a very expensive operation. Finally, the membership of the two groups differed. Of the participants in the discussions leading to ESRO, Sweden, Switzerland and Spain were not involved in the co-operative launcher discussions, whereas Australia as owner of the potential launch site, took part in the launcher discussions but not in those concerning scientific co-operation.

A consequence of this state of affairs was that the international negotiations leading to ESRO and ELDO proceeded to a large extent independently, and two quite different organizations resulted. It was nevertheless recognized by many at the time that this dichotomy was less than ideal if the objective was solely to seek the most efficient overall programme of activities in space.

The governments of the UK and France took the lead in encouraging European co-operation in launcher development. The initial plan envisaged the adaptation of the British Blue Streak as the first stage of the rocket system. Development was already well advanced at British expense, the design was technologically up to date with American experience, and the overall launching system would have well suited likely European needs for orbiting scientific and application satellites. It would have been capable of further development for more ambitious projects. As described on p. 124 agreement was reached in October 1961 at a conference in London between representatives of Belgium, France, West Germany, Holland, Italy, Britain and Australia.

Unfortunately the arrangements made for the execution of the project had serious shortcomings. For instance there was no nominated prime contractor for the overall launching system, and a tenuous management chain in a project of such complexity could not but aggravate the problems of overall system design. Whilst in Britain the responsible authority, the Ministry of Aviation, and the main Contractors all had much experience in this type of work, the same could not be said of all the participating countries. We only touch briefly on the ensuing events.

There was a perturbation in 1965 when the French proposed that the initial design for the first system, known as ELDO A, should be abandoned in favour of a more ambitious ELDO B which would use liquid oxygen and liquid hydrogen as fuel in the upper stages. It was decided that ELDO A should continue, with studies to be made of the ELDO B proposal. Henceforth these types became known as Europa I and Europa III. Prolonged delays and the escalation of overall costs led in 1966 to a major crisis when the British Government indicated a wish to withdraw from the Organization. After considerable debate, the UK contribution was reduced from 38% to 27% and a budget ceiling was agreed for the future programme in which a modified form of Europa I, designated Europa II, was included.

An even greater upheaval occurred in 1968–69. The agency in Britain now responsible for participation in ELDO was the Ministry of Technology. An overall review of the space activities with which the UK was concerned led to a policy decision that the Government would in future not support the development and production of satellite launching rockets, but would rely on purchase, in the first place from the USA. The Government would continue to support scientific space research and the applications of space technology. This policy was strongly opposed by the French Government, and others in Europe, who believed that a launching capability independent of the USA would be essential if freedom of action in the future was to be assured. However, the British view was not changed, and the UK withdrawal from ELDO was complete by the end of 1969.

ELDO itself, without Britain, continued until 1973, when the programme was finally abandoned and the organization subsumed in the European Space Agency. During its life, ELDO launched eleven rockets, ten being of the Europa I type from Woomera, and one Europa II from the French range at Kourou in Guiana. Although the Blue Streak first stage operated successfully in every case, each attempt to launch a satellite was marred by malfunctions elsewhere in the system.

2 Problems for ESRO

Prior to the withdrawal of the UK from ELDO, the British space science programme was relatively insulated from these events, but it was inevitable that the repercussions of the withdrawal would be felt in ESRO. Many European countries had a more unified organization for all space activities than the somewhat fragmented arrangements which persisted in the UK, and were less inclined to separate the individual crises which occurred in both ESRO and ELDO. This will be apparent as we describe the unsteady progress of ESRO in the late 1960s and early 1970s.

By 1966, the scientific programme of ESRO was well under way as we have recounted in Chapter 6, with sounding rocket launchings in progress and design and development work on the satellites ESRO I and II, HEOS A and TD-1 and TD-2. However, financially all was far from well, with project costs escalating and capital works slipping behind schedule, with consequent underspending. Governments refused to sanction the carry-over of unspent funds from one financial year to the next and ESRO Council members were unable to agree on the future level of resources. There were increasing pressures in Europe for the better co-ordination, or even amalgamation of all European space organizations into a single body charged with executing a single predetermined space policy. These issues were clearly beyond the remit of the ESRO Council itself, and were considered at the second meeting of the European Space Conference in July 1967. The assembled Ministers set up an Advisory Committee on Programmes under the Chairmanship of J.P. Causse, a French engineer who had been responsible for several successful French national space programmes (see Chapter 6, p. 160).

The report issued by this Committee in January 1968 called for the fusion of ELDO and ESRO. The name of European Space Agency was suggested for the combined organization. It would carry forward a European space programme containing satellites for both scientific and applied purposes, it would develop launch vehicles and support an appropriate research and development programme.

It was November 1968 before the European Space Conference considered the recommendations of the Advisory Committee on Programmes, when the Ministers concerned met in Bonn. The way was cleared for ESRO to plan its budget for 1969–71, with a growth of about 6% per annum, but unfortunately not even provisional planning figures were agreed for 1972–74. The Conference did resolve that ESRO should be allocated funds and authorized to carry out economic and technical studies of application satellites, and subject to the confirmation of governments, should be entrusted with the execution of such application satellite programmes as might be undertaken collectively in Europe.

The Conference also agreed in principle that European space activities and organizations should be integrated. A Senior Officials Committee with Working Groups on Programmes and Institutions was established to examine the matters in depth. The Committee first met under the Chairmanship of G. Puppi, of Italy, on 27 March 1969.

The Conference of Ministers envisaged a single European space programme which would include elements of scientific research, applications of space technology, and the development of satellite launching rockets. To allow for the possibility that some countries might not be prepared to participate in a relatively expensive launcher development programme, it was proposed that all member states should contribute to a minimum basic programme, with an option to contribute to the launcher programme. It was one thing to agree on the principle of such an arrangement, quite another to achieve agreement in detail when the interests and financial capabilities of the member states differed so widely. Professor Puppi's Committee faced a formidable task.

3   The British view

Naturally in the UK the Science Research Council, with its responsibility to foster the programmes and aspirations of British space scientists, gave increasing attention to these moves on the European scene. The Council, at this time under the Chairmanship of Sir Brian Flowers, had welcomed the mandate given to ESRO to include the study of application satellites in its activities, believing that this would be to the advantage of both application and purely scientific satellites. The Council also supported the concept of a single European organization for all space activities, subject always to the condition that the arrangements provided adequate safe-guards for an acceptable space science programme. There was nevertheless a strong body of opinion in British scientific circles that the best interests of science alone would be served by a withdrawal from ESRO and the deployment of the funds so released in a national programme and in bilateral co-operation with NASA.

4   The growing crisis in European space policy

Throughout 1969 and into 1970, officials, prominent amongst whom was J.F. Hosie of the SRC, struggled with the issues referred to them by the ESC. The problems of the ESRO budget and programme were serious enough in themselves. They were aggravated by the withdrawal of Britain from ELDO, and the tendency shown by some countries to recoup the extra costs which thereby fell on the other ELDO members, by reducing the financial resources allocated to ESRO. It must be to the credit of the ESRO staff that in the midst of so much uncertainty, the scientific and engineering programmes continued to be pursued with such a large measure of success.

Probably the most critical point in the whole affair was reached in November 1970, when the fourth meeting of the Ministerial European Space Conference failed to reach even broad agreement on programmes and institutional arrangements. Confusion about the way forward reached a new peak. At the following meeting of the ESRO Council, France and Denmark denounced the ESRO Convention, thus preparing for the possibi-lity of withdrawal at the end of 1971, following the first eight-year period of ESRO's existence. Other member states took counter-action through a Resolution which opened the way for other withdrawals to become effective at the end of 1971.

5   The international policy of NASA relevant to Europe

We have referred to the reaction of the ELDO difficulties upon European Space Policy and thus on the affairs of space scientists; at least these difficulties and reactions were almost entirely within the control of the European Governments involved. The post-Apollo programme of NASA, which at this time included the concept and development of the reusable space shuttle launching vehicle, presented opportunities and problems for Europe of a different kind.

NASA had always shown a willingness to assist other countries and other organizations such as ESRO on a bilateral basis by the provision of rocket launches for scientific satellites on very favourable terms. This had led to widespread co-operation in the scientific and related technological fields prior to and during the massive effort which NASA devoted to the Apollo programme, ending in 1972. 'Post-Apollo' marked a watershed in the affairs of NASA as new and ambitious programmes were conceived and studied. These included both the scientific exploration of the planets and the development of new and hopefully cheaper means of injecting satellites into orbit. Towards the end of 1969 the USA invited participation in some of the proposed post Apollo programmes, including the space shuttle (a reusable launcher), a large permanent space station, and Spacelab (a module carried by the shuttle and designed to carry experimental equipment and scientist–astronauts).

European governments and agencies were naturally interested in these proposals, and studies of the space station and space lab concepts were undertaken by ESRO, whilst ELDO examined the proposed shuttle launching system. The undoubted attractions of taking part in such advanced scientific and technological projects were tempered by the likely costs which would fall on European space budgets, perhaps some 10% of the total costs of the projects, and by the possible restrictive conditions which might be attached to the use of American launching systems. The American attitude, apparently, was that an Agreement between NASA and the European Agencies for the use of NASA launching systems for European payloads would guarantee no discrimination against such payloads provided that there was no conflict of interest with other American commitments. The most obvious possible conflict could arise if the Europeans wished to have a commercially competitive communications satellite launched. Thus two extreme points of view were taken on the one hand by the French, who held tenaciously to the belief that an independent European launcher facility was essential for the future well being of European space activities, and the British, who believed equally strongly that the wasteful duplication of American technology was unnecessary and that purchase of NASA launchers for all reasonable purposes could be relied upon.

6   The Puppi proposals

We can now summarize some of the constraints facing the new Chairman of the ESRO Council, Puppi, when the Council invited him to try and break the deadlock following the abortive meetings of the European Space Conference at the end of 1970. There was no agreement on the future funding and programme for ESRO, and the announced preparations of France and Denmark to withdraw threatened the very existence of the organization. Belgium, France and Germany favoured a balanced overall space programme for Europe, including satellites for scientific research, applications, and the development of satellite launching systems. Britain, on the other hand, while in favour of scientific space research and application satellite programmes, thought the development of European launching systems unnecessary and wasteful in view of the availability of American systems. The future of ELDO was uncertain; although set up largely at British initiative, Britain had withdrawn from the organization. The original timetable for the development of a European launching system had slipped badly and estimated costs to completion had risen by a factor of some eight times. There was no unanimity in Europe about the response to be made to the NASA offer of co-operation in the post Apollo programmes. There was widespread dissatisfaction amongst space scientists, particularly in Britain, about the high cost of the subscription to ESRO in relation to the scientific opportunities and benefits offered, and many believed that on scientific grounds alone, withdrawal from ESRO in favour of bilateral co-operation with NASA and an expanded national programme, would be the preferable course of action.

Puppi faced these difficult problems with a small international group of advisers and succeeded in making proposals for ESRO which enabled the deadlock to be broken. These proposals emerged in May 1971, and formed the basis for a 'Package Deal' the main elements of which were:

> The predominantly scientific nature of ESRO to be reoriented so that considerable emphasis would be placed on programmes of space applications in the fields of aeronautics, meteorology and telecommunications.
>
> The scientific programme and basic activities would remain mandatory, but would be reduced in size and scope.
>
> The theoretical and laboratory studies at ESRIN would be terminated by the end of 1973.
>
> The provision for scientific satellites would be phased down by

about 20% to a level estimated to permit the launch of one medium-sized satellite every one-and-a-half to two years.

The sounding rocket programme would be terminated during 1972 and ESRANGE disposed of to Sweden, with optional arrangements for its use as a special project when desired.

The scale of financial provision for the various activities proposed by Puppi is shown in Table 11.1, where the figures are in millions of Accounting Units (MAU) 1 MAU being approximately 1 United States dollar. Development and launching costs are included, but not the development costs of launching systems.

These proposals were extensively discussed and considered by the SRC in Britain, particularly with respect to the opportunities which British space scientists might have in relation to the costs to be borne by the SRC. The UK was at this time contributing about 20% of the ESRO funds and if each scientific satellite carried five or six experiments and all the member states participated, one British experiment could be expected to be included about every eighteen months. Such *pro rata* selection was very unlikely to be followed, but it gave an idea of the scale of opportunities. It convinced many in Britain that UK participation could hardly be justified if opportunities were appreciably less than this low figure suggested. However the scientific programme of ESRO was increasingly becoming subordinate to the proposed applications programme; within the UK the applications of space technology now fell within the remit of the Ministry of Technology, and ESRO activities in this field were neither funded nor controlled by the SRC.

Table 11.1.

Programme element	1971	1972	1973	1974	1977	
1. Basic activities and proportion of common costs	16	15	15	15	16	MAU
2. Scientific satellites	38	43	38	35	35	
3. Sub-total	54	58	53	50	51	
Telecommunication satellites	4	22	40	58	64	
Aeronautical satellites	5	7	18	23	4	
Meteorological satellites and new applications		1	5	9	21	
Sounding rockets and ESRANGE	9	9	8	8	8	
ESRIN	2	2	2	2	2	
Grand total	74	99	126	150	150	MAU

We are also referring to a period when co-operation with countries of Western Europe was a topic of very lively political interest, prior to the accession of Britain to the EEC. The matter of continued British membership, or withdrawal from ESRO was therefore far from being a question to be decided only on scientific grounds. In the event, the Puppi proposals, with minor changes, were accepted by all the member states at the ESRO Council meeting in December 1971. France and Denmark withdrew their denunciation of the Convention and the immediate crisis was over. The main effects on the space science activities of ESRO were to be:

1. The sounding rocket programme phased out by mid-1972.
2. The current activities of ESRIN to be ended in 1973.
3. ESRANGE to be transferred to Sweden in July 1972, but maintained as a special project for use by member states which so desired.
4. The level of support for the scientific satellite programme to be not less than 27 MAU per annum from 1975 onwards (compared with 35 MAU in the original Puppi proposals).

The implications of this change in emphasis in ESRO for British space scientists were commented upon by P.A. Sheppard, Chairman of the Space Policy and Grants Committee of the SRC. He foresaw that the level of activity now planned for the ESRO science programme after 1973 would be relatively so low that UK scientists would need to look predominantly to NASA for future opportunities in spacecraft. The end of the ESRO sounding rocket programme would cause some problems in the short term because of commitments which could no longer be fulfilled, but in the longer term the UK would again need to become self sufficient. On the financial side it was expected that the net result of the changed ESRO programme would be a significant reduction in the share of the UK contribution for which SRC would be responsible. Indications were that the SRC payments would fall from about £4.7 million in 1971–72 to about £2.4 million in 1974–75, the major part of the UK contribution falling to the Ministry of Technology. Subsequent events showed that Sheppard's predictions were largely fulfilled.

7   Changes in the ESRO organization following the 1971 package deal

We have mentioned in Chapter 6 the changes in senior staff and in the executive organization which took place in 1967 and 1968. The advent of the 1971 package deal, with the new optional programmes of space applications and the changed emphasis in the overall objectives of the organization, required modifications to the convention and to the committee structure below the Council.

The last meeting of the Scientific and Technical Committee, chaired by Puppi, took place in 1972. The control of the scientific programme was henceforth to be invested in a Scientific Programme Board. Since the scientific programme was a mandatory part of the whole, all member states were represented on this Board.

Other Programme Boards controlled the optional programmes in Aeronautical, Meteorological and Telecommunication applications, and Space Lab activities, with membership drawn from those countries choosing to participate in the relevant programmes. The UK was a participant in the first three of the above mentioned application programmes.

The wider responsibilities of the ESRO Council meant that it was no longer appropriate for the UK delegation to represent only the British scientific interests. Delegates from the Department of Trade and Industry were also appointed, and the full UK membership was: J.F. Hosie and H.H. Atkinson from the SRC, and D. Cavanagh and A. Goodson from the DTI.

The UK members of the Science Programme Board were H. Elliot from Imperial College London, and H.H. Atkinson and W.B.D. Greening from the SRC.

The year 1972 was a year of relative stability for ESRO, with the successful package deal behind it, and with three successful scientific spacecraft launched, in January (HEOS A-2), in March (TD-1) and in November (ESRO 4), it was also a year of achievement. However, the same could not be said about other aspects of European space policy and activities.

8 The genesis of the European Space Agency

We have referred to the NASA offer of co-operation with Europe in various parts of the proposed post-Apollo programme. As the feasibility and cost studies proceeded and there appeared to be no end in sight for the general arguments and indecisions in Europe, NASA offers were reduced until finally by the end of 1971 only the Space Lab project on which ESRO was working, remained.

November 1971 saw the last and unsuccessful attempt by ELDO to launch a satellite. This was the first launch of the Europa II design of rocket system and the first from the French equatorial launch site at Kourou in French Guiana, South America. The Europa II was basically the same as Europa I, but had an additional solid propellant fourth-stage motor of French design and manufacture. The first stage British Blue Streak again functioned successfully, and although the vehicle broke up just before the end of first-stage burning, this was eventually traced to a failure in the guidance computer due to electrical interference in the German third-stage system.

This unhappy affair increased if anything the disarray in Europe, the Germans being by now particularly disenchanted with the ELDO concept. The French, however, persisted in their unwavering support for the idea of an independent launching capability for Europe. They produced a design for a new launching system code named the L3S. This was later to become known as ARIANE. The L3S used a new liquid propellant first-stage rocket of French design, a second stage of similar design and a third stage using liquid oxygen and liquid hydrogen. The performance was expected to be at least comparable with the ELDO designs for the Europa series, and it was claimed to have a capability eventually of placing a 750 kg satellite in geostationary orbit. It is unnecessary to follow the complex discussions which were soon to lead to the abandonment of ELDO. Instead, we may note that a renewed interest and initiative seemed to emerge in Britain with the advent of a new Minister, Mr Heseltine, in the Department of Trade and Industry.

There had by now been so many years of discussion about European Space Policy, or perhaps the lack of it, that governments and organizations had clearer ideas about areas where an identity of views was sufficient to support fully co-operative programmes and those where diverse views would lead to different optional programmes.

The fifth meeting of the European Space Conference of Ministers took place in December 1972 in Brussels. This was a decisive meeting at which agreement in principle was reached on all the outstanding issues which had been contentious for so long. These included future co-operation with NASA in developing Space Lab, the inclusion of ARIANE as a European launcher, and of MAROTS a maritime communication satellite of special interest to Britain, and the establishment of a European Space Agency. This meeting of the ESC was reconvened in July 1973, by which time officials in the member states had transformed the agreements in principle into a detailed plan of action which was to become known as the 'Second Package Deal'.

The main items in the agreement were:

1. The French proposals for the rocket system L3S (ARIANE) to be adopted as a European satellite launcher in preference to the Europa series of ELDO.
2. Space Lab to be developed in co-operation with NASA for carriage into orbit by the Space Shuttle.
3. A maritime satellite (MAROTS) to be developed and placed in geostationary orbit as part of the applications programme.
4. A European Space Agency (ESA) to be established, in which all functions of ESRO and ELDO would be subsumed.

The agreement to establish ESA brings us to the point at which we need no longer follow the fortunes of the launcher, space lab and application satellite programmes, but rather concentrate attention on the way in which the new Agency concerned itself with space science in Europe.

9 The ESA Convention and space science

The terms of the Convention to establish ESA were agreed at the sixth and final meeting of Ministers of the European Space Conference in April 1975, and the Convention was signed on behalf of the Member States on 30 May 1975. The Agency began operations *de facto* in April 1975, but several years were to pass before the procedures of ratification by the Governments of the Member States were completed, and the Convention did not formally enter into force until 30 October 1980. However, for the purpose of this narrative we take April 1975 as our starting point. The advent of ESA implied the demise of ESRO, although the well-established structure of ESRO, its Council, programme Boards and Committees, its Director General and his staff, and the various laboratories and facilities clearly would form the basis on which the new Agency would be built.

The last ESRO Council was chaired by M. Levy of France; he was also Chairman of the Science Programme Board. Reporting to this Board were the Launching Programmes Advisory Committee (Chairman, H.C. Van de Hulst), the Astrophysics Working Group (Chairman, C. de Jager) and the Solar System Working Group (Chairman, A. Dollfus). As we shall note, it was necessary to expand this structure to deal with the wider scientific interests developed by ESA.

After the *de facto* beginning of ESA, Levy soon relinquished the Chairmanship of the Council which he had carried forward from ESRO, and he was succeeded in July 1975 by W. Finke of the Federal Republic of Germany.

It is useful now to refer to some of the Articles of the ESA Convention relevant to space science, particularly in those areas where significant changes from the original ESRO Convention have been made.

Article 1 requires that all member states should participate in the mandatory activities referred to in Article V and should contribute to the fixed common costs.

Article II describes the purpose of the Agency as providing for and promoting peaceful co-operation amongst member states in space research and technology, and their space applications – with a view to their use for scientific purposes and operational space applications. This is to be done by:

(a) implementing a long-term European Space Policy – concerting the policies of member states with respect to other national and international organizations.
(c) co-ordinating the European space programme and national programmes and by integrating the latter progressively and as completely as possible into the European space programme – in particular as regards application satellites.

Article III refers to Information and Data – the Agency shall ensure that any scientific results shall be published or otherwise made widely available after prior use by the scientists responsible for the experiments. The resulting reduced data shall be the property of the Agency.

Article V refers to the activities and programmes of the Agency. Activities will include some which are mandatory, in which all members participate, and some which are optional. All will participate in the latter except those who formally opt out.
   The mandatory activities are to include:
The execution of basic programmes such as education, documentation studies of future systems and technological research;
a scientific programme, including the use of satellites and other space systems;
collection of information and dissemination to members – and advice and assistance in the harmonization of international and national programmes.
   Regarding the co-ordination and integration of programmes referred to in Article II (c), the Agency shall receive in good time from members, information on projects relating to new space programmes, and will evaluate and formulate appropriate rules to be adopted for the internationalization of such programmes.

Article VIII refers to launchers and other space transport systems. It is particularly important and part is quoted verbatim:
When defining its missions, the Agency shall take into account the launchers or other space transport systems developed within the framework of its programmes, or by a member state, or with a significant Agency contribution, and shall grant preference to their utilization for appropriate payloads if this does not present an unreasonable disadvantage compared with other launchers or space transport means available at the envisaged time, in respect of cost, reliability and mission suitability.

Article IX concerns the use of Agency facilities by member states: Provided that their use for its own activities and programmes is not thereby prejudiced, the Agency shall make its facilities available, at the cost of the State concerned, to any member state that asks to use them for its own programmes.

Article XI defines the functions and responsibilities of the ESA Council, and in a particular reference to the science programme:
The Council shall establish a Science Programme Committee[3] to which it shall refer any matter relating to the mandatory science programme. It shall authorize the Committee to take decisions regarding that programme, subject to the Council's function of determining the level of resources and of adopting the Annual Budget.

Apart from the fact that Space Science only represented a fraction of the total ESA programme (financially about 15% in 1977), perhaps the two most important changes from the ESRO framework concerned the internationalization of programmes and preference for the use of ESA or member states' launching systems.

The ESA Convention calls for a positive approach to internationalization, with an obligation on member states to offer co-operation in programmes which they initiate. There is no doubt that the rapidly increasing costs of worthwhile scientific programmes would have influenced member states in this direction anyway, but the formal requirement gives an additional impetus to the process. The substance of Article VIII concerning the preferential choice of European launching systems where possible gave cause for considerable concern amongst space scientists when the matter was in the discussion stage. Whilst the majority no doubt accepted that the wider interests of European co-operation would best be served by using European systems whenever possible, there was understandable unease that judgements as to what constituted an 'unreasonable disadvantage' might need to be made on the basis of rather scanty evidence. No one who devoted years of work to the design and production of a satellite payload, scientific or not, would relish having it launched by any system less than one of proven reliability and suitability. It could be foreseen that this feature of ESA policy could be a potential source of friction should the development and proving of ESA launching systems take significantly longer or prove less successful than everyone hoped.

10 The ESA organization

The new ESA Council, which was in the first instance largely an expanded version of the ESRO Council, took up its duties in mid-1975 under the Chairmanship of W. Finke. The ESRO system of a Programme Board for each of the major application projects was maintained, but the mandatory science programme became the responsibility of a Science Programme Committee (SPC) reporting to the Council in accordance with Article XI of the Convention. The UK members of the SPC were H.H. Atkinson (SRC) and H. Elliot from Imperial College London. With such a broad field of science within its remit, the SPC required specialist advice on the main disciplines with which it was concerned, so it set up a Scientific Advisory Committee under the Chairmanship of M.J. Rees (Cambridge University). This Committee in turn was served by four Working Groups:

Astronomy	Chairman G. Setti (Italy)
Solar System	J. Geiss (Switzerland)
Life Sciences	H. Bjorstedt (Sweden)
Material Sciences	H. Ahlborn (Germany)

A fifth Working Group was added in 1977:

Space Telescope	Chairman F. Pacini (European Southern Observatory)

It will be noted that the earlier ESRO Working Groups on Astrophysics and the Solar System had been augmented by three more, representing the expanding boundaries of space science of concern to ESA.

In parallel with the development of the ESA Council structure and its delegate bodies, the executive structure of the former ESRO was expanded and modified to suit the new situation. The Director General of ESA at this time was R. Gibson, from the UK, who had been the Director General of ESRO and prior to that the Director of Administration of ESRO. Figure 11.1 indicates the main lines of responsibility for the science programme, and the institutions and projects related to that programme in 1975.

So the transformation of ESRO and ELDO into ESA was effected, although as we noted earlier, it was not until October 1980 that all the legal formalities were completed. Austria became an Associate Member, followed in 1981 by Norway. The initial organization of ESA sufficed for a time, but in due course changes and a general streamlining took place. However, these would more properly be described in a history of ESA rather than in this narrative.

11   New space science programmes

Before closing a chapter which has been almost exclusively concerned with matters of policy and administration, manoeuvre and compromise, we should note that despite the political and financial uncertainties, no doubt magnified by incomplete information at the time, ESRO and ESA staff and their counterparts in the member states developed new programmes in space science which were at the forefront of research.

We have already mentioned the three successful spacecraft launched in 1972, HEOS-2, TD-1 and ESRO 4. The next ESA spacecraft was successfully launched in August 1975 – this was COS B which had been planned by ESRO and was the first ESRO/ESA satellite in which there was no direct British scientific participation (see Chapter 15, p. 389). The single experiment carried by COS B detected gamma radiation over the energy range from 30 to 5000 MeV, and was still working well six years after launch.

Following the launch of COS B there was a gap until 1977, when on 20 April, the satellite GEOS was launched by a Thor-Delta rocket from the Cape Canaveral site in the USA. GEOS was a magnetospheric observatory satellite, but unfortunately partial failure of the second stage of the NASA rocket system resulted in the spacecraft failing to attain the intended geostationary orbit. Instead, only a low apogee of 11,752 km was achieved rather than the nominal 35,786 km. This resulted, amongst other things, in a serious degradation of the solar cell array and hence of the power supplies, as the spacecraft dipped into intense regions of radiation (the Van Allen belts). Some very rapid and resourceful reassessment of the situation by ESA and

Fig. 11.1 ESA executive organization 1975.

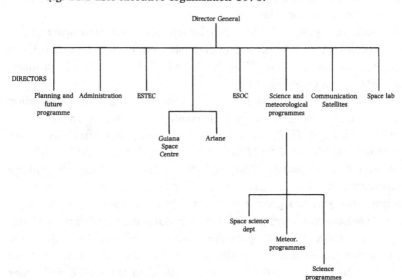

NASA staff resulted in a revised operations plan. The satellite was manoeuvred into an orbit which minimized the problems with the power supply and enabled command signals to be sent and scientific data to be received, albeit from a rather different orbit from that planned in the first place. The spacecraft carried an experiment devised by the Mullard Space Science Laboratory of UCL consisting of two electrostatic analysers which measured electron and proton fluxes up to energies of 500 eV. These performed well, and in 14 months of operations a variety of low energy plasma populations were sampled, the new orbit allowing a full local time scan of the low latitude magnetosphere to be made. Thus a considerable degree of success was rescued from apparent failure.

A second ESA spacecraft was launched in 1977 when on October 22 the European satellite code named ISEE-2, being the contribution to a three-satellite project constituting the International Sun Earth Explorer programme for magnetospheric research, was put into orbit by NASA together with its NASA companion satellite ISEE-1. This pair of satellites orbited together with a carefully controlled separation so that it was possible to distinguish between spatial and temporal variations in magnetospheric properties. The third satellite of the trio, ISEE-3 was successfully injected into orbit about the L1 libration point on 20 November 1977. This spacecraft, permanently located between the sun and the earth, can measure changes in the solar wind. The Group at Imperial College London, played a significant part in this project by providing magnetometer equipment for ISEE-2, and jointly with ESTEC and the Sterrekundig Instituut Utrecht, a low energy proton experiment for ISEE-3.

The following year, 1978, was also the year of launch of two spacecraft in which both ESA and British space scientists were heavily involved. These were the International Ultra-Violet Explorer (IUE) launched on 26 January, and GEOS-2, launched on 14 July.

The remarkable history of IUE which involved a tripartite co-operation between NASA, the SRC and ESA is recounted in Chapter 15.

After the launch of IUE, ESA had but two science projects remaining from the programme intiated by ESRO. These were GEOS-2 and EXOSAT. We have referred to the partial failure of GEOS-1 due to a faulty launch resulting in an unscheduled but still useful orbit. Its replacement, GEOS-2, successfully achieved the planned geosynchronous orbit, and again included in its payload plasma analysers provided by the Mullard Space Science Laboratory. The satellite has, with gaps due to the limited capacity of the ESOC ground station, provided a wealth of data the value of which has been enhanced by the correlations possible with data from ISEE and other spacecraft, and from the European Incoherent Scatter facility (EISCAT) based in Scandinavia.

EXOSAT, a satellite to be placed in a highly eccentric orbit for X-ray astronomical observations, was approved by the ESRO Council in 1973. Because of the 'Observatory' nature of the project, ESRO undertook to finance and manage the development of the scientific package, which included instrumental contributions from the Mullard Space Science Laboratory and from the X-ray group at the University of Leicester. EXOSAT was eventually launched by a NASA rocket system in 1983.

Meanwhile, UK space scientists can expect to play through ESA a significant and rewarding part in four more major projects in the 1980s. These are the Space Telescope (see Chapter 12, p. 249). GIOTTO (a spacecraft to rendezvous with Halley's Comet), HIPPARCOS (an astronomy satellite) and the International Solar Polar Mission space project. The transformation from ESRO to ESA can indeed be regarded as complete.

12

The Space Science Committee for Europe

1 The formation of the Provisional Space Science Board for Europe

In Chapter 4 we discussed the important role played by the Space Science Board (SSB) of the US National Academy of Sciences in relation to the space science programme of the National Aeronautics and Space Administration (NASA). The Board gives expert advice on the long-term programme and generally acts as a body to which members of the space science community may turn to air any dissatisfaction they may have. On the other hand, the Board does not interfere in the day-to-day running of NASA and its short term programme.

In 1973 the SSB, then under the Chairmanship of R.M. Goody, expressed their concern about the need for an international advisory body for space science. In a letter to Sir David Martin, the Executive Secretary of the Royal Society, H. Friedman, a member of the SSB, stated 'our concern is related to the growing interest in joint development projects for space exploration . . . much of the interest can be directly attributed to the increasing national budgets and growing political emphasis placed on sharing the costs of major projects internationally. Without an independent and respected source of scientific advice we believe it is possible that the science content and planning for joint projects may be overly influenced by aerospace industry requirements . . . .' He then suggested that he would welcome the opportunity to discuss the idea of an international advisory agency during his forthcoming visit to London to attend a meeting of the Bureau of SCOSTEP.[1] He also hoped that H.S.W. Massey, C. de Jager and B. Peters, who would be in London at the time, 28–29 November, would also take part in the discussions.

This initiative was welcomed without reservation by Martin and Massey who were well aware that, on the European scene, let alone the fully international one, there was no even remotely similar body to the SSB.

This had not been serious while ESRO was primarily concerned with space science but the forthcoming assimilation within ESA, an organization with a much greater resemblance to NASA in its overall responsibilities, had changed the situation.

At the discussion with Friedman on 28–29 November the Europeans suggested that, in the first instance, it would be wise to concentrate on the establishment of some independent advisory body for space science in Western Europe before attempting anything on the wider scene. Massey suggested that, as Physical Secretary of the Royal Society, he would be happy to call an *ad hoc* meeting of about 17–20 senior European space scientists to discuss the matter. Three US space scientists including Friedman would also be invited to take part.

This proposal was welcomed by R. Goody, the SSB Chairman. After discussion with other European colleagues, including J. Blamont, H.C. van de Hulst and R. Lüst and with the acting Director-General of ESRO, R, Gibson, Massey issued an invitation to the meeting couched in the following terms: 'After discussion with H. Friedman, C. de Jager and B. Peters I wish to invite you to a closed discussion meeting at the Royal Society to consider how to shape the future of space science programmes involving USA and Western Europe. The meeting might make a recommendation for some sort of continuing independent and influential advisory body. It is proposed that the discussion meeting begins at 10 am on 10 and 11 April 1974. It is hoped that each participant might give a 10-minute talk on their own personal views. The first part of the meeting will be devoted to broad discussion and the second to organisational matters.'

There was a very favourable response to the invitation and a fully representative group accepted. The SSB Chairman attended himself as one of the three for the USA, together with H. Friedman and F.S. Johnson. Apart from Massey, de Jager and Peters who attended as convenors, Western Europe was represented by P. Swings (Belgium) B. Stromgren (Denmark) J. Blamont (France) E. Amaldi and G. Occhialini (Italy) H.C. van de Hulst (The Netherlands) N. Herlofson (Sweden) R.L.F. Boyd, H. Elliot and S.K. Runcorn (UK) and R. Lüst (W. Germany). Neither H. Alfvèn (Sweden) J. Geiss (Switzerland) nor R. Monod (France) were able to accept.

After preparatory accounts of the present position in the USA and in Europe the desirability of establishing a space science advisory committee within W. Europe was considered in some detail. There was general agreement that this would be highly desirable. Among other things it was important to secure the position of space science in W. Europe now that ESRO was being absorbed within ESA while at the same time there were great advantages to be gained by closer contacts with the US space science community. One outstanding difference between the USA and W. European situations was the absence of any W. European organization comparable with the National Academy of Sciences to which an advisory committee would report and make its presence felt. As it happened, however, events were in train which helped partially to fill the gap in this respect.

On 1 December 1972 the Royal Society convened an informal meeting of leading scientists from Western European countries to discuss proposals made by the EEC about a common policy for scientific research and technological development. The meeting considered in particular the establishment of a European Science Foundation (ESF) of which academies and research councils within W. Europe, rather than of just the EEC, would be members. It was decided to take the matter further at a second meeting convened by the Max Planck Society in Munich on 13 and 14 April 1973, which was attended by representatives of sixteen national science organizations. As a result a preparatory group was formed to prepare for a further meeting in Paris. This meeting definitely decided to proceed with the formation of a European Science Foundation, appointed a steering group and arranged for further meetings to be held in Stockholm to complete the task.

Referring back now to the *ad hoc* meeting in April, Massey suggested that the proposed Foundation, being more nearly equivalent to the National Academy of Sciences than any other European body, might be just the one to which a European space science advisory committee should report. It was agreed that the scientists present at the meeting should constitute themselves a Provisional Space Science Advisory Board (PSSAB) for Europe, with Massey as Chairman and Lüst as Deputy Chairman. Furthermore, if informal approaches to be made by Amaldi and Lüst at the May meeting of the provisional ESF should be received favourably a letter would be sent without further delay to the Chairman Elect of the ESF prior to its inaugural meeting in November 1974, informing him of the existence of the PSSB, its aims and the hope that it could be incorporated as a committee of the ESF.

The suggested terms of reference of the proposed committee were as follows: 'To advise the European Science Foundation on

   (i) the role and significance of studies carried out by the use of spacecraft in the interest of basic science as a whole,
  (ii) the level of space activities appropriate in basic science in the context of the total space activity in Europe, and

(iii) the balance within space science of programmes in which the nations of Europe might be involved.

It will further be the duty of the proposed committee to collaborate with national academies or equivalent bodies in Europe, with the United States Space Science Board and any other comparable body with a view to promoting the best opportunities for reviewing, co-ordinating, proposing and supporting the most suitable scientific experiments in space.'

2 Incorporation in the ESF – the Space Science Committee

The European Science Foundation came into formal existence on 15 November 1974 at a plenary meeting in Strasbourg. Sir Brian Flowers was elected President and F. Schneider as Secretary General. Before long Flowers wrote to Massey requesting information about the activities, intentions and method of the PSSAB. Following discussion at a PSSAB meeting on 15 March 1975 a suitable reply was sent in time for the meeting of the ESF Council on 6 April 1975. As a result the ESF Council agreed that the PSSAB should be accepted in the first instance as an *ad hoc* Committee of the Foundation – its Space Science Committee (SSC) – subject to some small modifications in the membership. After two years the position would be reviewed in the light of experience, taking account of the fact that the ESF was establishing an Advisory Committee on Astronomy under Lüst's chairmanship.

It was obvious that these arrangements depended for their success on the agreement of ESA to co-operate with the ESF through the SSC. This presented no real difficulty because the PSSAB was in contact with ESRO and then ESA from the outset. In a document formally agreed by ESA and ESF in September 1976 it was stated that 'The SSC will normally concern itself with long-term issues related to the future of space science in Europe and is, according to its mandate, advisor to the ESF on the particular role of space science in the pattern of European scientific activities. Its reports and recommendations to the Executive Council and the Assembly of the ESF will as a matter of course be forwarded also to ESA with any observations from the Executive Council or from the Assembly. ESA will receive those recommendations on an annual basis or at other times, by mutual agreement. Within ESA the recommendations will be examined by the Director General, the Council and the Science Programme Committee.' Provision was also made for representatives of ESA to attend meetings of the SSC on the invitation of the Chairman. The final agreed terms of reference differed only in detail from those proposed initially.

During the two-year trial period the Secretariat of the Committee[2] would continue to be provided by the Royal Society in London.

While in principle being very similar to the US Space Science Board, in practice the position was rather different because, in contrast to NASA, ESA with its smaller resources and hence longer time between the larger space missions, did not work to a long-term scientific programme. As shorter term studies were already carried out within ESA by groups whose membership consisted largely of independent European scientists there was much less need for the SSC to undertake detailed scientific studies of the type which formed a large part of the task of the SSB in the United States. Nevertheless, as we shall see, the existence of the PSSAB and later of the SSC proved valuable at a quite early stage.

Before discussing these important early activities it is worth recording the stage finally reached in maintaining liaison between the SSB and SSC. This was very close from the earliest stages, thanks particularly to the SSB Chairman Goody and his colleagues Friedman and Johnson. However, in the course of time more formal relations were set up. The SSB now appoints one of its members as liaison member of the SSC. In all but purely local European affairs he is regarded in fact as a member of the SSC and attends all its meetings. As suggested in 1977 by A.G. Cameron, the then Space Science Chairman, who followed R. Goody in 1976 a full day is devoted by the SSB, normally at its Oct.–Nov. meeting, to discussion with members of the SSC of space science affairs of mutual interest. Usually four or so Europeans attend in Washington for this occasion and are able at the same time to acquire an on-the-spot picture of the present state of space science in the USA as viewed by the SSB. These arrangements have proved very satisfactory but they do not exclude other aspects of co-operation. Thus the SSC appoints liaison members to maintain contact with the work of the different subject sub-committees of the SSB and to attend meetings thereof, if appropriate. Again joint working parties on different aspects of space science, sponsored by both the SSB and SSC, have proved to be of great importance.

3 Early activities of the PSSAB and SSC – the ESA Convention

The convention establishing ESA was due to be signed on 30 May 1975. Certain members of the PSSAB were horrified to find that it contained no provision for the inclusion of a Scientific Committee within ESA. Without delay all members were consulted by telephone and all agreed that the matter must be rectified. The unanimous view of the PSSAB was conveyed to the acting Director-General and to the National Delegates of ESA and this had the desired effect, the following statement being included in the convention:

'The Council shall establish a Science Committee to which it shall refer any matter relating to the mandatory science programme; it shall authorize the Committee to take decisions on the programme subject to the Council's function of approving the level of resources and the annual budget. The terms of reference of the Science Committee shall be determined by a two-thirds majority of all Member States and in accordance with this article.'

If the PSSAB had not been in existence at the time it would have been much more difficult to achieve the desired aims so quickly.

4 The space telescope

The first major applications of the techniques of space research to astronomy took advantage of the fact that, from a space vehicle, observation could be made of radiations from extraterrestrial sources which could not be observed from the ground because of absorption in the earth's atmosphere. In this way ultra-violet and X-ray astronomy were born and developed quite rapidly, especially as they were soon providing new unexpected results of great importance.

Nevertheless, optical astronomy which is concerned mainly with observations of visible light from cosmic sources, remains the core of the subject. Any technique affording very substantial improvement in the 'seeing' of astronomical sources in the visible is bound to open up exciting scientific prospects. Such a technique becomes available if an optical telescope with a mirror of an aperture 2 m or so can be launched into space. As with earlier space experiments such a system could observe far down into the ultra-violet but its unique feature would be the great improvement in seeing in the visible owing to the elimination of the disturbing effect of atmospheric motions which limit the performance of a ground-based telescope. Additional advantages would be the great reduction in background radiation such as light from the night sky. It would of course be necessary that the telescope mirror could be fashioned to sufficiently high accuracy so that advantage could be taken of the avoidance of atmospheric motion.

The idea of a project of this kind dates back to quite early times in the history of space research. A thorough study of the scientific potentialities was made in 1969 by an *ad hoc* Committee of the SSB. Feasibility studies based on a 3.0 m telescope were completed by NASA in 1972 and contracts let for the definition and development of the telescope portion and the spacecraft. The Large Space Telescope, as it was then called, first appeared as a separate line item in the NASA Budget Submission for the fiscal year 1975. Only half the sum requested for that year was received and it was accompanied by a requirement that a less expensive programme be pursued and international co-operation sought.

Accordingly, NASA, through its contractors, examined the effect on the scientific performance and on the cost of the project resulting from reduction in mirror size.

It was found that a good compromise could be achieved by reduction of the primary mirror diameter to 2.4 m. As by June 1974 the figuring of a mirror of 1.8 m diameter had been achieved to an accuracy of 1/60 of a wavelength at 632.8 nm, there seemed every reason to expect that a 2.4 m mirror could be fashioned so that its performance would be diffraction-limited at this wavelength. Accordingly, all further design and development studies were based on the 2.4 m aperture telescope.

At the same time, NASA, mindful of the encouragement to seek international participation in the project, began discussions with ESRO about European involvement, in February 1974. These were essentially concerned with the possibility of ESRO providing one of the focal plane instruments for the telescope system and for this purpose the faint object camera (FOC) was selected as likely to be the most suitable. ESRO would also probably contribute through the provision of the solar cell arrays for the power supply.

By this time it was being realized, particularly by members of the SSB, that the LST programme would have a much wider effect on the astronomical community than any space project hitherto. For this time the 'classical' astronomers brought up to observe in the visible from the ground would be thoroughly involved. The advent of the LST could not fail to expand greatly the nature of their work. On the other hand, it would be essential that these scientists, who were far from being space science 'buffs', would participate fully, with enthusiasm, in the exciting new prospects opened up. This applied *a fortiori* to Europe where it was only in the UK that astronomy was beginning to be accepted as a science to which space research cannot fail to continue to make major contributions. Again the project was unique in space science in that it was planned to operate the telescope for 10–15 years at least. Advantage would be taken of the space shuttle, which should be fully operational in time to launch the LST, not only for this purpose but to refurbish, or even replace, instruments aboard at 2–3 year intervals. Bound up with all this was the problem of data processing and analysis which would result from the very rapid flow of data from the telescope. It was clearly necessary to work out means to encourage astronomers to use the LST in ways as near to their familiar ground based observatory practice as possible.

In human terms the problem was highly complicated, the mutual suspicions of different agencies and groups being aroused when something as big as the LST is involved. Thus there were the space science groups in the USA and in Europe, the ground-based astronomers in USA and Europe and the two space agencies NASA and ESA. It was necessary for all of these communities to work together for the project to be successful.

Equally, in view of the large scale of the project it was important to convince those concerned with budgetary decisions, primarily in the USA but also in W. Europe, that the LST programme was one of great scientific importance in the view of the international scientific community.

4.1 The Williamsburg meeting

In discussion with Massey, Goody suggested that the best way to proceed at this stage would be for the SSB and the PSSAB to organize jointly a meeting to study the relationship between the Large Space Telescope and the international community. This was further discussed between Goody, Johnson, Friedman and Massey at the COSPAR meeting in Varna in June 1975. It was suggested that there should be approximately 10 participants nominated by each organization, meeting some time during January 1976 in the USA. The aims would be (a) to make a statement on the relationship of the LST to European, US and international aspirations in astronomical research over the next 10–15 years (b) to study the most effective interface of the telescope with the international scientific community, including existing institutions with parallel problems and (c) to make recommendations as to the most effective means to optimise the scientific output from the LST.

This proposal was well received by NASA and by the SSB and PSSAB (now the SSC) and at its meeting on 22 September the latter body set up a small committee consisting of Blamont, Lüst, van de Hulst and Massey to represent it in discussions with the SSB. In particular Lüst and van de Hulst attended a meeting of the SSB in October where the arrangements were discussed at some length. It was decided to call the study 'International Cooperation in Space Observatories', to be held in Williamsburg, Virginia for 4–5 days commencing at 9 am on Monday 26 January 1976. In Goody's words 'the resulting report should be short, being primarily aimed at setting the tone for future collaboration and indicating guiding principles'. It should not be released until it had been reviewed by both the SSB and SSC. Finally on the 14 November a meeting was held at the National Academy of Sciences between R. Goody, Margaret Burbidge, M. Rosen (Secretary of the SSB) representing the SSB, and H. Massey and E.B. Dorling (Acting Secretary of the SSC) representing the SSC, to finalise the arrangements.

The SSC made arrangements for the ESF astronomy committee to be fully informed and to be represented at the meeting. The ten European participants were chosen as F. Bertola (Italy), G. Courtes (France), E.B. Dorling (UK), M. Golay (Switzerland), F.G. Smith (UK), M. Grewing (W. Germany), H. Massey (UK), M.J. Rees (UK), H.C. van de Hulst (The Netherlands) and L. Woltjer (The Netherlands and European Southern Observatory). In addition, after some preliminary hesitation, F. Machetto and E. Peytreman attended on the part of ESA as observers. R. Wilson (UK) also attended as an observer. The full list of those present is given in Annex 14.

The meeting, which took place under Massey's chairmanship, began with presentations on the present status of the LST project and its scientific objectives, followed by descriptions of the LST configuration and instrumentation. The following day was taken up with discussion of institutional and related questions and to the important problems of data acquisition and dissemination. After a day devoted to the discussion of the issues raised, the morning of Thursday 29 January was devoted to the formulation of conclusions and recommendations and in the afternoon these were discussed informally with the Administrator of NASA, J.A. Fletcher, and the Director General of ESA, R. Gibson.

There was no doubt of the success of the meeting in disseminating among the astronomical communities the unique potential of the LST and its technological feasibility. By doing so it already removed some of the difficulties holding up its further progress, as we shall see.

The inclusions and recommendations were published in a Joint Report on the conference by the National Academy of Sciences (through the SSB) and the European Science Foundation (through the SSC). An announcement to the Press, a copy of which is shown in Annex 15, was released simultaneously in Washington and London on 26 February 1976.

The first conclusion concerned the scientific importance of the LST. It was agreed that the capabilities of the LST represent a natural and achievable progress in astronomical technique, of a magnitude and importance that exceed those of all other likely developments in optical astronomy over the next few decades.

It was next recommended that every effort should be made to ensure that the focal plane instruments were such as to exploit fully the capabilities of the telescope. To assist in achieving this aim it was recommended that an advisory committee, appointed by the National Academy of Sciences and the European Science Foundation, should review, prior to final selection, the design of the focal plane instruments to ensure the availability of the best space qualified equipment.

The next recommendation was concerned with the operation of the LST. To carry this out effectively where the astronomical communities, not only of the USA but of W. Europe, are involved it was agreed that a scientific organization, probably a science institute, would need to be set up. Among its activities would be the planning and scheduling of observing programmes in conjunction with an international programme committee, providing links between the instrumentation and guest observers, and data reduction. The scientific staff of the institution would assist guest observers to use the instruments, while themselves also engaging in research.

As to observing time on the LST it was agreed immediately that the major factor in time allocation should be the scientific merit of the proposals. It was accepted, however, that some account would have to be taken of any contributions made to the project by organizations in other countries, as for example ESA.

Finally, attention was drawn to the need for study of the problem of handling the very large data flow, including processes of data distribution and requirements for manpower in data handling and scientific analysis.

Although Congressional approval for the ST as it was now called, the L being dropped, did not come till 1977 when $36 million were voted as start-up funds for the fiscal year 1978, there was a general feeling after the Williamsburg meeting that the project would go ahead. Not only did work proceed in the USA on this assumption but there was increased interest in ESA in participating in the project. An ESA Space Telescope Group was set up under the Chairmanship of F. Pacini and an ESA ST Instrument Science Team with H.C. van de Hulst as Chairman, while F. Machetto was chosen as ESA Project Scientist. A great deal of attention was devoted to the selection of the instrumentation for the Faint Object Camera as also the detailed drafting of a Memorandum of Understanding with NASA which was finally agreed on 7 October 1977.

4.2 The Science Institute and related matters

The concept of the science institute was a novel one in the space science context. While financed by NASA it was thought of as an essentially independent scientific body. While it was accepted by the SSB that in the event that ESA would contribute to the ST project, there would be European representation in the Institute at all levels, it was not at all clear whether ESA would be responsible for the selection of all of these. The first step towards formation of the institute was taken by the National Academy of Sciences, through the SSB, setting up a group under F.O. Hornig to study the whole problem. At the meeting of the SSC on 1 March 1976 the Chairman of the

SSB agreed that two members of the SSC would be invited to join the deliberations of this group. Unfortunately, however, this was not realized because ESA had replied quite properly in the negative to a query from NASA as to whether they would wish to take part in the study in any way. This was taken as an indication that the Europeans did not wish to take part and the SSC did not hear about this until it was too late. However, at the COSPAR meeting in Philadelphia in 1976 a discussion took place between Goody, Gibson, Rasool (NASA) and Massey and the confusion was cleared up sufficiently to allow for Woltjer and Dorling to represent the SSC as observers.

During the following year steps were taken to ensure that contact between the SSC and ESA was close enough to avoid any repetition of this misunderstanding. In particular on 5 October 1977 a joint meeting on the Space Telescope project was held at the Royal Society's rooms in London attended by representatives of the Space Telescope Working Group of ESA, the SSC and the Astronomy Committee of the ESF. The Chairman of the SSB A.G.W. Cameron also attended as well as F.S. Johnson who was at the time the SSB liaison member to the SSC.

The timing of the meeting was very appropriate because the Memorandum of Understanding between NASA and ESA concerning the ST had been agreed and was to be formally signed on October 7. Cameron was able to give an account of developments concerning the ST institute. He reported that NASA had set up an internal committee to consider the recommendations made by the study carried out under the Chairmanship of D.F. Hornig. A preliminary report from this Committee agreed with these recommendations apart from some minor matters so that it was very probable that an institute would be established which would be operated by an independent consortium of universities. Responsibility for policy-making would reside within a Board of Trustees on which there would be room for European membership. The Board would designate the Director of the Institute. There would be a Users' Committee which would include all users of the ST, operating through open meetings and a Steering Committee having access to the Director of the Institute. It was suggested that the Institute should start to operate in 1979. Arrangements were made at the meeting to ensure through common membership that the SSC, the ESF Astronomy Committee and the Space Telescope Working Group should keep closely in touch with events especially with regard to European involvement in the Institute. Considerable discussion also took place about the desirability of a European Space Telescope Centre and what form it might take.

A further matter reported at the meeting by the Chairman concerned the progress made in setting up the Joint Advisory Committee on focal plane instrumentation referred to on p. 252. Two meetings to plan the committee procedure had been held, one in Washington on 25 July 1977 and one in London on 9 August 1977. It was agreed that action should be taken at the request of NASA and ESA and that the instrument review should begin after the instrument selection by NASA was completed. Meanwhile NASA and ESA would keep the Committee informed of the progress of instrument development.

This Committee finally met in Washington 29–31 March 1978. The members were W.K. Ford, J. Mathe, E. Ney and D. Williamson nominated by the SSB and R. Wilson, R. Caysel, R.J. van Duimen and D. Lemke by the SSC. Ford and Wilson acted as co-chairmen. They presented a very favourable report on the selection of the present complement of instruments.

Turning back now to the relations between the European astronomical community and the ST Institute it was decided to set up a European co-ordinating Facility at which the ST data archives could be stored in a way accessible to European users as well as data processing facilities provided for those European groups working on ST material who wished to use them. The UK submitted a proposal that the Facility should be located at the Royal Observatory Edinburgh but the European Southern Observatory establishment at Garching was chosen. The SSC nominated van de Hulst and Lüst to keep closely in touch with matters relating to the ST.

In March 1981 it was announced that the ST Institute would be operated by the American Association for Research in Astronomy (AURA) and would be located at John Hopkins University in Baltimore, Maryland. R. Giacconi who had made such fundamental contributions to X-ray astronomy was chosen as Director. Pacini was nominated as a European member of the Board of Directors of AURA, and together with van de Hulst as a member of the Science Institute Council. European representation on the Visiting Committee, the Users Committee and the whole position regarding the Time Allocation Committee was still under consideration. In addition to European membership of three overseeing bodies, ESA would employ about 15 people, about half of whom would be astronomers, to work at the Institute normally for four year periods.

5 Other activities of the Provisional Committee – the Spitzingsee Conference

In addition to the various issues concerned with the space telescope the Provisional SCC began to examine its role in long-term planning of European space research in relation to the Science Advisory Committee of

ESA. It also discussed the use of Spacelab and of the Space Shuttle generally from the point of view of space science. The future of the Committee within the ESF soon became an urgent matter as the two year 'probationary' period for the Provisional Committee ended in late 1977. A further activity which had to be planned and which would take place before that date was a second joint conference sponsored by the SSB and the Provisional SSC.

The proposal for this conference arose because of the success of the Williamsburg Conference in bringing together the astronomical communities of the USA and Europe. It was suggested that a similar conference should be held on high energy astronomy as a natural complement. Some reservations were felt about this because the Williamsburg Conference was built around the core of the development of a specific spacecraft, the space telescope, whereas the proposed successor meeting would be dealing with a variety of possibilities in various stages of development, many well into the future. Nevertheless there was general support for the idea. It would be the turn of the SSC to be hosts for the conference and at an early stage the Max Planck Gesellschaft offered accommodation at Spitzingsee near Munich which was gratefully accepted. A provisional organizing group was set up consisting of Pinkau and de Jager from the SSC to be joined by two representatives of the SSB. In fact most of the organization was planned by this group somewhat augmented at the COSPAR meeting in Innsbruck in June 1977.

The meeting took place over five days beginning 9 October 1978. Attendance was restricted to 15 members from each community including L. Culhane, Massey, K. Pounds, A.P. Willmore and A. Wolfendale from the UK. It proved to be very successful and achieved the aim of establishing complete agreement as to the way the high energy astronomy should develop and bringing the USA and European scientists into an even more intimate collaboration than hitherto.

A 50-page document reporting the proceedings was published jointly by the SSB and the SSC as an 'advocacy input' document the preamble stating that 'It is a document signed by its authors as individuals but it is being made public so that those who wish to obtain guidance from their deliberations may do so'. One extract from the recommendations which was most relevant for the further activities of the SSC ran as follows.

'We have noted that the funding for space science in Europe is about 1/10 of that in the USA and not at all in proportion to either resources or population. There are a host of essential and exciting programmes waiting to be carried out but limited by funding. In order to realize an optimum scientific programme in X-ray and γ-ray astronomy in which European scientists can remain competitive, can continue and improve the development of scientific and technological talent and are able to co-operate as equals with Americans in joint missions which may lead to 'world class' projects funding must be increased substantially in Europe. Ideally, funding for NASA must also grow at a rate slightly more than the inflation rate in order to realize the great potential in X-ray and γ-ray astronomy!'

We shall return to the question of financial resources for European space science on p. 258 below. Meanwhile the success of the Spitzingsee meeting ensured that regular joint advisory meetings of a similar kind would be organized covering other scientific disciplines.

6 The future of the SSC – establishment as a Standing Committee of ESF

A document presenting the case for establishment of the SSC as a Standing Committee of the ESF was agreed and submitted to the ESF in May 1978. It was considered by the Executive Council of that body at its meeting on 16–17 October 1978. Although there was some suggestion that the SSC might become a subcommittee of the Astronomy Committee[3] based on the fact that much of the SSC's previous concerns had been with astronomy it was made clear that the SSC covered also a wide range of subjects in geophysics. The proposal was submitted to the General Assembly of the ESF on 7–8 November 1978 and accepted.

The new standing committee took over on 1 January 1979 with J. Geiss of Switzerland who was very experienced in European space science both as a practitioner and on policy issues, as Chairman, and a provisional membership is listed in Annex 16. The terms of reference, shown in Annex 17 differed little from those of the provisional committee and continuity was maintained in all arrangements with the SSB.

One of the earliest major activities of the new Committee was to carry forward the consideration of the long term future of European space science. Intensive study of the situation was initiated and a draft document prepared under the title 'Space Science in Europe' which was considered at two meetings of the SSC from which a special report emerged which was submitted to the Executive Council of the ESF on 18 September 1979. It discussed in particular the inadequacy of funding, a specific instance of which was referred to in the report of the Spitzingsee Conference (see p. 255). A very relevant paragraph reads as follows:

'The weakest link in the development of European space research is the lack of adequate flight opportunities. We have to achieve a launch frequency of about one satellite per year, fully instrumented by Europe, to develop our areas of excellence. The expectation that Spacelab would help in this context has not been fulfilled because of its high cost. Therefore a significant increase in ESA's fund for scientific projects is requested. At the same time ESA should strive to achieve a more economic use of its funds. We argue that there is now in fact an opportunity to recover the setback in the scientific programme incurred in 1972 and to improve the situation beyond this, because the large expenses required for the development of Spacelab and Ariane are coming to an end.'

The document went on to recommend that the mandatory scientific budget be increased. Although the document has been widely circulated the progress made up to the time of writing in achieving its aims has been small. In this respect the UK is in a specially difficult position because, as explained in Chapter 10, the cost of British space science must be borne from the overall scientific budget and is in direct competition with other scientific programmes and projects. On the other hand, in most continental European countries, expenditure on space activities including both the pure science and the applications and the launching systems comes from a single source so there is no direct competition between space science and other science. The advantage of the UK system in making it possible to assign priorities for pure science on scientific grounds alone has already been stressed in Chapter 10, p. 216.

6.1 Planetary Science in Europe

In 1978, Runcorn drew to the attention of the President of the ESF, Sir Brian Flowers, the desirability of a European meeting to discuss lunar and planetary science in Europe. This matter was referred to the SSC who sponsored a Workshop on the subject held in Strasbourg on 15–19 September 1980, Fechtig and Runcorn being the co-Chairmen. The terms of reference were to take stock of ongoing activities in Europe in the field of planetary research, to make a number of practical recommendations which should seek to improve planetary research in Europe and to submit a report in which the conclusions from these studies are spelt out.

The report, a local European advocacy report as it were, duly appeared in 1982 under the title 'Planetary Science in Europe – the Present State and Outlook for the Future'. After consideration of the report the suggestion was made at an SSC meeting that discussions should be initiated with the USA about possibilities for collaborative projects in planetary studies using space vehicles. Following correspondence initiated by J. Geiss between F. Press, President of the US National Academy of Sciences, H. Curien, President of the ESF, H. Levy, Chairman of the SSB sub-committee on lunar and planetary sciences a joint US–European lunar and planetary workshop was set up and, at the time of writing (April 1983) is still deliberating.

It would seem appropriate to end this chapter at this stage with the SSC actively involved in many important matters and with well-established relations with ESA, the SSB and the Space Telescope Institute.

13

Scientific studies by British space scientists – figure of the earth and the neutral atmosphere

In this chapter and the following two chapters we attempt to convey an impression of the scale and effectiveness of research in the different branches of space science carried out by British scientists from the beginning. The sheer volume and variety of this work is immense and it would be both impracticable and pointless to attempt anything like a comprehensive account. Instead we shall concentrate attention on a number of areas in which the UK contribution has been especially impressive while at the same time saying something in much less detail about the remaining work. It is in any case somewhat invidious in describing scientific research to select work done by scientists from one country as all the work is truly international and we shall try to avoid any trace of chauvinism in our account.

In this chapter we concentrate on research work directed towards the study of the earth, including the solid earth and the lower, middle and outer neutral atmosphere. Some reference will also be made to lunar and planetary studies. Although much of the earth's environment is under solar control, it is convenient to deal with the work done in solar physics, including X-ray and ultra-violet solar astronomy, in Chapter 15, which is devoted to astronomy.

Of the subjects covered in this chapter, major UK contributions have been made in four areas – the analysis of satellite orbits to obtain information about the figure of the earth and atmospheric density and winds, the use of grenade and glow release tracking and of Skua rockets to determine the temperature and wind profiles in the lower and middle atmosphere, the development and highly successful operation of infra-red devices for determining temperature, pressure and composition height profiles in the lower and middle atmosphere from satellites and the exploration of the properties of the ionosphere and magnetosphere. In addition, a number of observations have been made of the intensities of spectral lines in day and night airglow emissions. We now discuss the contributions under these heads.

1 The analysis of satellite orbits

1.1 Orbit determination

We have already described in Chapter 3 how UK observers were quickly at work tracking the first satellites, Sputniks 1, 2 and 3 by radio, radar and optical methods. They enjoyed an initial advantage in that they were well placed geographically to observe these satellites and so were able quickly to lay a foundation on which to build effectively for the future. Already by 1958 their results established the great value of accurate satellite orbit data for obtaining new information about atmospheric density and temperature, atmospheric winds and the figure of the earth. There was clearly seen to be a completely open-ended programme ahead and plans for effective long-term involvement were soon being implemented.

Thus the Radio and Space Research Station, later to become the Appleton Laboratory, assumed responsibility for supplying information for orbit predictions which made it possible for observers using their different techniques to make accurate observations of satellite transits from which accurate orbits could be calculated. Computer programmes developed at the Royal Aircraft Establishment carried out these accurate orbit determinations. Indeed, on the initiative of D.G. King-Hele, the RAE became the UK centre for orbit analysis and has maintained a leading position in this subject from the outset.

The most accurate observations used in orbit determination are those made with the Hewitt Camera which is in fact the most accurate device available for this purpose. It was designed in the late 1950s by J. Hewitt working at the Royal Radar Establishment (RRE) Malvern, to record the track of the Blue Streak ballistic missile. When the Blue Streak defence programme was cancelled in 1962 the cameras became available for satellite observation for which it was well suited. Now under the ownership of the Ordnance Survey, one was installed at Malvern and the other at the Earlyburn station of the Royal Observatory Edinburgh where they were used for satellite observations.

The internal arrangement of a Hewitt camera is shown in Figure 13.1. It is a Super-Schmidt f1 of 61 cm (24 inch) aperture with a field view 10° in diameter. Light enters through the Schmidt corrector plate and is focused by the 86 cm diameter mirror on to the plate. The accuracy of measurement is markedly increased by ensuring, through introduction of the lens in front of the photographic plate that the rays converge on to a plane rather than a curved surface. For satellite tracking the main shutters are opened for 10-second periods during which a rotating chopper produces breaks in the trail at intervals of 1 s which act as time markers.

The directional accuracy of the camera is 1″ of arc. A satellite of magnitude 10 can be recorded even at 4000 km distance if it is moving at 0.1° per second but if it is moving ten times faster the limiting magnitude is 7½.

For some years the Hewitt Cameras at Malvern and Edinburgh were owned and operated by the Ordnance Survey but in 1978 this responsibility was taken over by the University of Aston. In 1979 the second camera was transferred from Edinburgh to Siding Spring in New South Wales at the site of the UK Schmidt telescope. It is now operating successfully and is of particular value because of its location in the Southern Hemisphere. Previously the only accurate optical observations in that hemisphere were made by the kinetheodolite transferred from the Royal Greenwich Observatory to the Royal Observatory, Cape of Good Hope, in 1966 which became operational two years later. Because of its location observations made with this instrument were of especial value – until it closed down in 1981 it made about 40,000 observations on about 15,000 satellite transits. Previous to 1968 the most southerly kinetheodolite was that operated by the Meteorological Office in Malta (see Chapter 4, p. 64) which recorded 950 transits up to its closure in May 1968. The first Hewitt Camera, located at Malvern since

Fig. 13.1. The layout of the Hewitt camera.

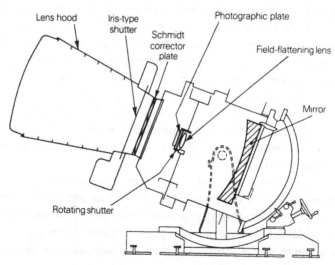

1962, was moved in 1982 to an excellent new site at the Royal Greenwich Observatory Herstmonceux where it is operated by a team from the University of Aston.

Radar tracking is available under all weather conditions, both day and night, but is very expensive. However, regular radar surveillance of satellites is carried out by military organizations and these data are made available to civilian scientists through the US Navy. They make an important contribution to the orbit determinations made at RAE.

In addition, a valuable input comes from a group of about 50 volunteer observers, mainly in Britain, who literally make observations from their back gardens, as many as 20,000 in any one year. Their devotion to this task is remarkable as many of them have been at work for more than 10 years. The world's leading visual observer is R. Eberst of the Royal Observatory, Edinburgh, who has in 25 years made 90,000 observations to an accuracy of about 100" of arc, as high as that attained by radar.[1] Although these visual observations[2] are much less accurate than those made with the Hewitt Camera, they provide the essential 'weight of numbers' needed to eliminate any bias due to the small arc of the orbit covered by the accurate observations.

The activities of the volunteer observers are co-ordinated by the Royal Society through the Sub-Committee on Optical Tracking of the National Committee on Space Research, D.G. King-Hele having been the Chairman since 1972. The Sub-Committee issues regularly a priority list of satellites to be observed, selected primarily for their interest in connection with geophysical research. It also arranges for selected observers to be equipped with binoculars, star atlases and stop watches. Until recently the British Astronomical Association has operated a simplified prediction service to train amateur observers and has prepared an instruction manual for their use. When judged proficient an observer's name is added to the address list of the Satellite Prediction Service, now operated by the University of Aston.

1.2 Air density

We have already explained in Chapter 3, p. 51 how the air density close to the perigee of a satellite can be determined from observation of the rate of decrease of orbital period with time. The group at RAE under D.G. King-Hele which made the first determinations of air density from observation of the first Sputniks continued very actively to utilize data from other satellites to study air density. Before long many other groups[3] world-wide joined in this work but the RAE group continued to play an important role.

It was found that the air density exhibits a strong degree of solar control manifest in day-to-night variations, a marked dependence on epoch in the solar cycles and on solar activity. Despite the complexity of these variations Jacchia succeeded in incorporating them in a semi-empirical model which proved very satisfactory except that it did not properly allow for a semi-annual variation which is observed and proves to be quite complicated, both to predict and to understand. As an example of the many contributions made by the RAE we describe briefly the results obtained about air density, and particularly the semi-annual variation by D.M.C. Walker[4] in a detailed study based on observations of the orbit of a particular object over the period December 1971 to April 1975.

The object, known in standard satellite classification as 1971-106A, was one of a pair of satellites launched by the Soviet Union to test high-speed interception. It was the hunter satellite which exploded after passing the target satellite. The largest piece which remained in orbit after the explosion was 1971-106A, pursuing an orbit with a period of about 105 min, perigee height 230 km, apogee height 1800 km and inclination 65.7° It remained in orbit for 40 months to decay on 4 April 1975. It was selected for high priority observing by the Appleton Laboratory–RAE system and, from 5400 observations, accurate orbits were computed at RAE at 85 epochs. The calculated perigee heights were correct to about 200 m throughout the object's lifetime.

Figure 13.2 shows the air density at a height of 245 km throughout 1972 which Walker derived. Comparison is made with the variation of certain solar indicators over the same period, the intensity of the 10.7 cm radiation from the sun, the local solar time, and the geomagnetic index A_p which is a measure of geomagnetic activity. Using the Jacchia model the solar control effects can be eliminated giving rise to the curve labelled Density Index D which shows clearly a semi-annual variation. The difficulty in incorporating this variation to any atmospheric model arises because it varies in phase and amplitude from year to year as shown in Figure 13.3 for 1972–5 according to Walker.

The atmospheric density scale height H at any altitude h is defined by

$$\frac{1}{H} = -\frac{1}{\rho}\frac{d\rho}{dh} \tag{1}$$

where ρ is the atmospheric density. It is an important quantity as it is directly proportional to the air temperature divided by the mean molecular weight and it is very desirable to check atmospheric models by determining it from orbit analysis. In fact it may be obtained to quite good accuracy from the change of perigee height Δh_p while the eccentricity decreases from e_0 to

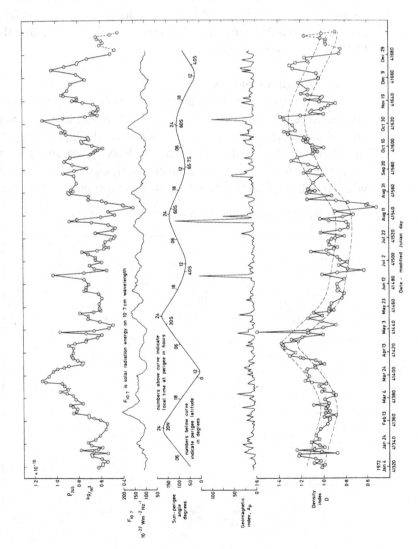

Fig. 13.2. Values of the atmospheric density of 245 km altitude in 1972, and density index D with solar radiation index $F_{10.7}$, Sun–perigee angle and geomagnetic index Ap.

e_1 through the relation $\Delta h_p = H \ln (e_0/e_1)$. However Δh_p is likely to be small unless the satellite is near the end of its life, in which case special care is necessary to obtain orbits of sufficiently high accuracy. As a further example of work at RAE we refer to the analysis carried out by H. Hiller[5] on an object in the last sixteen days of its life.

The object was the second-stage rocket (1972-05B) used in the launching of the ESRO satellite HEOS 2 (see Chapter 6, p. 146). This entered a polar orbit of eccentricity 0.06 with a lifetime of about $6\frac{1}{2}$ years. Using 1360 observations orbits were determined daily with an accuracy of 30–70 m both radially and transversely. Eleven values of the scale height were obtained, with a 2% error. The RMS value was about 4% higher than the CIRA reference atmosphere.

1.3   Atmospheric rotation

In Chapter 3, p. 52 reference was made to the interpretation by Merson, King-Hele and Plimmer[6] of the small change in inclination of the orbit of Sputnik II with time as due to rotation of the high atmosphere at a considerably faster rate than that of the earth on its axis, giving rise to W–E zonal winds. Since that time many further analyses have been carried out in which the RAE group has played the predominant part and there is now a

Fig. 13.3. Variation of $\bar{D}$ and the standard $\bar{D}$ during the year.

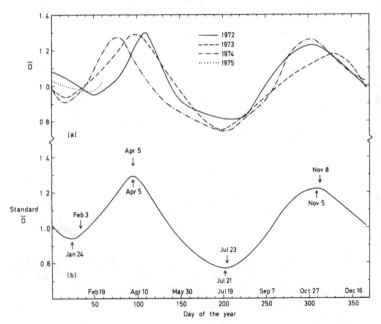

considerable body of information available about high altitude zonal winds and their variation with time of day, season etc.

Figure 13.4 shows the results obtained by King-Hele and Walker[7] from analysis of 85 orbits up to 1982. The diagram gives the atmospheric rotation rate and the zonal wind near latitude 30°. The diagram shows that the average rotation rate of the atmosphere at heights of 250–400 km is considerably greater than the earth's rotation rate. But there are wide variations with local time, the maximum west-to-east winds being in the evening and of order 100 m/s, while in the morning the winds are mostly from east to west but not so strong. The diagram is for average season, but there are also substantial variations with season, the rotation rate being about 0.1 higher than average in winter and 0.1 lower in summer.[8] It is of interest to know that Hiller[9] from analysis of the orbit of 1972-05B referred to previously, obtained a rotation rate as high as 1.40 ± 0.05 rev day^{-1} in conditions of high solar activity.

1.4 Figure of the earth

In Chapter 3, p. 51 we described how King-Hele and his colleagues obtained a corrected value for the polar flattening of the earth from the rate of precession of the orbital plane of Sputnik 2 about the earth's axis. The value they obtained, $1/298.21 \pm 0.03$, is quite close to the most accurate results available at the time of writing ($1/298.257$).[10]

However, since that time the determination of the earth's gravitational

Fig. 13.4. Variation of atmospheric rate Λ with height and local time for an average season.

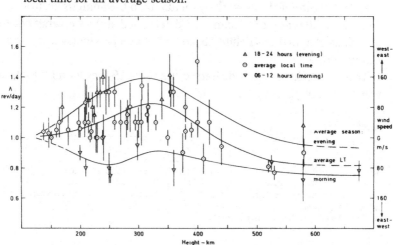

potential from analysis of satellite orbit data has been carried out to a remarkable degree of accuracy.

If the potential is averaged over longitude it can be expanded in the form

$$V_z = \frac{GM}{R} \left\{ \frac{R}{r} - \sum_{n=2}^{\infty} J_n \left(\frac{R}{r}\right)^{n+1} P_n (\cos \psi) \right\}, \quad r > R \qquad (2)$$

where ψ is the co-latitude angle and r the distance from the centre of the earth, G is the constant of gravitation, M the mass of the earth, R its radius and the J_n are numerical coefficients, $P_n (\cos \psi)$ is a zonal harmonic (Legendre polynomial) of degree n. The polar flattening is determined by J_2. The even harmonics determine the rate of orbital precession about the earth's axis and the rotation of the orbit within its plane. This causes the perigee to move from the northern to the southern hemisphere and back. Combined with the effect of the odd harmonics, an oscillation is produced in the perigee distance from the earth's centre of the order of 10 km, depending on the orbital inclination i. If accurate orbits could be determined for a wide range of values of i, accurate values of the coefficients J_3, J_5, J_7 up to high degrees could be obtained. In practice the accuracy is limited because there are still wide gaps in the inclinations covered.

The British programme has aimed particularly at using optical tracking with Hewitt cameras to obtain accurate orbits useful for determining the odd coefficients J_3, J_5 etc. The first results, in 1965[11] were improved in 1974[12] by the addition of eight new orbits, including two at inclinations 65° and 62° close to the critical inclination of 63.4° at which the oscillation in perigee becomes especially large. In a later paper, D.G. King-Hele, C.J. Brookes and G.E. Cook[13] extended this work by including in the analysis accurate orbits for two satellites at 62.9° inclination determined at the University of Aston as well as of a further satellite, Explorer 46, at a previously unrepresented inclination of 38°.

Truncating the series of odd harmonics at $n = 19$ they found the following values for $10^9 J_n$, $n = 3.5 \ldots 19$.

Table 13.1.

n	3	5	7	9	11	13	15	17	19
$10^9 J_n$ (a)	-2530 ± 4	-245 ± 5	-336 ± 6	-90 ± 7	159 ± 9	-158 ± 15	-20 ± 15	-236 ± 14	-27 ± 19
(b)	-2536	-226	-363	-117	229	-223	-14	-93	-11

(a) Derived as above.
(b) From the GEM 10 B model (see p. 271).

We shall discuss the comparison of these values with those obtained by a very comprehensive analysis of the full gravitational potential in 1.5 which are also given in Table 13.1 under (b)

1.5 Variation of the gravitational potential with longitude

The full gravitational potential including terms which depend on the longitude can be written as

$$V = V_z + \frac{GM}{R} \sum_{n=2}^{\infty} \sum_{m=1}^{n} \left(\frac{R}{r}\right)^{n+1} P_n^m (\cos \psi) \left\{ C_n^m \cos m \phi + S_n^m \sin m \phi \right\} \quad (3)$$

where $P_n^m (\cos \psi)$ is a tesseral harmonic. The longitude-dependent terms are all oscillatory so that the perturbations they produce are relatively small under most conditions. Under certain circumstances, however, they may be enhanced by resonance effects which make possible the determination of sets of the coefficients C_n^m, S_n^m for fixed m with reasonable accuracy.

Thus a primary orbital resonance occurs when a satellite repeats its track over the earth after one day. The most frequent example which occurs in practice is that of 15th order ($m = 15$) in which the satellite repeats its track after 15 revolutions while the earth makes one revolution. In general, in a $\beta : \alpha$ resonance the track repeats after β revolutions of the satellite and α of the earth so that the effect of the longitude terms of order β build up day after day until the resulting perturbations may be measured with sufficient accuracy to enable a linear combination of terms involving C_n^β and S_n^β to be determined. The coefficients in this linear combination, which is usually referred to as a lumped coefficient of order β, depend on the satellite inclination so that, with results for a sufficient number of inclinations, individual coefficients C_n^β, S_n^β with different n may be obtained.

The RAE group has devoted special attention to the determination of these coefficients. The first analysis of a satellite orbit at the most common resonance, that of 15th order, was carried out in 1971 by Gooding[14] for the orbit of Ariel 3. It was based on analysis of 281 orbits which he had determined from Minitrack observations. Somewhat later (1975) King-Hele, Walker and Gooding[15] were able to analyse data from satellites at different inclinations to obtain values for individual coefficients of order 15. These have been improved by subsequent calculations. At the time of writing the most detailed and accurate analysis is that carried out by King-Hele and Walker[16] who used the observed perturbations in 23 resonant orbits at various inclinations to determine C_n^β, S_n^β with $\beta = 15$ and $n = 15, 16, 17 \ldots 33$. The coefficients for odd and even values of n were obtained from analysis of the perturbations in the inclination and eccentricity respectively. The accuracy of the analysis is greater the longer the time the satellite is moving under resonance conditions. This depends largely on the magnitude of the air drag. Among the cases considered, the longest resonance lasted for about five years but for some inclinations it was necessary to use resonances which were effective for only a few months.

Table 13.2 gives the values of the coefficients obtained by King-Hele and Walker.

Resonances of lower order can be analysed in principle at least as accurately as for 15th order but in such cases if the orbit is nearly circular as is desirable for accurate analysis, resonance is only likely to be reached after periods of 20 years or more. However, more results have been obtained from 14th order resonances.[17]

A further possibility is to analyse the 29:2 and 31:2 resonances but the effects are much smaller as the results are less accurate. Nevertheless, some good values have been obtained[18] from such analysis.

In recent years comprehensive models of the earth's gravitational field have been derived in which all terms in the expansion (3) have been included up to some chosen limit and equations obtained for the various coefficients from accurate optical tracking data, laser observations of high accuracy, surface gravimetry and radar altimetry. These equations are then solved by computer to obtain the coefficients and hence the field.

The GEM 10 B model[19] uses 840,000 observations including 21,300 laser ranges 150,000 optical observations and 270,000 US Navy Doppler measurements. 1654 surface gravity measurements were also included. The geopotential field was represented by the series (3) up to terms of degree and order 36. Even more elaborate models are in preparation.

Table 13.2. *Values of the coefficients C_n^{15}, S_n^{15} derived by King-Hele and Walker.*

n	15	16	17	18	19	20	21
$10^9 C_n^{15}$	-22.7 ± 0.6	-11.0 ± 2.7	11.3 ± 1.0	-43.0 ± 1.8	-13.3 ± 0.8	-24.3 ± 2.3	15.9 ± 0.7
$10^9 S_n^{15}$	-7.4 ± 0.6	-21.5 ± 1.7	6.7 ± 1.2	-22.5 ± 1.2	-11.8 ± 0.9	-6.2 ± 1.6	8.7 ± 0.8

n	22	23	24	25	26	27	28
$10^9 C_n^{15}$	24.1 ± 2.0	14.3 ± 1.6	1.4 ± 3.8	-12.7 ± 2.0	-13.3 ± 5.8	-6.8 ± 1.4	-15.4 ± 6.4
$10^9 S_n^{15}$	10.2 ± 1.6	-1.3 ± 1.9	-21.8 ± 3.3	0.6 ± 2.4	14.4 ± 5.5	12.7 ± 2.0	-8.4 ± 6.3

n	29	30	31	32	33	34	35
$10^9 C_n^{15}$	-2.2 ± 1.8	-4.0 ± 6.8	27.9 ± 2.9	7.8 ± 6.2	6.5 ± 2.9	9.6 ± 6.3	-6.8 ± 4.1
$10^9 S_n^{15}$	0.3 ± 1.9	-16.0 ± 6.3	-2.0 ± 3.9	2.5 ± 5.1	-12.0 ± 3.8	5.6 ± 5.2	3.3 ± 4.6

While the GEM 10 B model probably gives the geopotential field accurate to 1 or 2 m it does not follow that the higher order coefficients in the series representation are accurate. As the values of the coefficients are obtained by solution of more than a million equations for more than 1200 unknowns, very different sets of coefficients might lead to nearly the same field. Accurate values of the coefficients are of importance in supplying an indication of the mass distribution in the interior of the earth. It is therefore desirable to check as many coefficients as possible against those derived by the orbit analysis methods described previously.

In Table 13.1 such a comparison is carried out for odd zonal harmonic coefficients and it is seen how the agreement is quite good for the lower harmonics and it becomes progressively less satisfactory for the higher ones.

Table 13.3 compares values of the coefficients C_n^{15}, S_n^{15} derived from 15th order resonance (Table 13.2) with those given by GEM 10 B up to $n = 24$. On the whole, in view of the immense complexity of the calculations, the comparison is quite good and shows what enormous progress has been made in which UK scientists especially those under D.G. King-Hele at RAE, have contributed in a major way to a fully international co-operative activity.

2 The neutral atmosphere studied by rocket techniques

2.1 The atmosphere below 100 km

In Chapter 3, p. 28 we described how the grenade experiment was one of the first to be carried out in the British Skylark rocket research programmes. This experiment was designed to measure the altitude profile of atmospheric temperature and horizontal wind components up to an

Table 13.3. *Comparison of coefficients C_n^{15}, S_n^{15} derived from analysis of 15th order resonance with those of GEM 10 B*

n	$10^9\ C_n^{15}$		$10^9\ S_n^{15}$	
	15th order res.	GEM 10 B	15th order res.	GEM 10 B
15	-22.7 ± 0.6	-19.7	-7.4 ± 0.6	-6.4
16	-11.0 ± 2.7	-14.4	-21.5 ± 1.7	-27.8
17	11.3 ± 1.0	2.5	6.7 ± 1.2	4.8
18	-43.0 ± 1.8	-48.3	-22.5 ± 1.2	-18.7
19	-13.3 ± 0.8	-20.6	-11.8 ± 0.9	-15.3
20	-24.3 ± 2.3	-23.9	-6.2 ± 1.6	4.8
21	15.9 ± 0.7	15.2	8.7 ± 0.9	9.5
22	24.1 ± 1.6	24.1	10.2 ± 1.6	-1.3
23	14.6 ± 1.6	15.4	-1.3 ± 1.9	4.1
24	1.4 ± 3.8	3.1	-21.8 ± 3.3	-5.1

altitude of about 85 km. The principle of the experiment has been described in Chapter 3 – it is basically a sound ranging method based on the observation of the series of explosions ejected at regular intervals from a rocket launched into a vertical trajectory. Ground-based equipment which must be employed consists of ballistic cameras to photograph the flashes from the explosion against the star background, an array of flash detectors to read the time of arrival of the light from each flash and an array of microphones to record that of the sound pulse. With suitable analysis it is possible to determine the mean atmospheric temperature and the mean components of wind speed between each explosion. The upper limit of 80–85 km to the altitude for which data could be obtained was determined by the attenuation of the sound in propagating in an increasingly rare atmosphere. It turned out in practice that data could be obtained to higher altitudes because above about 100 km a grenade explosion produced a bright glow which persisted for some time. This arose through chemical reactions with atomic oxygen which is the most important oxygen allotrope above 110–120 km. From the rate of expansion of the glow cloud the diffusion coefficient of the constituent gas in the ambient atmosphere could be determined and hence information obtained about the atmospheric pressure. Observation of the drift of the cloud as a whole gave information on the wind speeds. At a later stage spectra of the glow cloud were taken. Prominent features are bands of AlO from the intensity distribution in which the temperature can be derived (see p. 278).

We have already referred in Chapter 3 to experiments in which sodium is vaporized from a rocket in local twilight to glow brightly by fluorescence in the sunlight. Once again atmospheric density and winds may be determined as with the grenade glows. This was originally suggested by D.R. Bates and was the first active experiment carried out in the high atmosphere. Bates and his group at Belfast planned and carried out experiments very early in the Skylark programme but these are subject to the limitation of only being operative at local twilight in the upper air.

Other techniques were developed for study of parameters of the neutral atmosphere at altitudes above 100 km but in this section we will confine ourselves to describing results obtained below this altitude especially because atmospheric tidal effects are most prominent towards the upper end of this region. Also, only a little higher, atmospheric turbulence becomes unimportant presenting a new aspect to study.

In addition to the grenade experiment, winds may also be measured by following, through radar reflection, the tracks of ribbons of metallic strip ('window' or 'chaff') ejected from the rocket. Again, many observations of this kind were made in the British Skylark programme.

Turning back now to the grenade experiments they were pursued with great determination and energy from the outset. While the first developments involved co-operation between almost the entire staff of the Department of Physics concerned with space research at UCL[20] in 1962, Groves took over responsibility for the whole programme which benefited very much from his skill both in theory and in the analysis of complicated and extensive data, as well as his ability to plan and carry through the technological side of the work. 28 grenade experiments using Skylark rockets were carried out at Woomera where the generous and skilful co-operation of many members of the Australian Weapons Research Establishment, including among others A. Blight, R. Cartwright and P. Malcolm, was a vital factor in their success.

By 1962 sufficient successful firings had taken place to determine the seasonal variations in the temperature, pressure, density and wind speeds at heights up to 85 km above Woomera. Figure 13.5 shows the seasonal variation in density derived in this way.[21] In Figure 13.6 temperature–height profiles obtained in seven grenade flights in 1962 are shown.[22] A striking feature of these results is the presence of oscillations which are particularly marked at the higher altitudes and which were also present in the wind speed profiles. These have been averaged out in the studies of seasonal variations but are of great interest in connection with diurnal and semi-diurnal variation associated with atmospheric tidal phenomena.

Fig. 13.5. Seasonal variations of atmospheric density ρ with altitude over Woomera derived from grenade experiments.

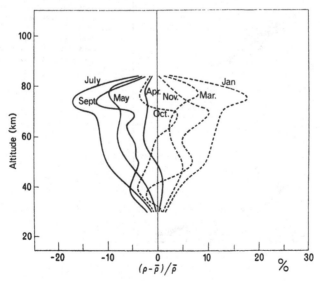

To investigate this aspect, grenade experiments must be undertaken at much closer intervals in time and with the spacing between successive explosions as small as possible. The first step in this direction was taken in 1962 when, of the six successful grenade firings, two were in daylight but it was necessary to face the considerable problems involved in launching a number of Skylark rockets with grenade payloads during one night. Those of us who had witnessed rocket firings on occasional visits were astonished at the courage of those who would attempt such a task but, during the night of 15–16 October 1963, four rockets were successfully launched at local times of 19.21, 00.39, 21.16 and 04.52 hr.[23] These rockets carried more grenades than in the past so the average spacing between explosions was reduced to about 4 km. Figure 13.7 shows the height profile of the components of wind speed derived from these four flights. The oscillatory character of the results above 40 km is clear and exhibits a regular change with local time. Finally, on the night of the 29–30 April 1965 an attempt was made to launch no less than seven Skylark grenade rounds,[24] While the first and last of these, launched at local twilight and dawn were not successful, good results were obtained from the five remaining. Figure 13.8 shows the temperature profiles obtained which, apart from the second flight,

Fig. 13.6. Temperature–height profiles derived from grenade experiments in 1962 as indicated.

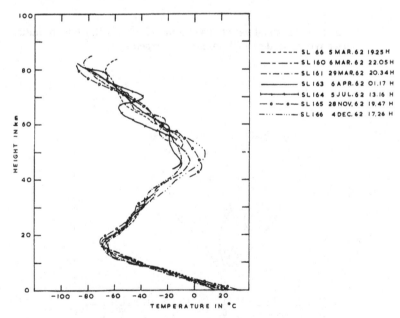

Fig. 13.7. Height profiles of horizontal components of wind speed obtained from four grenade flights during the night of 15–16 October 1963 at Woomera.

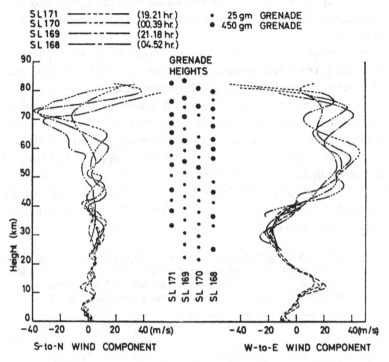

Fig. 13.8. Temperature profiles obtained from five grenade flights during the night of 29–30 April 1965 at Woomera.

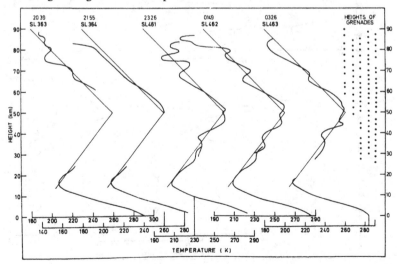

are all of the characteristic form, showing oscillations which become more marked at the higher altitudes and which vary with time in at least a semi-regular way. Perhaps even more striking are the wind data as may be seen from Figure 13.9 which shows the height profiles of the S–N component of wind speed.

These results pointed clearly to the influence of atmospheric tidal oscillations. To study this aspect many more data at different geographical locations were required. Groves and his group made a direct contribution by carrying out grenade experiments with Petrel rockets at Sonmiani in Pakistan in a collaborative programme between NASA, the Royal Society through the British National Committee for Space Research (BNCSR) and the SUPARCO space agency in Pakistan (see Chapter 7, p. 170) and at Thumba in India as part of the Commonwealth Collaboration (see Chapter 7, p. 175) as well as with Skylark rockets at the ESRO ranges in Kiruna (see Chapter 6, p. 142) and Sardinia (see Chapter 6, p. 142). However, important as this work was, it could only supply a small fraction of the data required.

Fig. 13.9. Height profiles of the S–N component of wind speed obtained from five grenade flights during the night of 29–30 April 1965 at Woomera. The lower five profiles were derived from grenade ranging observations, the upper five from observation of grenade glows.

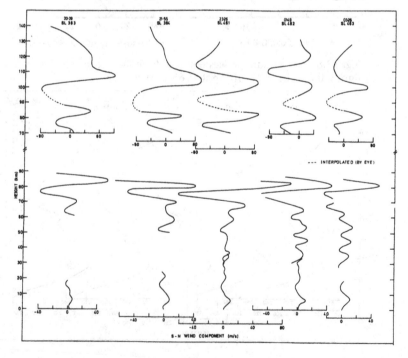

Groves[24] then undertook the task of analysing data from co-ordinated series of grenade flights carried out by NASA at Wallops Island (37.8°N 75.5°W) Natal, Brazil (5.9°S 35.1°W) and Kourou, French Guiana (5.1°N 52.6°W) with particular reference to tidal theory. He also included in his analysis the much greater volume of data available from observations of wind speeds made up to altitudes of 60 km above North America and adjacent ocean areas through the Meteorological Rocket Network. At the same time he has worked in detail with the so called classical tidal theory,[25] including the effect of heat sources as well as gravity. As a result, a much more detailed and reliable understanding of the phenomena which, though studied in some detail as early as 1898, presented difficult problems of interpretation which still remained until information from space techniques has been forthcoming.

2.2   Structure of the neutral atmosphere above 100 km

The most effective means of observing the temperature density and wind structure of the atmosphere above 100 km has been the use of various means to produce glow clouds at chosen altitudes whose motions and optical properties can be studied. Winds near 100 km can also be determined from ground-based observation of meteor trails.

The first artificially produced glows were those of sodium vapour clouds fluorescing in sunlight. These suffered from the disadvantage of only being effective at local twilight when the glow could be observed from the ground. Also, while it was in principle possible to determine the temperature at the location of the glow from observation of the width of the sodium D lines, in practice this was difficult because of strong self absorption.

To overcome the latter problem, the group at Queen's University Belfast used a lithium vapour cloud – the infra-red resonance line of Li at 8126 Å does not show self-absorption.

It was shown in 1972 that a lithium glow cloud could also be used to measure the winds distribution in sunlight. These experiments were carried out at Woomera[26] in a collaboration between UCL (D. Rees and M.P. Neal), RSRS (K. Burrows), the Weapons Research Establishment, Salisbury, Australia (C.H. Law and A.D. Hind), and the University of Brisbane (R.S. Fitten). The lithium was released from a Skylark rocket at an altitude near 200 km and daylight observations of the glow were made by two techniques, one using a differential photometer and the other a Faby-Perot plate with a resolution of 1.0 Å together with an interference filter with a 3.0 Å band pass around the lithium resonance line. The glow cloud was observed for 20 minutes and from its motion the wind speed at 202 km was measured as

$32 \pm 2 \, \text{km s}^{-1}$ in a direction specified by an azimuthal angle of $221 \pm 2°$. After this successful flight the lithium glow technique was frequently used by the UCL and other groups.

We have already described how the glow from a grenade explosion above 120 km may be used not only at twilight but also throughout the night.

Finally trimethyl aluminium (TMA) was introduced as a substance which produced a trail at night through chemical reaction in the atmosphere. For many purposes it superseded the earlier methods of glow production.

The first region of special interest occurring near 100 km in the atmosphere is the turbopause which marks the transition between turbulent mixing and laminar flow in which diffusive separation of atmospheric constituents occurs. The earliest glow experiments carried out by the Belfast group with sodium glows showed that the transition is a very sharp one and occurs quite close to 100 km altitude – below the turbopause the trail is obviously turbulent.

Figure 13.10 shows the onset of turbulence in a TMA trail at 102.7 km altitude, released at Woomera from a Skylark in the morning of 31 May 1963. The time sequence photographs were taken in a joint experiment on atmospheric structure carried out between D. Rees of the UCL group and R.G. Roper, K.H. Lloyd and C.H. Low[27] of the Weapons Research Establishment at Salisbury, Australia. Use was made of the Baker–Nunn satellite tracking camera located 160 km from the Skylark launcher. Isodensity scans of part of the trail are shown. It will be seen that laminar flow in the trail persists at least until 30 s after release but breaks down into turbulence by 50 s. Similar observations in the evening of the same day obtained from a second rocket firing showed that, at 106 km, the transition from laminar to turbulent structure takes place between 23 and 61 s after release while at 108 km the structure remains laminar throughout the time of observation. Reasons for this behaviour were discussed.

Turning now to observations at much greater altitudes we show in Figure 13.9 the S–N components of wind speed from 94 to 140 km obtained by Groves from tracking grenade glows released at 3 km intervals during the course of the campaign of Skylark launchings carried out at Woomera on the night of the 29/30 April 1965 described on p. 274. The oscillatory character of the results was also a feature of the observations of the E–W component. In both cases the amplitude of the oscillations is much greater than at altitudes below 90 km.

Rees, Roper, Floyd and Low, in the course of the experiments discussed above on 31 May 1968, extended wind observations at Woomera to an altitude of 240 km using TMA release between 80 and 140 km and grenade glows between 150 and 240 km. Figure 13.11 shows results they obtained during the morning flight, again exhibiting large amplitude oscillations which are probably mainly due to tidal effects. The results at the highest altitudes are not inconsistent with those obtained from satellite tracking (see p. 267).

Fig. 13.10. Photomontage of position of the morning trail produced by TMA release from a Skylark rocket at Woomera. The kink in the trail occurs at an altitude of 102.7 km. The trail is laminar at + 30 s.

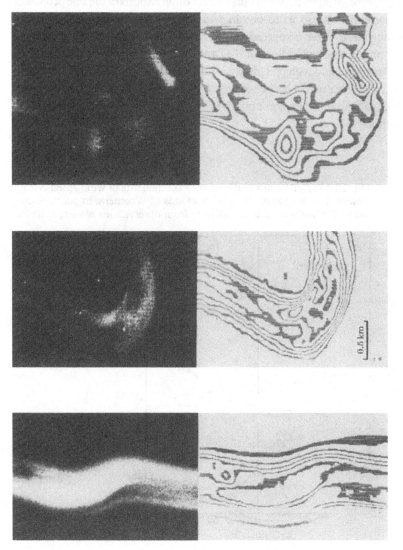

To extend wind observations to altitudes of 300 km the UCL group in collaboration with the Appleton Laboratory developed a tuned dye laser radar system for tracking sodium clouds which was operated successfully on several occasions, particularly in rocket flights at high latitudes. It was first tested satisfactorily during the 1973 High Latitude Rocket Campaigns and flown with other payloads during the 1976–77 Campaign. Useful results were obtained.

Meanwhile at UCL Rees was developing an ultrastable single etalon Fabry–Perot interferometer to make direct observations of the line profiles of airglow emissions so as to obtain wind speeds and temperatures in the thermosphere. After a successful trial in a balloon experiment to observe atmospheric absorption lines of H_2O and O_2 the etalons were incorporated in an instrument developed jointly by UCL and the University of Michigan which was successfully launched on 3 August 1981 as part of the payload of the NASA Dynamics Explorer mission. The technique is clearly one of great potential.

Fig. 13.11. Northward and eastward components of wind speed △, ○ observed by Rees *et al.* using glow clouds at Woomera, in comparison with observations made at Adelaide from observations of meteor trails.

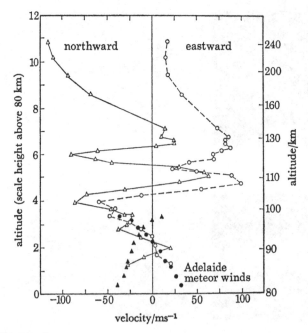

2.3 Other middle-atmosphere measurements

The UCL group first under Groves and since 1974 under Rees has taken part in many combined experiments, to measure winds at high latitudes during auroral displays, to study the S_q current system at mid latitudes and to help in the investigation of the dynamical condition of the atmosphere during winter anomaly conditions in the D region of the ionosphere (see Chapter 14, p. 323). Many interesting and useful results have been obtained in this way which are too numerous to be reported here. It must suffice to mention the launch of four Nike-Cajun rockets between 16 and 25 January 1969 during a period of stratospheric warming (see p. 323). These rockets carried grenade and TMA payloads prepared in co-operation with the Meteorological Institute of the University of Stockholm and the Swedish Space Technology Group. Their launch times were co-ordinated with six other grenade launches from northern hemisphere locations (Wallops Island, Fort Churchill and Point Barrow) as well as a meteorological sonde from S. Uist.

A cooling of about 45 K was observed at an altitude of 45 km whereas at 70 km a warming of about 50 K was found showing that a mesospheric warming followed the stratospheric. During these flights, wind speeds of $220\,\mathrm{m\,s^{-1}}$ were observed at 49 km on 19 January, among the highest ever recorded in the stratosphere.

3 Remote sounding from satellites with infra-red sensors

In the preceding section we described how measurements have been made of the temperature, density and wind distribution in the neutral atmosphere using sounding rockets. Important as these measurements are they refer to particular locations and times and it is very difficult to obtain a global picture from them (see, however, p. 276), especially as for meteorological applications the global picture available needs to be as clearly continuous in time as possible. It would therefore be of great value to develop techniques which enabled remote sounding of the atmosphere below to be carried out from earth satellites. We describe now how infra-red sensors have been developed for this purpose in the UK and flown with great success in several satellites as well as in the Pioneer Venus mission (see p. 293). At an early stage the Meteorology Sub-committee of the British National Committee for Space Research, under the Chairmanship of P.A. Sheppard noted the work of Kaplan[28] who pointed out how it would be possible to make observations of the vertical profile of atmospheric temperature below a balloon or satellite. Very soon, two physicists experienced in infra-red technology, J. Houghton at the Clarendon Laboratory, Oxford and S.D. Smith at the physics department of the University of Reading, who had worked together at RAE during the 1950s, began a collaboration which was to prove most fruitful. This was directed towards the practical implementation of Kaplan's suggestions with the development of suitable sensors for operation in an earth satellite. This would make possible for the first time the global sounding of atmospheric temperature, of great value for meteorology.

In very general terms the idea of the remote sounding is quite simple. If the sensor aboard a satellite observes radiation of a frequency which is very weakly absorbed in the atmosphere beneath, then this radiation will be emitted from the surface, or close to it, with an intensity determined in particular by the surface temperature. On the other hand, if the radiation is strongly absorbed, the sensor will observe only that which is emitted by the atmosphere close to the space vehicle. If the emitting molecules are still in thermodynamic equilibrium at this altitude the observed intensity will be determined by the atmospheric temperature near the vehicle. Radiation with intermediate absorption will come mainly from some intermediate level so that its observed intensity will be determined by the temperature at that level. In this way the temperature profile could be obtained by suitable choice of a range of frequencies of infra-red radiation.

To express this in more precise terms, if k_v is the absorption coefficient for radiation of frequency v then the intensity of radiation emitted in the vertical direction by a slab of atmosphere of thickness δh at an altitude h where the temperature is $T(h)$, which is received by a sensor at an altitude h_0, is given by

$$k_v \, B_v \, (T) \, \tau_v \, \delta h \tag{4}$$

where

$$\tau_v = \exp \left\{ - \int_h^{h_0} k_v \, d \, h \right\} \tag{5}$$

and $B(T)$ is the Planck radiation function. Integrating over all such slabs the total intensity received in the vertical direction from the atmosphere below h_0 will be

$$I_v = \int_0^{h_0} k_v \, B_v \, (T) \, \tau_v \, d \, h \tag{6}$$

$$= \int_{\tau_0}^{\tau} B_v \, (T) \, d \, \tau_v \tag{7}$$

where

$$\tau_v = \exp \left\{ - \int_0^{h_0} k_v \, d \, h \right\} \tag{8}$$

This may be transformed to

$$I_v = \int B_v \, (T) \, \kappa \, (y) \, dy \tag{9}$$

where $y = - \ln p$, p being the pressure at height h, and

$$\kappa(y) = \frac{d}{dy} (\exp - \int k_v \, dh) = d \, \tau_v/dy.$$

According to (9) the observed intensity is a weighted average of the black body intensity, the weighting factor being $\kappa \, (y)$. (10)

For a single collision-broadened spectral line of strength s centred at a frequency v_0 for a path at pressure p,

$$k_v = s\gamma_0 \, p/\pi \left\{ (v-v_0)^2 + \gamma_0{}^2 p^2 \right\} \tag{11}$$

where γ_0 is its width at 1 atmosphere pressure. In the outer wings of the line $v-v_0 > \gamma_0{}^2 p^2$. Also, if the absorbing constituent is uniformly mixed throughout the height range considered, dh is proportional to dp so that τ_v is equal to $\exp(-\beta p^2)$ where β is a constant and

$$\kappa(y) = 2 \, (p/p_0)^2 \, \exp \, (-p^2/p_0{}^2), \tag{12}$$

p_0 being the pressure at which $\kappa \, (y)$ is a maximum. In terms of $\ln p$ the half width is approximately constant or, since $d(\ln p)$ is approximately proportional to $d \, h$, it may be expressed in height units, as close to $10 \, \text{km}$.

This example assumes that the radiation detected is completely monochromatic whereas in practice several lines will be observed and the resolution of the detector will be finite. For the best results this resolution should be as high as possible, consistent with sufficient intensity of radiation being observed to make accurate measurement practicable. It was in dealing with this requirement that Houghton and Smith made a most important contribution.[29]

Remembering that it is necessary that the radiation should arise from a single constituent which is well mixed and that the excited states concerned should be in thermodynamical equilibrium with the ground state over the height range to be explored, the best choice for the radiation is the $15 \, \mu m$ band of CO_2 which is strongly absorbed in the atmosphere.

For meaurements at low altitudes, Houghton and Smith introduced a tube containing CO_2 at a suitable pressure in front of the detector. This absorbed the radiation from the line centres so only that at the wings of the lines penetrated and the detector was observing only weakly absorbed radiation coming from low altitudes.

To measure at relatively high altitudes it is the strongly absorbed radiation at the line centres which must be detected. To achieve this Houghton and Smith introduced a chopping system in which the detector was switched at a controlled frequency between observing through the CO_2 filled cell and through an exactly similar empty tube. The difference signal at the chopping frequency was then that due to the radiation around the line centres.

With this arrangement the effective resolution is the line width but the intensity observed is that from all the lines admitted by the filter placed before the absorbing tubes and is large enough for measurements.

In a variant of this system, the CO_2 pressure in the tube is modulated by a piston so that the effective absorption of the detected signal may be varied continuously between low and high values.

So far we have only considered looking vertically downwards but an alternative which has some advantage, particularly for measurements at high altitudes, is to look tangentially towards the earth with a narrow field of view so that height resolution is obtained by scanning the limb instead of spectroscopically. The loss of intensity due to the narrow field of view may then be partially offset by using a wide spectral bandwidth. With limb scanning the weighting functions are very sharp so that higher vertical resolution is obtainable. This is at the cost of somewhat poorer horizontal resolution. Using the pressure modulated system and limb scanning temperature sounding may be carried out up to nearly 100 km at which altitude some correction is necessary owing to departure from thermodynamic equilibrium. Limb scanning is also advantageous for measurements of atmospheric composition which can be determined more precisely and definitely.

The first selective chopping infra-red radiometer designed and constructed by Houghton and Smith and their collaborators consisted of six independent filter radiometers with spectral band pass varying between 3.5 and $10 \, cm^{-1}$. Four channels were used to observe below 30 km and two which employed the principle of selective chopping described above observed the region between 70 and 80 km altitude above up to 40 to 50 km. The detector was a thermistor bolometer. It was flown in the Nimbus 4 experimental meteorological satellite launched by NASA in April 1970. From the outset its performance was excellent and it continued in operation until superseded by an improved version launched in Nimbus 5 in December 1972. Instead of chopping between two cells the radiation was measured in turn between four cells, one empty and the other three containing CO_2 at different pressures. It used a pyroelectric detector making it possible to work with a much narrower field of view. Again this instrument proved to be very satisfactory and continued to provide data regularly for a period of nearly 10 years.

A pressure modulator radiometer was flown on Nimbus 6 launched in June 1975 and a limb scanning instrument on Nimbus 7 launched in October 1978.

The performance of all these instruments has been remarkably good – in fact continuous temperature data on a global basis have been obtained for over twelve years from the four infra-red instruments designed and developed by the Oxford and Reading (transferred to Heriot-Watt University since 1970) groups and flown in four successive Nimbus satellites. In 1981 the instruments in Nimbus 5 and 6 were still in working order and have been occasionally operated to check data obtained with the limb scanning instrument.

It goes without saying that this extensive data bank has been of great value in meteorological research, especially in connection with atmospheric circulation in the stratosphere and ionosphere. We can only show here one or two examples by way of illustration.

Figure 13.12 shows a temperature cross-section of the atmosphere up to 90 km over the latitude range 80°N to 80°S for 4 August 1975, derived solely from measurements made by the selective chopper radiometer on Nimbus 5 and the pressure modulated radiometer on Nimbus 6. Of special interest are the very cold stratosphere in the southern winter, the warm polar stratopause and the very cold polar mesopause in the northern summer.

Some remarkable effects arise from the interaction between the normal zonal circulation and planetary waves propagating upwards from the troposphere. Among these are the stratospheric warmings which occur in winter in the northern hemisphere. During such occasions the temperature in polar regions at the stratopause may rise by as much as 50 K or more, exceeding, for a few days, that at midsummer. These major warmings occur in late December or January, in which case the high polar temperatures are followed by rapid cooling and very low temperatures, or in March after which the temperature remains as in normal summer conditions. Figure 13.13 shows observations made with the pressure modulated radiometer on Nimbus 6 for the period November 1976 to January 1977. These are radiances originating near 80 km and 45 km averaged around the latitude circle at 80° N and around the equator. Observations made with the selective chopper radiometer on Nimbus 5 originating near 20 km are also shown.

Three periods of stratospheric warming are clearly seen in the observations near 56 km where the observed radiance at 80° N is as great or greater than that at the equator. Variations in the mesosphere near 80 km are in antiphase to those at the 45 km level while those at 20 km are in phase but much less marked. After the third and most intense warming the stratospheric temperature fell very rapidly while that in the mesosphere rose, so that at the end of January it was warmer than the stratosphere.

In the southern hemisphere the variability in temperature is much less than in the northern and more gradual. During a northern stratosphere warming the temperature falls in the southern stratosphere. Figure 13.14 shows observed radiances near 45 km averaged over circles of latitude at 80° and 50°S.

Figure 14.15 of Chapter 14 illustrates temperature observations taken

Fig. 13.12. Temperature cross-section of the atmosphere from 80°N to 80°S as deduced from measurements made by the selective chopper radiometer on Nimbus 5 and the pressure modulation radiometer on Nimbus 6, for 4 August 1975.

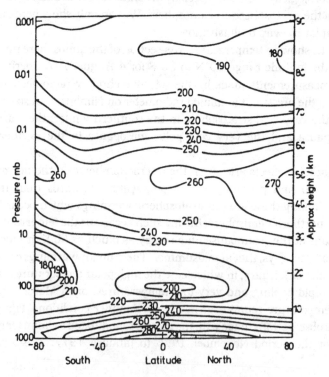

with the pressure modulation instrument in Nimbus 6 in connection with the interpretation of winter anomaly effects in the D region of the ionosphere.

It is interesting to note that following on the Oxford work, the UK Meteorological Office has built ten Pressure Modulation Radiometers for the US TIROS series of satellites (their operational weather satellites) beginning in 1979, to provide routine sounding of stratospheric temperatures.

4 Atmospheric composition

The Meteorological Office has carried out a number of measurements from rockets and satellites of the concentrations of ozone (O_3) and molecular oxygen (O_2) in the atmosphere.

Fig. 13.13. Radiances for the period November 1976 to January 1977 averaged around the 80°N latitude curve and the equator. The curves labelled PMR refer to results obtained with the pressure modulation radiometer on Nimbus 6, CHAN 3000 referring to altitudes near 80 km, CHAN 2115 near 45 km. The curves labelled SCR were obtained with the selective chopping radiometer on Nimbus 5 and refer to altitudes near 20 km.

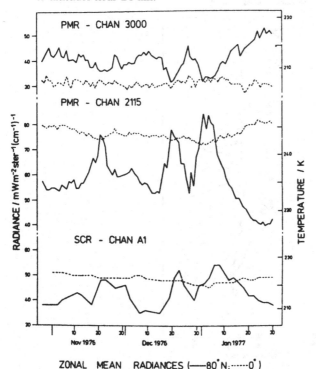

ZONAL MEAN RADIANCES (——80°N;······0°)

The principle of the method employed, from a satellite for example, was to observe the solar radiation at a suitable wavelength as it passed through the earth's atmosphere at times near sunrise and sunset at the satellite. The wave length is chosen to be one which is strongly and selectively absorbed by the atmospheric constituent under study. For O_2 a suitable choice is 145 nm. From the attenuation of the radiation as it passes through different thicknesses of atmosphere the concentration of the particular constituent can be obtained as a function of altitude.

We have already described some of the results obtained for O_3 (see Chapter 5, p. 95) from observations made from the satellite Ariel 2. Many hundreds of altitude profiles of O_3 concentration were obtained covering a large part of the earth's surface.

A typical altitude profile for O_2 obtained from instruments aboard Ariel 3 has been shown in Chapter 5, Figure 5.17. The results are of special interest

Fig. 13.14. Mean radiance measurements for the southern hemisphere averaged around latitude circles at 80°, 50°S and the equator as a function of time, an equivalent temperature scale being shown on the right hand side. These measurements were made with the selective chopping radiometer on Nimbus 4 and refer to altitudes near 45 km.

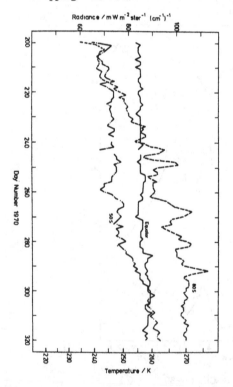

as they are for altitude ranges in which the proportion of atmospheric oxygen in the bimolecular form is decreasing rapidly with height the monatomic form becoming increasingly dominant.

Concentration of O_3 and O_2 have also been observed from a number of rocket flights including two from a high latitude at Fort Churchill using a Black Brant rocket. Petrel rockets have also been used to study the diurnal variation of O_3 concentration.

A very important development initiated and developed by the Appleton Laboratory in collaboration with the Department of Physics at the University College of Wales Aberystwyth was that for determining the concentration of atomic oxygen in the mesosphere and lower thermosphere by *in situ* observations from rockets. Knowledge of these concentrations is essential for the interpretation of many phenomena occurring at these

Fig. 13.15. Concentrations of atomic oxygen on three winter days at S. Uist, ——, – – –, ... compared with mass spectrometer measurements at El Arenosillo, Spain △, □.

winter days

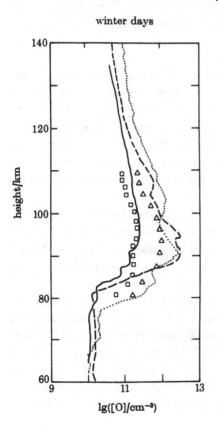

altitudes (see p. 291) and it is very difficult to make any accurate predictions when the monatomic form is becoming an increasingly minor constituent. The method depends on the observation of resonance fluorescence and absorption when the ambient atmosphere is irradiated by light of wavelength 130 nm, the ultra-violet resonance line of atomic oxygen. The first rocket flights including this payload were carried out in 1974 and gave good results. Figure 13.15 shows results obtained on a number of Petrel flights from S. Uist, during the day. Data such as these have been extensively used for interpreting the behaviour of the D region of the ionosphere (see Chapter 14, p. 320) and the intensity of the main night airglow (see p. 291). Observations have also been made at high latitudes in connection with noctilucent clouds and other phenomena.

Much information regarding the global distribution of CH_4, N_2O and H_2O, which is assisting in studies of ozone chemistry and stratospheric dynamics, has been obtained from the US satellite Nimbus 7. This was launched in 1978 as a Stratospheric and Mesospheric Sounder, and contained a limb sounding instrument built by the Oxford group using Pressure Modulation Radiometer techniques for composition measurements.

5   Airglow studies

In Chapter 3 we described briefly the different experiments planned for the early Skylark flights. Included in these were measurements of the intensity of day and night air glow emissions initially planned by D.R. Bates and E.B. Armstrong of Queen's University, Belfast. This was a very appropriate choice of programme because Bates had been particularly interested in the problem of determining what reactions are responsible for the night airglow emissions and already in the early 1960s was a world authority on the subject.

As referred to in Chapter 3, p. 31, special interest attached to observations of the variation of emitted intensity of the various radiations as a function of altitude because the results obtained by the ground-based techniques were in strong disagreement with theoretical prediction. Early rocket observations gave results supporting the theory.

The first successful flight in which measurements of the intensity and altitude profiles of night air glow emission were made, was in a Skylark from Woomera on 18 October 1965. It was followed by two Skylark flights to make similar measurements specifically for the O_2 lines at 7619 and 8645 Å. These were fully successful and it was found that the former arose from a layer at an altitude between 45 and 66 km and was 20 times more intense than the latter which was emitted from a layer at an altitude between 80 and 95 km.

Similar measurements were carried out for the dayglow emission of lines at 6300, 5890, 5577, 5199 and 3914 Å. Because the ratio of the intensity of the sunlight to that of the dayglow is $> 10^7$ an attitude stabilized rocket equipped with special baffles had to be launched, with the sun low in the sky.

A successful experiment to observe the resonant scattering of sunlight by the atoms and ions of Mg, Fe and Ca was flown successfully in an unstabilized Skylark rocket on 17 October 1972. Good signals were observed from all the detectors and, although the data analysis was very complex, evidence for the existence of layers of metallic atoms and ions at an altitude near 100 km was found.

Perhaps the most interesting observations are those which have provided new evidence about the source of the most prominent night airglow emission, that of the green line at 5577 Å resulting from a transition between the metastable ^1S of O and the ground ^3P state. Chapman[30] in 1931 suggested that the metastable atoms are produced through the reaction

$$O + O + O \rightarrow O_2 + O\,(^1S) \tag{13}$$

the importance of which was first suggested two years earlier by Franck. For many years this mechanism was accepted although there was no evidence about its adequacy as no measurement of the reaction rate had been made. In 1964 in order to explain certain laboratory results Barth[31] suggested the alternative possibility of a two-stage process

$$O + O + M \rightarrow O^1_2 + M \tag{14a}$$

$$O^1_2 + O\,(^3P) \rightarrow O_2 + O\,(^1S) \tag{14b}$$

where M is a third body, probably either or both O_2 and N_2 and O^1_2 is an intermediate excited state of O_2. However, in the absence of any direct evidence about the rates of (14a or 14b) this proposal did not find favour initially. A somewhat confused situation then developed in that the first attempts to make laboratory measurements of the relevant reaction rates, not only of the processes of production of O (^1S) but also of its destruction under atmospheric conditions which were necessary to determine its equilibrium concentration, seemed to favour the reaction (13). It was later realized that the measured production rates could well have referred to production by the Barth mechanism so the choice between (13) and (14) remained undecided.

An important step forward was taken when it became possible to make nearly coincident measurements of the green line emission and of the atomic oxygen concentration.[32] On 9 September 1975 a Petrel rocket was launched from S. Uist containing photometers to measure the intensity and altitude profile of the green line and of the Herzberg bands of O_2. Twenty-six minutes later this was followed by a second Petrel carrying the resonance lamp experiment developed by the Appleton Laboratory (see p. 290) to measure atomic oxygen concentrations. Figure 13.16 shows the observed altitude profiles. From analysis of these observations, using laboratory measurements of the rates of quenching of $O(^1S)$ by collision with O_2 and $O(^3P)$ respectively it was found that, if (13) is the dominant production process for $O(^1S)$ the rate must vary very rapidly with temperature T, as T^{-7}. This would not be consistent with observation of the effect of gravity waves near 95 km which produce temperature variations of nearly 100 K (see Figure 13.8) but cause little change in green line intensity. No such difficulty arises if the two-stage mechanism (14) is the dominant source of O (^1S). Similar evidence has come from experiments by American scientists. To clinch the matter finally, the reliability of laboratory measurements of the relevant quenching reaction rates must be established. The situation as of 1981 has been thoroughly reviewed by Bates.[33]

6   Micrometeoroids

On 28 October 1971 the Prospero satellite was launched into an orbit with a perigee height of 447 km and apogee 1582 km by a British Black

Fig. 13.16. Variation with altitude of the intensity of the green line (557.7 nm) and Herzberg bands in the night airglow observed using a Petrel rocket launched at S. Uist on 9 September 1975. The concentration of atomic oxygen observed in a rocket launched 26 min later is also shown.

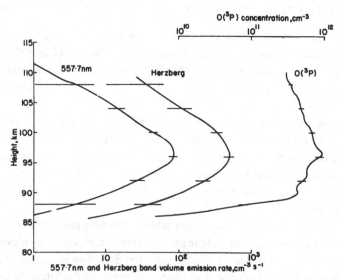

Arrow rocket from Woomera.[34] This was the first and, up to the time of writing, the only satellite to be launched by an all-British launching system. The payload was technological in character except for one experiment designed to study the flux of micrometeoroids in the near neighbourhood of the earth.

The problem of determining this flux had proved to be a difficult one. Thus in Chapter 3 the negative result of such an experiment carried in Ariel 2 was reported. Results in other experiments seemed to err in the other direction in observing too great a flux. The detector on Prospero was designed, developed and constructed at Birmingham University[35,36] and depended on observation of the plasma generated by impact of a high velocity particle on a solid surface. From laboratory experiments in which the positive and negative charges produced are separated, it is found that the total charge of either sign produced in the impact is proportional to $m^\alpha v^\beta$ where m is the mass of the impinging particle, v its velocity. Over the velocity range of these experiments ($50\,\text{m s}^{-1}$ to $5\,\text{km s}^{-1}$) $\alpha \sim 1$ and $\beta \sim 3$. In the case of micrometeoroid impact on a surface carried by a satellite, the impact velocities are higher (50–$80\,\text{km s}^{-1}$) and it is necessary to assume that the laboratory results can be extrapolated to such velocities.

The Prospero detector had an exceptionally high sensitivity, the minimum detectable mass being as low as $10^{-15}\,\text{g}$, two orders of magnitude lower than for any other detector used up to that time and in a region in which solar radiation effects are becoming important. Observations were made over 285 days during which 94 micrometeoroid particles were recorded. The mean particle flux [37] was 0.21 particles $\text{m}^{-2}\,\text{s}^{-1}\,(2\pi\,\text{sr})^{-1}$. However, many events were strongly clustered within two periods of 9 and 15 days respectively. If these were excluded as due to some special event the average was reduced to $0.12\,\text{m}^{-2}\,\text{s}^{-1}$. These results are consistent with the values obtained for particles with mass below $10^{-13}\,\text{g}$ derived from observations of zodiacal light.[38,39]

7 Lunar and planetary observations

7.1 *The Venus temperature probe*

While the UK scientists have played a considerable part in the analysis of data from lunar and planetary missions including the laboratory study of samples of lunar materials provided from both the USA and USSR, only one experiment involving major UK participation has been flown, up to the time of writing, in a planetary mission. This was, most appropriately, an infra-red temperature sounder very similar to those used in terrestrial studies (see Chapter 13, p. 281), carried on board the Pioneer Venus orbiter launched on 20 May 1978. This was designed to sound the atmosphere of Venus from the cloud tops at about 65 km above the surface to an altitude of about 150 km with a vertical resolution from 10 to 40 km. As for the earth the method was based on observation of the CO_2 emission in a number of wavelength channels, CO_2 being the major constituent of the atmosphere of Venus. Seven spectral channels were used, in three of which the pressure modulation technique (see p. 284) was used for spectral discrimination. As well as temperature sounding the instrument included observing channels through which information about the cloud structure and water vapour content could be obtained. The design and development of the instrument[40] involved a collaboration between the group at Oxford University under J.T. Houghton and one at the Jet Propulsion Laboratory, Pasadena.

The instrument performed very well and has obtained a great amount of data about the atmosphere of Venus – in the first 72 orbits about Venus, 800,000 or so temperature soundings were made together with the associated measurements of albido and humidity. The zonal mean equivalent temperature in the northern hemisphere derived from the radiance measurements at 80 km increases from 205 K at 0° latitude to 224 K at 90° and varies little between day and night. At 100 km, however, while a similar situation prevails at latitudes higher than 60° N the night time temperature falls below the daytime at lower latitudes, the respective values being 169 and 186 K at 0°. This is but a small sample of the great amount of data obtained from which wind distributions and cloud cover are being derived.

7.2   Lunar samples analysis

When samples of lunar material from the Apollo 11 and 12 missions became available near the end of 1969, some 15 research groups in Universities in the UK were all prepared to make experimental studies of the material using a wide variety of techniques.

Several groups began mineralogical and petrological investigations of the samples using such techniques as electron and optical microscopy, electro microprobe and chemical analysis, X-ray fluorescence, Mossbauer spectroscopy and powder and single crystal X-ray diffraction. The groups involved were headed by S. Agrell, J.H. Scoon and P. Gray at Cambridge University, M. Brown at Durham University, S.H.U. Bowie at the Institute of Geological Sciences, S. Tolansky at the University of London Royal Holloway College, J. Zussman at Manchester University and M. O'Hara at Edinburgh University.

Under the general heading of geochemistry, G. Turner at Sheffield University developed the $^{40}Ar/^{39}Ar$ activation method for dating rocks. A.A.

Smales and his group at the Atomic Energy Research Establishment, Harwell, used spark source mass spectrometry, emission spectrography and X-ray fluorescence spectrometry to carry out time element analysis and the group at Bristol University under G. Eglinton was concerned with the detection of organic compounds.

A number of groups carried out physical measurements on the samples. Measurement of infra-red absorption, specific heat and thermal conductivity of lunar rocks and fines were made by J. Bastin and J. Clegg of Queen Mary College London (QMC) and by G. Fielder of UCL. A collaboration between J.C. Geake of the Manchester Institute of Science and Technology (UMIST), G.F.J. Garlick at the University of Hull and A. Dollfus at the Paris Observatory studied a number of optical properties including luminescence of the samples induced by ultraviolet radiation and by proton bombardment. Induced luminescence and thermoluminescence were also studied by J.A. Edgington of QMC and I.M. Blair at Harwell.

Magnetic properties of the samples which are very important for interpretation of the magnetic history of the moon were studied by the group under K. Runcorn at the University of Newcastle on Tyne.

At a somewhat later stage a group at the University of Birmingham under S.A. Durrani began the study of charged-particle track phenomena in lunar crystals and glasses.

Further samples became available from later Apollo flights and in 1971 somewhat larger samples became available for collaborative study between groups at Cambridge, Sheffield, Paris and Zurich. These included a 25 g slab from Apollo 14 and 150 g from Apollo 15. Samples also became available from Apollo 16. In addition, a gift of 0.5 g from each of the samples returned automatically in the Russian Luna 16 and Luna 20 missions was received by the Royal Society from the Soviet Academy of Sciences. These were fractionated and distributed to the research groups.

These investigations proceeded with much enthusiasm and skill culminating in a Discussion Meeting at the Royal Society on the 'The Moon – a new appraisal from space missions and laboratory analysis' organized by Runcorn, Eglinton, Brown and H.C. Urey, it was held between 9 and 12 June 1975. Over 70 papers were presented from a wide international range of speakers, including several from the UK and the proceedings were published as a separate volume[41] by the Royal Society.

7.3 Ground-based planetary studies

Although there has been direct participation in planetary probing by a UK group in only one experiment (see 7.1) there is strong interest in the

UK in planetary studies. It would be out of place here to refer to much of the ground-based work but it is appropriate to point out that UK scientists have been otherwise closely involved in US planetary projects. Thus J.E. Guest of UCL was a member of the Viking Orbiter Imaging Team and G. Hunt, then at the Meteorological Office, of the Imaging Science Team for the Mariner/Jupiter/Saturn mission. Hunt also participated in the Viking programme as a Guest Investigator and member of the Infra-Red Thermal Mapping Team.

Following his transfer to UCL[42] in 1978 Hunt extended the range of his planetary studies including use of the IUE (see Chapter 15, p. 358) to obtain the first resolved ultra-violet spectra of Uranus and Neptune. His group also developed and put into application a system for imaging data concerning planetary atmospheres known as the Interactive Planetary Image Processing System.

While Hunt concentrated on planetary atmospheres, Guest, at the University of London Observatory has concerned himself particularly with the study of surface features appearing in planetary images.

14

Scientific studies by British space scientists – the ionosphere, the magnetosphere and cosmic rays

As we have explained in Chapter 1, the British scientists who were most interested initially in the use of rockets for scientific research were those concerned with the ionosphere so it is not surprising that, since the inception of a rocket research programme, a great deal of attention has been directed towards ionospheric studies. This has involved the use of a wide variety of techniques ranging from ground-based radio tracking of satellites to *in situ* measurements of ionospheric properties from space vehicles. In fact, as we shall see, much of the work has been concerned with the topside ionosphere and this has naturally connected up with studies of the magnetosphere to which many ionospheric physicists have made major contributions.

1 Ways and means

At the time when these programmes were planned, most knowledge of the ionosphere was confined to the information obtained from ground-based sounding which basically provided the height profile of electron concentration above the sounding station. Even this was incomplete as explained in Chapter 1, p. 2, especially as it could provide no information about the region above the F layer maximum – the topside ionosphere. No direct method of determining other important ionospheric parameters such as the electron temperature or the positive ion composition was available. By the early 1950s the impact of the new rocket techniques was making itself felt in that exploratory measurements had been made, particularly of the ion composition and of the electron concentration in the gaps in the altitude profile. The likely availability of satellite vehicles suitably instrumented offered a promise of making *in situ* measurements of such quantities as the electron and ion concentrations and temperatures, as well as ion composition, on a world-wide basis over considerable period of time. This latter feature was especially important because the behaviour of the ionosphere depends on local time and season, geographic and geomagnetic location, magnetic activity, solar activity and epoch in the 11-year solar cycle. A few years after space techniques became practicable, the incoherent back-scatter technique for ionospheric sounding was developed. With this technique, the vertical profile of electron concentration, of electron and ion temperatures and, in certain circumstances, of the ion composition, above the sounding station could be measured. As compared with measurements from satellites the back-scatter technique had two advantages. First of all, it gave direct vertical profiles of the quantities measured whereas for satellites account must be taken of the quite rapid orbital motion which makes it difficult to obtain such profiles. The second arises from the fact that, unless fitted with special onboard propulsion a satellite will not circulate for long in an orbit at an altitude less than 200 km. This generally limits the *in situ* coverage of the ionosphere to the F_2 region and above whereas the back-scatter technique operates quite effectively at present down to about 80 km. On the other hand the number of back-scatter stations is necessarily limited (there are six in operation at the time of writing) and it is only possible to obtain a global pattern for the different ionospheric parameters from satellite observations. In other words the satellite and back-scatter techniques are complementary.

The choice of suitable instruments for *in situ* ionospheric probing is rendered difficult by the fact that, even for plasmas on a laboratory scale, as in electric discharges, reliable methods for measuring the basic parameters have not been easy to develop. Special care had to be taken in interpreting the data obtained with Langmuir probes in terms of the probe geometry, the discharge conditions etc. The effective use of the probes in satellites posed still further problems and it is not surprising that little use had been made of such techniques for *in situ* measurement of ionospheric parameters even by the early 1960s. It so happened that Boyd and Sayers had had extensive experience of using Langmuir probes in laboratory plasmas so that from the beginning of the British Space Science programme it was most appropriate that they chose to work on the application of these probes to *in situ* measurements in the ionosphere. As a result, Ariel 1 was one of the first satellites to make such measurements effectively. It was so successful that the probe techniques employed were flown in several later satellites as well as in many sounding rockets. An important part of the work concerned the optimization of the techniques and the checking of their operation against results obtained by other methods such as the incoherent back-scatter technique. As we shall see, data were obtained on a global scale for such quantities as electron temperature and ion composition under a wide variety of conditions, providing an essential background to a thorough understanding of the ionosphere above 260 km, mainly in fact the topside ionosphere.

A further very effective method for study of this region, combining ground-based and space techniques, was the topside sounder. It observed the topside ionosphere in very much the same way as the classical ionosonde used in ground-based studies, observes the lower ionosphere, except that the transmitting and receiving equipment was contained in a satellite circulating high above the F_2 maximum. We have already referred in Chapter 7 to the way in which the first topside sounder named Alouette 1, was designed and built in Canada and launched on 29 September 1962 into an approximately circular orbit about 1000 km above the earth's surface. The data rate from operation of this device was very great and the success of the enterprise depended on efficient data processing and analysis. From the outset, J.W. King and his colleagues at the Radio Research Station played a major part in this side of the work[1] and contributed much to the success not only of Alouette 1 but of its successors Alouette 2 and the ISIS satellites.

We have already described in Chapter 3, p. 39, how British scientists quickly took advantage of the launching of the first Sputniks with radio transmitters aboard to use the radio signals both for satellite tracking and, by comparison with accurate optical observations, to disentangle effects on the transmissions from the satellite due to the ionosphere. The comparatively low frequency, 20 and 40 MHz, of the Sputnik beacons helped in this respect. At a little later stage beacon satellites with transmitters in this frequency range were launched specially in the USA to facilitate ionospheric research. The Radio Research Station (later RSRS and later still the Appleton Laboratory) with its receiving stations at Winkfield, Singapore and Port Stanley in the Falkland I., took a major part in this work, which has continued to the present time. A large contribution has also been made by L. Kersley and other members of Beynon's group at the University College of Wales, Aberystwyth. From these studies the quantity which emerges from analysis of satellite Doppler effects or from Faraday rotation of the plane of polarization of radio waves, is the integrated electron concentration along the path from the point of reception on the ground to the satellite. This will include contributions from both the top and bottom side ionospheres.

Similar techniques may be used to measure electron concentrations using vertical sounding rockets by observing either the difference in time taken for two radio pulses of different suitably chosen frequencies to propagate from a ground transmitter to a transceiver in the rocket and back to ground or by the Doppler or Faraday effect in continuous wave transmission at suitable frequencies from the rocket. In each case the refractive index of the ionosphere close to the rocket and hence the electron concentration may be obtained. The group at Aberystwyth under Beynon introduced these techniques into the Skylark programme as an alternative to the *in situ* probe methods described above.

The pulse technique involved the reception at the rocket of hf and vhf signals radiated from ground-based transmitters. Throughout the flight pulsed radio signals on one of two preselected fixed frequencies in the range 2–5 MHz, together with pulsed signals on 104 MHz were received at the rocket every 1/50 second, and the received pulse patterns telemetered back to ground-based receivers. Comparison of the observed time delays of the hf (2–5 MHz) signals (which are refracted and reflected by the ionosphere) and those of the vhf (104 MHz) signals (which are effectively uninfluenced by the ionosphere) enables the electron density/height profile in the ionosphere to be deduced.

Figure 14.1 shows the photographic record of echo-range versus flight time (or rocket height) for Skylark SL 120 launched at 1456 local time on 17 September 1964[2] using this pulse technique. The thin black trace refers to the 104 MHz pulse signal up to the top of the rocket trajectory and down again. The other traces refer to the recorded pulse patterns initially on 2.3 MHz and later in the flight on 3.6 MHz. For the first 200 seconds of the flight the lower frequency 2.3 MHz, appropriate for exploring the E-region was employed and for the remainder of the flight the higher frequency 3.6 MHz, appropriate for probing the F1 layer, was used.

The interpretation of the E-region echo traces, observed in the first 200 seconds of the flight, is given in Figure 14.2 (*a*) and Figure 14.2 (*b*) which respectively show a small scale line tracing of the observed echo traces and a sketch of how these various recorded multiple-reflected echo traces arise. Trace A represents the 104 MHz pulse signal. The coalescing of direct and indirect signals on 2.3 MHz as the rocket ascends through the top of the ray trajectory is clearly seen both on the first reflection (signals B and C), the second reflection (D and E) and the third (F and G).

The situation is seen to be more complicated at the higher altitudes where pulse patterns on 3.6 MHz were recorded. Thus the dense black patch between altitudes of 150 and 190 km represents a large spreading of echoes due either to ionospheric irregularities or some plasma resonance effect. On the down-leg the effect is still present but some long delayed echoes extending outward from the direct ray are now resolved. Many other detailed features

301

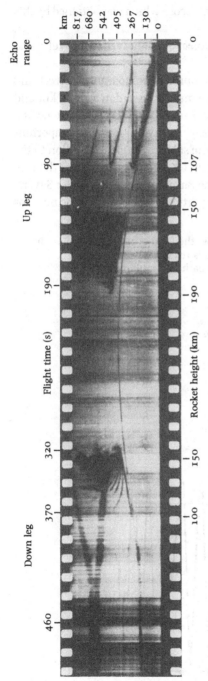

Fig. 14.1. Delay time record obtained on Skylark rocket flight SL 120 for ionospheric sounding using the pulse technique (1456 LMT 17 September 1964).

can be discerned on this record but these will not be discussed here since our aim has been only to illustrate the type of record which was obtained by this technique.

This rocket experiment was a joint project of the University of Wales and RAE Farnborough.

The ionosphere below 200 km is of course of comparable interest and importance. The back-scatter technique may be used down to 80 km and may be supplemented by use of vertical sounding rockets to extend the geographical coverage and to fly instruments to measure important quantities such as the spectral intensity of solar radiation in addition to the usual ionospheric parameters. Here again UK scientists, including the UCL group under Boyd and Willmore and the Birmingham group under Sayers, have made important contributions. In particular they realized the need for

Fig. 14.2. (*a*) Sketch showing how the form of the record shown in Fig. 14.1. before 150 km altitude is reached.
(*b*) Interpretation in terms of ray path and multiple reflections.

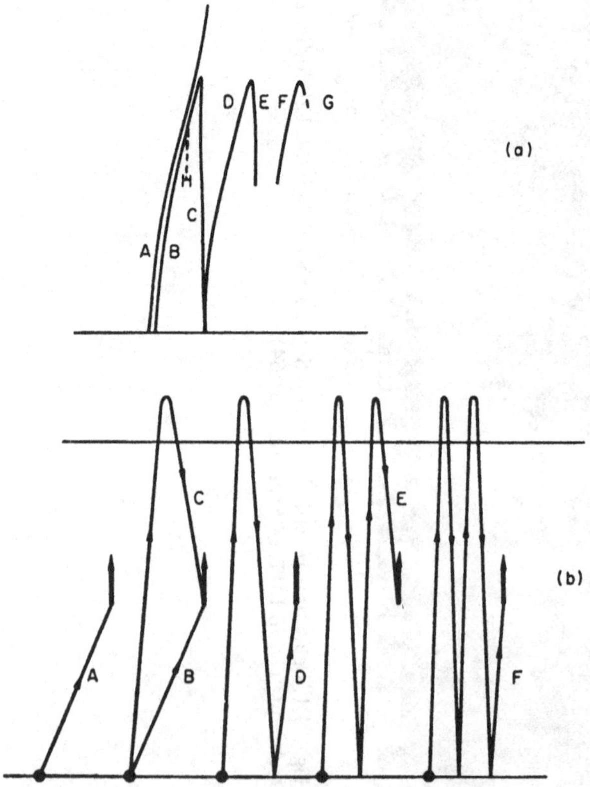

integrated rocket payloads to observe simultaneously sufficient parameters describing the ionosphere and the solar radiation to make possible a self-consistent check in terms of ionospheric theory. Their rocket measurements above 150 km were however overtaken by the US Atmospheric Explorer satellites which were fitted with auxiliary propulsion so as to circulate for long periods between 150 and 200 km altitude. The first Atmospheric Explorer (AEC) was launched in December 1973.

Below 150 km, the Petrel rocket was available in addition to Skylark for ionospheric studies. A great number of flights were made for this purpose including studies of the equatorial electrojet in the Commonwealth programme and of mid-latitude sporadic E ionization, an interest stimulated by the very first scientific results from the Skylark programme which observed a thin sporadic E layer over Woomera (see Chapter 3, p. 46).

The atmospheric ionization below the E region, say below 90 km, usually referred to as the D region has always been of interest to British scientists. As explained in Chapter 3, p. 32, it is important in radio transmission via the ionosphere since it causes absorption of the radio waves. This is because it is located at altitudes where density of the neutral atmosphere is much higher than in the main ionosphere. For the same reason the behaviour of the D region ionization is very complex and involves a wide range of interesting physics. The importance of simultaneous measurement of relevant quantities is even greater than for the main ionosphere and there are many more such quantities to measure – the behaviour of the D region depends strongly on the atmospheric composition, including that of minor constituents such as CO_2, O, O_3 and H_2O, while the ionic composition below about 75 km is very complicated, consisting mainly of clustered ions. It is clearly necessary for many groups to collaborate in these studies. As early as 1962 J.E. Hall of the Radio Research Station and K. Bullough of the University of Sheffield carried out the first rocket study[3,4] specifically of the D region using Skylark rounds SL108 and SL109 launched respectively on 15 August and 20 September 1962. A continuous wave of frequency 202 kHz was transmitted from the ground and the field strength measured using receivers connected to three orthogonal dipole aerials carried on the rocket. Very satisfactory results were obtained and the technique was used on many other occasions. In addition to this work at the RRS, L. Thomas and P.H.G. Dickinson have played an active part in D region studies both in experiment and theory. Beynon and his collaborators at Aberystwyth have also been involved. The UK scientists have also been involved in many collaborative projects with other European scientists.

The contribution of UK scientists in providing laboratory data on reaction rates of importance for understanding the D region has also been important in these studies.

Returning now to *in situ* observation of quantities of importance for the interpretation of ionospheric and magnetospheric phenomena, we have not so far mentioned the important measurement of magnetic and electric field strengths. Within the ionospheric magnetic field measurements may be used to investigate electric current systems such as the equatorial and auroral electrojets while their importance for study of the magnetosphere is obvious. In Chapter 3 we described the earliest experiments prepared for launch by Skylark rockets. Only a little while after these got under way a group from the Space Research Laboratory of the Department of Geophysics at Imperial College, under S.H. Hall, began work on the design of magnetometers to be flown in Skylark rockets. This group eventually flew rubidium magnetometers successfully and took part in collaborative experiments with other UK groups to investigate the S_q current system sporadic E, etc. Somewhat later, when Hall took up an overseas appointment, K. Burrows and colleagues at the Radio and Space Research Station assumed responsibility for the development of the magnetometers. These instruments have flown successfully in rocket flights both in mid-latitudes and in the auroral zone, again mainly in collaboration with other groups engaged in investigating simultaneously different aspects of a particular phenomenon. In collaboration with the UCL group under D. Rees, a vector rubidium magnetometer was developed and used effectively.

Magnetic field measurements in the magnetosphere and beyond have been made by the Cosmic Ray Group at IC by H. Elliot, using fluxgate magnetometers. In particular, instruments of this type were flown very successfully in the ESRO eccentric orbit satellites HEOS I and II (see Chapter 6, p. 146 and pp. 335–336, below).

The measurement of electric field strengths is intrinsically more difficult than that of magnetic field. One of the first methods used which still continues to be one of the most effective, is to release into sunlight a barium vapour cloud in the region at which the electric field is required. We have already referred to this technique and its applications to the study of the density and wind distribution in the neutral atmosphere. In sunlight some of the barium is photoionized. The ionized cloud can be distinguished from the neutral by its optical spectrum and its mass motion and expansion by diffusion can be followed. From the mass motion it is possible to disentangle the component arising from the effect of the ambient electric field on the motion of the ions in the presence of the ambient ionosphere and magnetic field and hence derive the magnitude of that field. This technique which was introduced by Föppe *et al.* (see Haerendel *et al.*[5]) was soon taken up by the group at UCL (see p. 317) and also extensively used, often in collaboration, by G. Martelli and colleagues at the University of Sussex and P. Rothwell and colleagues at the University of Southampton.

An onboard probe technique for measuring electric fields was developed and operated by Sayers and his group at Birmingham. It was an improved version of a method used earlier in which the potential difference between two probes at a fixed distance apart in the plasma is measured.

The study of the magnetosphere and its interaction with the ionosphere in the auroral zone and elsewhere requires, in addition, the development of onboard instruments for measuring the flux and energy distributions of the electrons, protons and heavier positive ions in the auroral regions, the magnetosphere and beyond. There was no lack of effort in developing such devices. D.A. Bryant and his colleagues at the Radio and Space Research Station began by using low-energy electron detectors consisting of an electrostatic energy analyser followed by a secondary emission multiplier, and progressed to much more sophisticated instruments. For example, in two ESRO Skylark rocket flights on 28 and 29 October 1970 from Kiruna to study auroral electrons they included instruments to measure electron densities in the energy range 1–14 keV. Three electrostatic analysers detecting electrons with energies of 4, 6 and 10 keV and one detector which counted all electrons with energies greater than 1 keV were mounted with their apertures along the spin axis of the rocket. Continuous measurements were taken with three detectors to give high time resolution. Energy spectra and pitch angle distributions were obtained from measurements made with two electrostatic analysers detecting electrons in energy steps from 2 to 14 keV, mounted with apertures along and at 30° to the rocket axis. Comparable equipment was flown by RSRS on later flights. Thus in the payload of the first Skylark 12 rocket (see Chapter 3, p. 34) launched, to study the particle streams and plasma waves associated with an auroral arc, detectors were flown which were sensitive to electrons of 0.5–25 keV energy and positive ions of 0.5–25 keV per unit charge.

At UCL (MSSL) a detector for suprathermal electrons and protons with energies between 5 and 500 eV has been developed which uses hemispherical electrostatic analysers and channel multiplier detectors. The energy resolution is ~ 0.1. It has been flown successfully on several rocket flights as well as on the GEOS 1 and GEOS 2 satellites.

The cosmic ray group at IC under Elliot, who had been concerned with cosmic ray studies in pre-satellite days, naturally possessed the experience to make measurements not only of magnetospheric particles but of energetic solar proton streams and cosmic rays. Thus to the ESRO II (p. 334) payload they contributed (a) two Geiger-Müller counters to measure the total flux of cosmic rays and of particles in the radiation belts, (b) a solid state proton range telescope to study the energy spectrum of solar flare particles and protons from the radiation belt and (c) combined proportional counters scintillation counter and Cerenkov detectors to observe the ratio of the proton to α-particle flux in cosmic rays and solar flare particles. Again in the eccentric orbit ESRO satellite HEOS I (p. 334), apart from a fluxgate magnetometer for measurement of interplanetary magnetic fields, they included two particle detectors for detection and measurement of anisotropies in the fluxes of cosmic rays and particles emitted from the sun.

The group from the University of Leeds under P.L. Marsden used a Cerenkov counter to measure the flux and energy spectra of primary cosmic ray electrons which was flown successfully in ESRO II (see p. 338).

Particle detectors in the energy range 1 keV to greater than 100 keV were also flown by the group at the University of Southampton under G.W. Hutchinson, in a number of rocket experiments.

A somewhat different kind of magnetospheric experiment is that of observing from a satellite very low frequency radio noise. This naturally includes man-made as well as natural noise, either atmospheric in origin such as spherics and whistlers arising from lightning in the lower atmosphere or hiss and chorus from the ionosphere. These waves are of special interest as they propagate along magnetic field lines in the plasmasphere and beyond. The first British observations of vlf noise from a satellite were made by the group from the University of Sheffield under T.R. Kaiser from the satellites Ariel 3 and Ariel 4. They continued this work using, in particular, Petrel rockets as vehicles, partly in collaboration with a group at the University of Southampton.

Related to these studies are those at somewhat higher frequency carried out from satellites primarily to observe radio signals due to lightning flashes in the atmosphere. Work of this kind began with equipment carried in the Ariel 3 satellite (see Chapter 5, p. 98) which was designed, constructed and operated by a group from the Radio Research Station (Appleton Laboratory) under F. Horner. It has led to very interesting results (see p. 331).

Observations of radio frequency noise were made in collaboration between the Nuffield Radio Astronomy Laboratories, University of Manchester (D. Walsh and A.P. Hayes) and the Appleton Laboratory (V.A.W. Harrison) using equipment on the Ariel 4 satellite (see Chapter 5, p. 103).

The Ariel 1 satellite carried a Cerenkov counter designed and operated by the Imperial College group to determine the energy spectrum of cosmic rays with charge $> 5e$ (see Chapter 5, p. 90), the main aim being to study the modulation of cosmic ray intensity in the earth's neighbourhood by the interplanetary magnetic field, a subject which they were able to pursue much further from the interplanetary magnetic field measurements which they made from the ESRO satellites HEOS 1 and HEOS 2 (see p. 335).

Finally we must refer to the equipment carried in Ariel 6 which was primarily a cosmic ray satellite directed towards measurement of the energy spectra and charge distribution of the heavier components of the cosmic radiation, i.e. those with nuclear charge $> 30e$. This involved the simultaneous measurement of the responses of Cerenkov and gas scintillation counters to a cosmic ray transit. The experiments were designed, constructed and operated by the group at Bristol University under P. Fowler that is very experienced in ground-based studies of the composition of cosmic radiation. Some of the results obtained are described on p. 337.

2 Results obtained in ionosphere studies
2.1 Topside ionosphere
The main British contributions to the study of the topside ionosphere using spacecraft has come from *in situ* probe measurements and from major contributions to the analysis of data from topside sounding and beacon satellites.

As explained earlier the probe measurements have been carried out by two groups, that at UCL under Boyd and Willmore and that at Birmingham University under Sayers. The probe techniques used have been somewhat different though with considerable overlap.

One of the important features of the probes used by Boyd and Willmore was that they operated with a small component added to the dc voltage sweep for the measurement of electron and ion temperatures T_e, and T_i respectively. In the usual dc mode of operation T_e, for example, is obtained by sweeping the voltage through the probe characteristics from floating to space potential and then deriving the temperature by use of the Boltzmann relation. In the ac mode additional small ac voltages of acoustic frequencies are applied to the probe and the cross modulation gives the temperature from the slope and second derivative of the current-voltage curve directly without depending on the detailed form of the probe characteristics. This was applied both to the measurements of electron and of ion temperatures. On the other hand to obtain electron concentrations the current/voltage derivative dj/dV of the probe must be measured at space potential V_s. This involves determining V_s, roughly the voltage at which $d^2j/dV^2 = 0$. The positive ion concentrations on the other hand are given in terms of the effectively constant value of dj/dV for ion retarding potentials $< \frac{1}{2}Mv_s^2$ where M is the ion mass and v_s the satellite velocity. With a mixture of different ions the technique may be employed to measure the concentrations and temperatures for each variety of ion.

The ac mode of operation has clearly many advantages for probes in satellites. Thus during a typical voltage sweep between V_s and the floating potential V_f a satellite will travel between 0.1 and 1 km so that the probe characteristics may be appreciably distorted by changes in the ambient plasma. It may also be distorted by modulation due to satellite spin. In the ac mode of operation the measurements are made in a considerably shorter time so that any distortion of the characteristics has much less effect.

As far as concentration measurements are concerned the same arguments no longer apply but the ion concentration, depending only on dj/dV at retarding potentials less than a well-known quantity, might be more accurate than that of the electrons which require the determination of the space potential.

The Boyd-Willmore probes have been flown in a number of satellites, Ariel 1, Explorer 20 (launched 25 August 1964), Explorer 31 (29 November 1965), OGO E (4 March 1968), ESRO 1 A (3 October 1968), ESRO 1 B (1 October 1969) and ESRO IV (22 November 1972) as well as in many sounding rockets. Comparison has been made from time to time when opportunity offered between measurements made with the probes and by incoherent back-scatter techniques. Table 14.1 shows such a comparison for probes carried on ESRO IV with back-scatter observations from the Royal Radar Establishment (RRE), Malvern (latitude $\sim 52°$N).

It will be seen that the comparison, though far from precise, does show that there is quite good agreement as far as T_e is concerned but considerable discrepancies exist for n_e. On the other hand under the conditions prevailing

Table 14.1.

								ESRO IV		RRE
Date	Time	Lat.	Long.	Height	n_e	$n(O^+)$	T_e	n_e		T_e
	T	deg	deg	km	10^{10}	m^{-3}	°K	10^{10}	m^{-3}	°K
19 Dec. 72	12.31	51.4	34.4	381	7.4	10.8	3454	11.0 ± 1.5		3300 *
	14.09	52.7	9.8	376	7.96	12.6	3196	12.0 ± 1.5		3000 *
8 Jan. 73	01.50	52.3	355	342	4.17	5.7	1773	10*		–
6 Jan. 73	14.05	51.8	355.5	890	0.9	1.05	5919	$1.1 \pm$		> 3500
	15.43	52.9	330.9	890	1.05	1.05	5180	$1 \pm$		> 3500
14 Jan. 73	3.47	52.3	320.6	456	1.2	2.08	2223	3.5		2400 ± 450
	3.48	53.2	320.5	462	3.16	–	2686	3.5		2400 ± 450

*Extrapolated from measurements beginning 30 min later.

O^+ was the dominant positive ion and $n(O^+)$ should be equal to n_e. It will be seen from the Table that in fact $n(O^+)$ is in better agreement with the back-scatter measurement than is the value given from the electron probe.

The comparisons which have been made between back-scatter and data from Ariel 1 and Explorer 21 in general support the conclusions from Table 14.1 that the probe temperature measurements are in agreement with the back-scatter results. This applies both to T_i and T_e. Also, in conditions under which one positive ion is dominant, n^+ appears to give a better measurement of n_e than that derived from the electron probe.

Figure 14.3 shows the dependence of T_e on latitude as determined[6] from the probe on ESRO 1. These apply to conditions in the quiet phase of a solar cycle. According to the theory, in which the energy input to the electrons arises from photoionization, T_e should fall rapidly to the neutral gas

Fig. 14.3. Contours of electron temperature for the altitude range 600–700 km during magnetically quiet times, obtained from observations made from the satellite ESRO 1.

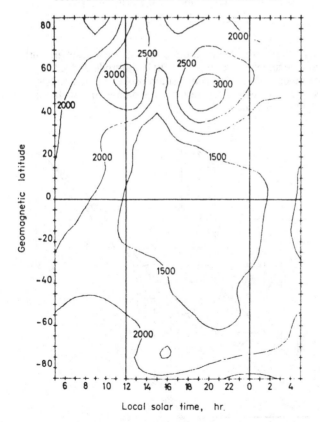

Local solar time, hr.

Fig. 14.4. Comparison of electron temperatures at 40° magnetic latitude derived from ESRO 4 and Ariel 1 data.

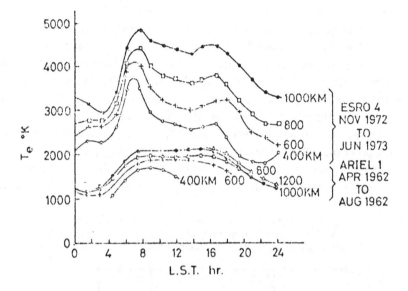

Fig. 14.5. Variation of the composition of the topside ionosphere with local time and geomagnetic latitude at 700 km altitude derived from observations made with the ion composition probe on Ariel 1. The heavy lines represent regions where there are data values within 1 h 5° lat. and 100 km altitude.

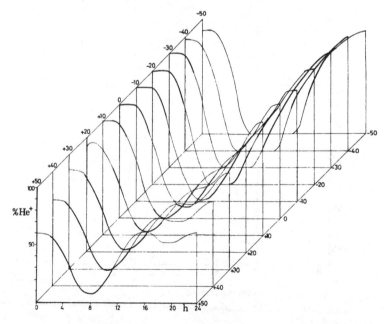

temperature soon after sunset. Back-scatter observations confirm that this is so at altitudes of 200 km but Figure 14.3 shows that it certainly does not apply at much greater altitudes. The existence of some source of energy for the electrons at night is implied.

Figure 14.4 gives a comparison between measurements of T_e made from the UCL (MSSL) equipment on ESRO IV with those from Ariel 1. Although both sets of measurements were made during the quiet phase of the Solar cycle they nevertheless differ quite considerably.

A great number of observations of T_e have been made using the Boyd–Willmore ac mode operated probes and these have contributed substantially to the understanding of the behaviour of the topside ionosphere.

Turning to ion composition and temperature, Boyd and Laflin analysed in detail the results obtained with the positive ion probe on Ariel 1. Figure 14.5 shows the variation of the composition of the topside ionosphere with local time and geomagnetic latitude at an altitude of 700 km. At these altitudes the only positive ions present of any significance are He^+ and O^+ and the percentage of He^+ is the variable plotted. A corresponding plot for the total ion content is shown in Figure 14.6. A great many results of this kind have been obtained from probes in the other satellites referred to above as well as in many sounding rockets.

Detailed world-wide information on the electron concentration in the

Fig. 14.6. As for Fig. 14.5, but now referring to the total ion content.

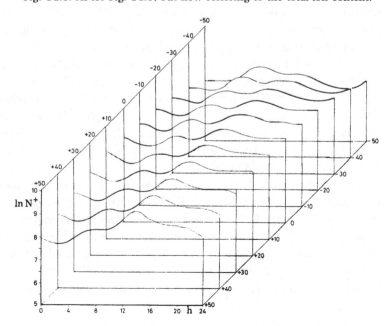

topside ionosphere has come from observations made by Sayers and his group using the capacity probe described in Chapter 3, p. 31 (see also Chapter 5, p. 87) and from the analysis of the great amount of data obtained from topside sounder satellites. As mentioned earlier, a major contribution to this analysis has been made by UK scientists at the Appleton Laboratory and at British universities.

The Sayers capacity probe has been flown in the satellites Ariel 1, Ariel 3, Ariel 4 and Fr 1, the last being a co-operative French–American satellite similar to the Anglo–American Ariel series, as well as in many sounding rockets. Checks made by comparison of results obtained by the probe with those obtained in other ways, such as ground-based sounding have shown that it gives reliable results.

The topside sounding satellites were a great success and provided a great amount of data calling for a collaborative effort in carrying out the analysis of the data to obtain in the first instance electron concentration as functions of altitude, geomagnetic and geographic location, time of day and magnetic activity. From such data scale heights (see Chaper 13) could be derived[7] and, in conjunction with observations by other techniques of electron and ion temperatures, used to derive mean ion masses or vice versa.

From observations of n_e with the capacity probe in Ariel 1, Sayers and his colleagues discovered ionization 'ledges' in the topside ionosphere aligned along certain magnetic field lines. King et al.[8] at the Appleton Laboratory found similar features from their analysis of data from Alouette 1. These authors used the data to make a thorough study of the so-called equatorial anomaly – the fact, under certain conditions, that the ion concentration as a function of latitude does not show a peak at the magnetic equator but at magnetic latitudes of 10° or so on either side. Figure 14.7 shows a typical set of results for n_e as a function of magnetic dip latitude at altitudes ranging from 440 to 710 km. The anomaly is clearly seen at 550 km and below. It will be seen also that the maxima of n_e lie along the magnetic field line. Figure 14.8 shows a similar set of results obtained by Thomas and Rycroft[9] from analysis of data obtained about six months later. The anomaly is less marked but has already appeared at 400 km.

Thomas and Rycroft[9] derived vertical scale heights H_v from their data on n_e. At magnitude dip latitudes of 25°, H_v for an isothermal plasma at temperature T is equal to $2kT/\bar{m}_i g$ where k is Boltzmann's constant, g the acceleration of gravity and $\bar{m}_i$ the mean ionic mass. A model for variation of T with dip altitude between 800 and 1000 km was constructed on the basis of measurements from Ariel 1 and other satellites, and using this $\bar{m}_i$ was derived as a function of dip latitude, two examples of which are shown in

Fig. 14.7. Contours of electron density at real heights above ground for Singapore at 10.00 local time, derived by King *et al.* from analysis of data obtained with the topside sounding satellite Alouette. The magnetic field line (MFL) calculated from the magnetic dip is drawn in.

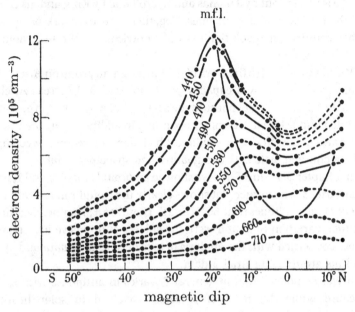

Fig. 14.8. Variation of electron density with latitude in the topside ionosphere on 7 March 1963, at local afternoon, derived by Thomas and Rycroft, from analysis of data obtained with the topside sounding satellite Alouette.

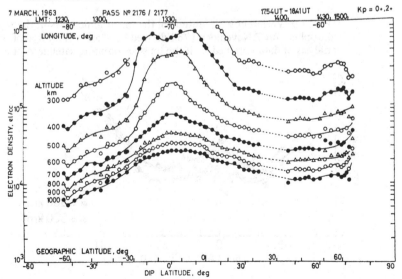

Figure 14.9. These results agreed well with those obtained from measurements with the positive ion probe on Ariel 1 (see Figure 14.5).

It is impossible to do justice to the vast amount of data processing and analysis carried out by Thomas and Rycroft and by King and his colleagues on the topside sounder satellites. Together with the work of scientists in other countries it revolutionised our knowledge of the behaviour of the topside ionosphere.

We have referred to the work of the Birmingham group on Ariel 1, but the same group also flew a capacity probe on Ariel 3. Whereas Ariel 1 was launched during a quiet phase in the solar cycle, Ariel 3 began operations at a period quite close to the solar maximum. In addition to n_e, measurements were also made of T_e by a technique which does not require sweeping of the whole characteristic between floating and space potential.

In Chapter 5, Figure 5.15, we showed an orbital plot of n_e and T_e obtained from Ariel 3. The former plot shows clearly the troughs in ionization which occur at mid-latitudes near 60° N and S. These are a prominent feature of the latitude variation of n_e and were the subject of intensive investigation by those concerned with the analysis of topside sounding data and by Sayers and his group using Ariel 3 data.

It will be noticed that in general T_e varies in antiphase with n_e. This is because while the rates of processes involved in solar heating are independent of n_e, cooling processes occur at rates proportional to n_e.

Many intense magnetic storms occurred during the observational lifetime of Ariel 3. Figure 14.10 shows the effect of a specially intense storm which

Fig. 14.9. Variation with latitude of the mean ionic mass in the topside ionosphere on 7 November 1962 derived by Thomas and Rycroft from analysis of data obtained with the topside sounding satellite Alouette.

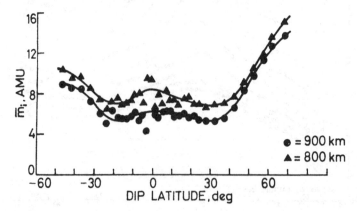

occurred on 25 and 26 May 1967, on the mid-latitude trough in the southern hemisphere.

The Birmingham group, apart from obtaining a wealth of data on n_e and T_e from Ariel 1, Ariel 3 and Ariel 4, also made a number of successful measurements from Fr 1.

2.2 *The ionosphere between 150 and 200 km*

As explained earlier, until the advent of the Atmospheric Explorer Satellites in 1972, the only space vehicle available for studies of the ionosphere below 200 km was the vertical sounding rocket. The UK groups, those at UCL and Birmingham, both of whom used *in situ* probing, and the

Fig. 1·4.10. Magnetic storm effects in the early evening ionosphere 24 to 27 May 1967 observed by Sayers *et al.* using the capacity probe on Ariel 3.

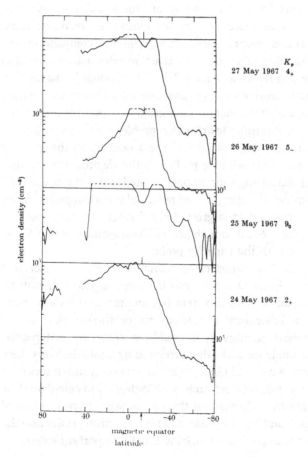

group at the University College of Wales, Aberystwyth, who used propagation methods, carried out an extensive programme using Skylark, Centaure, Black Knight, Nike-Apache, Dragon and Petrel rockets. Much attention has been devoted to studies in the auroral zone which we shall describe later on p. 323.

We can only select a few of these experiments for further discussion here. One of the most interesting was carried out on 3 April 1969 with a sun-stabilized Skylark rocket and invloved co-operation between UCL (MSSL) and Birmingham. The importance of simultaneous measurement of related ionospheric and solar parameters was clearly realized by that time but this experiment was the first one to attempt such measurements in relation to ionospheric theory in the 150–200 km range. At these altitudes diffusion effects are unimportant and the properties of the ionosphere are determined by the solar UV spectrum, the neutral atmospheric composition and the reaction cross-sections for a wide variety of atomic collision reactions including photoionization, electron impact ionization, recombination, charge transfer and other ionic reactions. Though far from complete, quite a considerable body of information on these reaction rates had developed at the time. Willmore and his colleagues[10] therefore developed a payload to make the following measurements: (a) absolute solar flux in wavelength range 80 Å–1050 Å with a wavelength resolution of 10 Å using a scanning EUV spectrometer, (b) atmospheric extinction profiles of five strong lines in the solar EUV spectrum using a fixed EUV spectrometer, (c) the ambient electron density using a capacitance probe, (d) the electron temperature using spherical and plane Langmuir probes operated in the ac mode and (e) the total positive ion density using a fixed potential spherical probe. As an additional check on the n_e data, a ground-based ionosonde was operated close to the launch site over the launch period. Good agreement was found with results obtained with the capacity probe.

The aim was to use the atmospheric extinction profiles to determine neutral atom and molecule densities and the neutral gas temperature. Thence using the absolute solar flux data (a), and photoionization cross-sections given from laboratory measurements or theory, the rate of production of different primary ions could be calculated. Moreover, allowance could be made for ionization produced by photoelectrons since their rate of production could be obtained in the same way and it was only necessary then to use data on electron impact ionization to calculate their contribution. Having then determined the rate of primary production of different ions O^+, O_2^+ and N_2^+ by photo and photoelectron production the equilibrium concentrations of each ion, including the important secondary ion NO^+, could be calculated from known rates of ionic reactions and electron recombination. If the input from laboratory experiment and theory is valid, then the sum of these concentrations should equal the measured electron concentration. In fact, rather indirect methods had to be used to determine the neutral composition and temperature but agreement within 30% was obtained for altitudes extending from 130 km to the rocket apogee at 270 km. In view of the complexity of the collision reaction scheme which is required, this represented quite good agreement although a much more severe test would have been possible if the ion composition had also been measured as in the Atmospheric Explorers which followed.

Willmore and his colleagues also used data on atomic reaction rates to calculate electron temperatures from the observed photoelectron spectra and neutral composition and obtained results which correlated with the direct measurements, at least above 140 km.

2.3 Mid-latitude sporadic E layers

We have remarked earlier that, following the observation of sporadic E in the first successful scientific flight of Skylark (see Chapter 3, p. 46), British scientists devoted much attention to exploring this phenomenon and in the end carried out quite elaborate co-operative rocket flights to check the proposed explanations due to Whitehead.[11] Through the effect of horizontal winds moving the ions across the horizontal component of the earth's magnetic field perpendicular to the wind direction, the ions are subjected to a vertical force. Across a region of strong wind shear the sense of this force will reverse in going through the region. Hence, if the force is upward just below the wind shear region it will be downward just above, so the ions will be compressed into a narrow layer which constitutes sporadic E. The relatively long lifetime of the ions before recombination is understood from the mass spectrometer measurements of Narcisi and his colleagues according to which the ions are metallic and hence atomic so that they recombine very slowly. To facilitate observations of sporadic E the UCL (MSSL) group introduced a simple fixed potential positive ion probe which was flown on many occasions. Two were flown in a collaborative payload[12] in a Skylark rocket launched on 2 March 1971 from Woomera. In addition there was included a rubidium vapour magnetometer (RSRS), two electric field probes (RSRS and Birmingham) and barium and TMA release canisters (UCL). The rocket was launched at twilight so that the drift of the vapour glow trail from the TMA could be photographed from the ground. Figure 14.11 shows the observed ion density profiles on the upleg and downleg of the flight, a sporadic E layer showing up clearly, with peak close to 107 km altitude. Figure 14.12 shows the vertical ion velocity profile calculated from the wind velocity observed from the TMA trail. It is seen to show a minimum close to 107 km but this would not be effective as the vertical velocity does not change sign. However, from the barium release observations and measurements made with the two electric field probes, which gave concordant results, electric fields were determined which contributed significantly to the vertical velocity of the ions as seen from the full line curve of Figure 14.12. With this curve the required conditions are met and we would expect a sporadic E layer located at the point of zero vertical velocity, 107.6 km, as observed. While confirming the theory in general terms this experiment shows that one must take account of electric fields as well as neutral wind effects.

Fig. 14.11. Sporadic E altitude profiles observed on the upleg and downleg of the flight of Skylark SL 922 at Woomera on 2 March 1971.

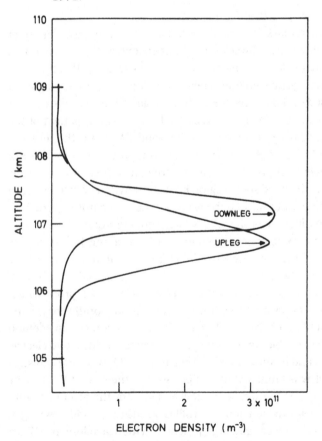

2.4 *The equatorial electrojet*

Observations of the equatorial electrojet were made in collaboration between Indian and British scientists, forming part of the Commonwealth co-operative programme. These involved magnetic field measurements carried out with magnetometers aboard rockets launched through the electrojet from the Thumba rocket range in South India (magnetic (dip) latitude 0.47°S) in February 1972.[13] Figure 14.13 shows current density altitude profiles obtained at four different times under quiet magnetic conditions. It will be seen that the general shape of the profiles is similar on all four occasions. The total current measured was found to vary in proportion to the ground magnetic variations. In the same campaign[14] a Petrel rocket carrying a rubidium magnetometer was launched within the first hour of the main phase of a world-wide magnetic storm. The contribution from the equatorial electrojet to the magnetic field depression at ground

Fig. 14.12. Vertical velocity of the ionospheric plasma as a function of altitude calculated from wind data only and from wind data with inclusion of electric field effects, for the flight of SL 922 at Woomera on 2 March 1971.

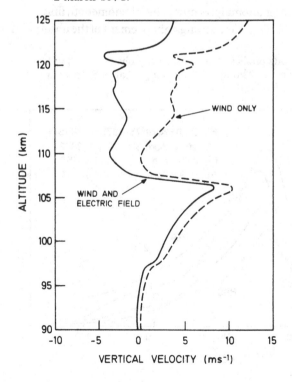

was found to be negligible. The same result was found by Burrows in a Skylark flight over Woomera (mag. latitude – 41°) during a weak magnetic storm. No significant perturbation of the normal S_q current was observed.

2.5   The D region

The D region of the ionosphere is highly variable. Under 'normal' conditions the primary source of ionization in the region is due mainly to solar Lyα radiation penetrating to the relatively low altitude (< 90 km) through an atmospheric absorption window and ionizing the minor constituent nitric oxide, with some contribution from solar X-rays ionizing the major constituents O_2 and N_2. When a solar flare occurs, the intensity of hard X-rays is markedly increased, thereby increasing the rate of electron production at quite low altitudes. Examples of this effect were observed from Ariel 1 (see Chapter 5, p. 90 and Figure 5.9). The disturbance of the D region by solar flares is not a latitude-dependent effect. However, at high latitudes, disturbances occur due to the precipitation of ionizing electrons into the atmosphere from the magnetosphere or to energetic bursts of protons from the sun. We shall say something about observations of these phenomena on p. 323. At middle latitudes a further disturbance of the D region, known as the winter anomaly, occurs. This phenomenon, first noted by Appleton in 1937, is observed as a strong enhancement of the absorption

Fig. 14.13. Altitude profiles of current density observed during four rocket flights from the Thumba Rocket Range (Lat. 8.52°N Long. 76.87°E, dip. lat. 0.47°S).

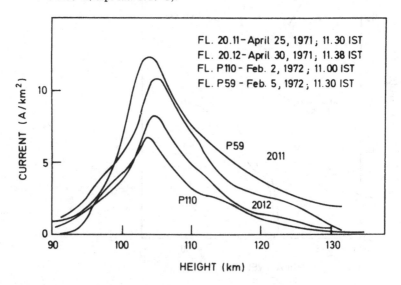

of radio waves at mid-latitudes in the winter for periods of one or two days, an effect which must be due to a marked increase in electron content in the D region.

Figure 14.14 shows profiles of electron concentration covering the D region, obtained from measurements made from Petrel rockets, launched from S. Uist by the group from the Appleton Laboratory[15] (see p. 322), which exhibit the seasonal variation between summer and winter. Even under these average conditions, the electron concentration below 85–90 km is substantially greater in winter, especially at lower altitudes. These features are more marked under winter anomaly conditions. As an example we show in Figure 14.14 electron concentration – altitude profiles obtained from the Petrel flights on days during which the absorption is very strong, corresponding to winter anomaly conditions, when it is less strong and when it is negligible. Many Petrel rockets to investigate the winter anomaly were also launched from South Uist for the University College of Wales.[16]

Fig. 14.14. Observed profiles of electron concentration in the D region at S. Uist obtained using Petrel rockets:
I 3 Dec, 1971 strong absorption, II 1 Dec. 1971 weak absorption, III 6 Dec. 1971 very weak absorption, IV 26 June 1971 weak absorption.

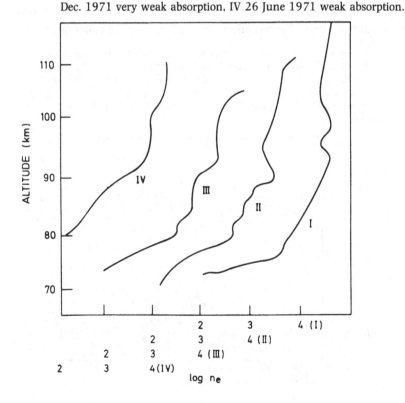

A great deal of research effort has been devoted to determining the dynamic and chemical factors which give rise to the anomaly. It seems likely that the increased ionization under these conditions arises from increased concentration of nitric oxide but it is necessary to understand how these increases occur. To investigate these questions it is essential to undertake organized rocket programmes in which as many relevant quantities as possible are measured in the same flight, and others in nearly simultaneous flights, over the periods before, during and after winter anomaly onset. At the same time complementary ground-based observations should be made.

For instance a co-ordinated launching of three Petrel rocket payloads from S. Uist took place in November 1974, instrumented by the Max Planck Institut Heidelberg, the Appleton Laboratory and the University College of Wales.[17] The results showed that throughout the height range 70–115 km, the profiles of enhanced electron density on the winter anomaly day were practically coincident with that of NO$^+$. They pointed strongly to the conclusion that winter anomaly absorption occurs as a result of a greatly enhanced concentration of NO, particularly at D region heights.

An extensive programme[18] was organized by scientists from Western Europe during the winter of 1975–76. Twenty-two scientific groups carried out 38 different kinds of observations from 47 rockets, 33 balloons and 12

Fig. 14.15. Comparison of winter anomaly activity and temperatures at 80 km. Upper full line curve – Infra-red flux measured by the pressure modulation radiometer of the Nimbus 6 satellite. The broken line indicates the long-term temperature trends. Lower full line curve is the strength of the anomaly measured daily by means of radio absorption.

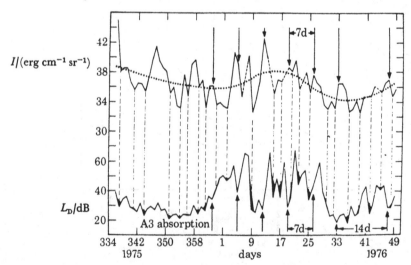

ground stations, the rockets being launched from El Arenosillo in Spain (37° 6′ N, 6° 44′ W). Measurements were made of the neutral atmospheric temperature winds and composition including minor constituents, electron and ion concentration and ion composition.

There is no doubt that atmospheric temperature and wind conditions are important for onset of anomaly conditions. Figure 14.15 shows a comparison between the observed ionospheric absorption and the atmospheric temperature near 80 km as measured with the pressure modulation radiometer on the satellite Nimbus 6 (see p. 285) over the period of the programme. In both cases, wave-like structures appear, the minima in which are well correlated. The mean variations are also very similar but the correlation is less marked between the maxima.

In addition, 31 rocket launches included equipment to measure wind profiles.[19] Of these, three carried lithium release payloads for measurements of wind speed over the altitude range from 90–130 km using the technique developed by the UCL group and D. Rees (see Chapter 13, p. 281) and were operated successfully by Rees and Scott. It would be out of place here to discuss the significance of the atmosphere structure measurements for the understanding of the winter anomaly but they showed clearly that dynamical and temperature effects are very important.

2.6 The aurora and the polar ionosphere

The polar aurora arises largely from the impact excitation of atmospheric atoms and molecules by electrons precipitated into the high-latitude atmosphere from the magnetosphere at locations within the so-called auroral oval. Some contribution also comes from excitation by solar protons with incident energy of some MeV – this would extend into the polar cap. A bewildering variety of auroral forms occur and the phenomena are highly variable. Under quiet magnetic and hence quite solar conditions they will be weak or absent, becoming much brighter when the sun is active.

Because of the great variability of auroral phenomena it is an advantage to be able to launch suitably instrumented rockets from ranges close to the auroral zone into auroral displays, to measure energy distributions of electrons and protons as functions of location and time, magnetic field to determine current distributions, electric fields and flow patterns of neutral wind and ionization. UK scientists have devoted much effort to studies of this kind taking advantage not only of the ESRO launching site at Kiruna in Sweden but also the Norwegian range at Andøya. Over the periods 1973–74 and 1976–78 special High-Latitude Rocket Campaigns were organized during which co-ordinated rocket flights took place but there have been many other rocket experiments carried out, involving co-operation between UK groups with others from the UK, or from other European countries.

On October 1, 1968 a group from the Radio and Space Research Station led by D.A. Bryant carried out a joint experiment with the Norwegian Defence Research Establishment and others in a Nike-Tomahawk rocket launched from Andøya over an auroral arc to measure fluxes of electrons and energies between 0.4 and 30 keV (see p. 305). It was found that the mean energy of the electrons decreased steadily as the rocket moved from the centre to the northern boundary of the arc. Further measurements of this kind were made by the RSRS group using the set of three electrostatic analysers at energies of 4, 6 and 10 keV together with a detector for electrons with energies greater than 1 keV (see p. 305), flown on an ESRO Skylark rocket from Kiruna on 29 October 1970. Again they observed large changes in the electron energy spectra as the rocket moved through an auroral arc, the mean energy being largest towards the centre of the band. It was deduced that the arc occurred at the boundary between the two different magnetospheric plasma.

Experiments were also carried out by the RSRS group in which the rockets were launched through pulsating aurorae. Thus in a flight on 6 March 1967, again in a joint experiment with the Norwegian Defence Research Establishment, a Nike-Apache rocket was launched into a pulsating aurora. A series of 4-second pulsations in the electron intensity were observed at two different electron energies (4 and 10 keV). It was noted that the higher-energy pulsations occurred 0.55 s earlier than those at lower energy.

One of the difficulties in studying variable phenomena from a space vehicle is to determine whether the changes observed are spatial or temporal in character. A major step forward in this respect was first taken in experiments carried out in two ESRO Skylark rockets launched from Kiruna in 1972 in a collaboration between the University of Toulouse, the Max Planck Institute, Lindau and the RSRS. In these rockets part of the payload was ejected forward at a speed as high as $20 \, \mathrm{m \, s^{-1}}$ so that the incoming stream of auroral particles could be sampled simultaneously at two points along almost the same trajectory.

The first rocket was launched in March 1972 into a pulsating aurora. Most changes in particle intensity were observed simultaneously by the main and ejected payload showing that they were temporal in nature. In agreement with earlier observations the time occurrence of the pulsations was earlier the higher the energy, consistent with an origin of the pulsations at about 4.5×10^4 km distance, locating it at the geomagnetic equator on the magnetic field line which enters the atmosphere in the auroral zone. Similar results were obtained by the RSRS in later flights. In one flight in 1974 it was found that the pulsations originated at distances varying between 4×10^4 and 9×10^4 km, the larger distances suggesting that the geomagnetic field lines between the equator and the auroral zone had been temporarily stretched to twice their normal length. During the 1976–77 High-Latitude Campaign, RSRS flew equipment to measure electron energy spectra in the range 0.5–25 keV and also of positive ions in the same energy range. In the second flight of a Skylark 12 rocket, to an altitude of 690 km into a pulsating aurora, it was found that pulsations in the positive ion flux originated at the same time and place as those of the electron flux – near the equatorial crossing of lines of force from the auroral zone.

Returning to observations made in steady auroral arcs, the second ESRO Skylark rocket with the ejectable payload was launched in September 1972 into such an auroral feature and in this case changes in intensity were not observed simultaneously, showing that they were spatial in character. Further flights during the 1973–74 campaign confirmed earlier observations and supported the conclusion that the arc appears at the boundary between plasmas of different density and temperature. The first Skylark 12 to be launched (Chapter 3, p. 34), from Andøya reaching a peak altitude of 715 km, carried the RSRS equipment as above to measure both electron and positive ion spectra and was fired through a quiet arc. It was found that in general when electrons were accelerated, positive ions were retarded suggesting the action of an electric field. However, on some occasions, both were accelerated.

A third Skylark 12 was launched on 13 October 1977, during the 1977–78 campaign, into a diffuse aurora exhibiting some pulsations, reaching an altitude of 717 km, followed a little later by a Fulmar rocket reaching 245 km which penetrated a bright, active aurora. Both carried the RSRS electron and proton-energy measuring equipment, among other experiments. Electrons with energies at and below a pronounced peak in the energy spectrum were strongly aligned along the geomagnetic field.

While we have concentrated in some detail on the RSRS experiments, a great number of other auroral investigations have been carried out by British scientists. For example, J.J. Sojka and W.J. Raitt of UCL (MSSL) flew their equipment for measuring the energy spectra of suprathermal electrons (5–500 eV) (see p. 305) in three rocket flights during the 1973–74 High-Latitude Campaign and detailed energy spectra were obtained. These were compared with values calculated from their production as secondaries by the energetic electrons observed in the same rocket flights by equipment flown by the RSRS and Southampton University. Good agreement was found for primary electrons with energy greater than 1 keV. Further flights of the MSSL equipment in the improved form similar to that[20] operated on GEOS I and II were carried out by J.F.E. Johnson during the 1976–77 campaign. Bursts of superthermal electrons were observed when the rocket was emerging from a stable auroral arc and on other occasions.

It is not possible to do justice to all the experimental groups who have carried out auroral experiments, such as the Birmingham group measuring n_e and T_e, the Sussex group with their wave experiments and the Southampton group measuring electron energy spectra as well as taking TV pictures of aurora which showed, for example, that the discrete bright auroral arcs into which rockets were flown are much narrower in the N–S direction than the regions of particle precipitation measured from rockets. We shall say something about observations of vlf noise in relation to aurorae in 3 while we have already referred to neutral wind observations in the auroral zone carried out in connection with High-Latitude Campaigns in Chapter 13, p. 280.

A great number of observations of the polar ionosphere have also been made by British scientists from satellites such as Ariel 3, Ariel 4, ESRO 1A, ESRO 1B, ESRO IV and GEOS 2, sometimes in conjunction with observations from rockets or from the ground.

3   Low-frequency radio noise

Low-frequency electromagnetic waves generated in the atmosphere propagate along magnetic field lines through the plasmasphere and under some conditions, the plasmapause. Observation of their propagation, apart from its interest in helping to determine the nature of the sources of the noise, is also likely to yield useful information about the behaviour of the plasmasphere and plasmapause. Extensive studies of both vlf and elf radiation using data from rockets and satellites as well as from ground-based equipment have been carried out by the group under T.R. Kaiser at the University of Sheffield, partly in collaboration with the group at the University of Southampton under M.J. Rycroft.

A very comprehensive survey of low-frequency noise above the ionosphere was carried out by the Sheffield group using equipment on the satellites Ariel 3 (see Chapter 5, p. 100)[21] and Ariel 4 (see Chapter 5, p. 102).[22] In the former they made observations at three frequencies 3.2, 9.6 and 16 kHz in the vlf range and this was extended in Ariel 4 to include two elf channels at 0.75 and 1.25 kHz and one vlf at 17.8 kHz.

The low-frequency noise observed arises from man-made sources such as the transmitters at 16 kHz at GBR (Rugby), at 15.5 kHz from NWC (Australia) and at 17.8 kHz from NAA (Maine, USA), from thunderstorms (spherics) and from disturbances due to entry of charged particles into the ionosphere.

The natural emissions are broadly classified as broad-band hiss, chorus and impulses. To assist in distinguishing these types of signal, the output from the broad-band channels gives the peak, mean and minimum signals observed in successive 28 s sampling intervals. An additional narrow-band, minimum-reading channel was included at 16 and 17.8 kHz to identify the emission from the Rugby and Maine transmitters. It was noted by comparison of Ariel 3 vlf observations with those of medium frequency made from the same satellite by Horner and Best (see p. 100) which were directed towards thunderstorm detection, that thunderstorms produce high-peak impulsive signals at low frequencies. For such a signal to be present it was taken that the peak/mean ratio of the signal should exceed 30 dB and the main signal should exceed noise at 3.2 kHz. Impulse counters provided by the Appleton Laboratory were included in the 3.2 and 9.2 kHz channels to count the spherics impulses above three chosen signal levels and hence obtain more information about the amplitude distributions in the pulse.

Figure 14.16 shows the wave field observed from the GBR transmitter during a north–south pass of Ariel 3, to the west of Rugby but quite close to the magnetic meridian. In terms of the invariant latitude Λ the signal rises steeply at $\Lambda = 60°$ N to a maximum at $\Lambda = 52°$ slightly south of Rugby, falls to a minimum at the magnetic equator $\Lambda = 0$ then rises to a low plateau in the southern hemisphere and thence to a well-defined maximum near $\Lambda = 45°$ S before falling sharply near $\Lambda = 60°$ S. The maximum in the southern hemisphere is due to one-hop whistler propagation. At night it is displaced by approximately 6° to the west of the geomagnetic conjugate point, a result also found for night-time propagation from the NAA transmitter. Following dawn however the displacement is to the east. These displacements are probably due to longitudinal gradients in the electron density in the magnetosphere. On 11/12 January 1972 the conjugate maximum from NAA transmissions occurred at invariant latitudes 5–10° higher than usual.

Evidence of the strong influence of man-made ground-based sources is also very clearly present in the geographical distribution of the observed emission at the lower frequencies of 3.2 and 9.6 kHz in the northern summer.[23] Figure 14.17 shows the global distribution of the frequency of occurrence of emission at 3.2 kHz, observed in 1972 during northern summer (days 117–213) with intensity greater than 4.8×10^{-16} w m^{-2} Hz^{-1}, the background intensity being normally 5 to 10 dB lower. Most striking features are the strong maxima over North America and its geomagnetic conjugate in the Southern hemisphere. There are also three other maxima over Eurasia indicated as N_1, N_2, N_3, a sharper, weaker maxima over the UK and one to the magnetic north of Japan. The Southern region of strong emission marked X is probably conjugate to N_1. All of the northern maxima are associated with regions of high industrialization involving strong generation and and consumption of electric power and all of the emission zone maxima in both the northern and southern hemisphere are located between the magnetic shells with $2 < L < 3$.

It seems very probable that the regions of strong emission arise from power line harmonic radiation propagating upwards which is amplified by interaction with electrons precipitating between the magnetic shells $L = 2$

Fig. 14.16. The magnetic component of the 16 kHz wavefield observed on an NS pass to the west of Rugby on 29 May 1967, 4.35 to 6.00 UT. Heavy line = minimum reading from 16 kHz narrow band channel. Fine line = minimum reading output from 16 kHz wide band channel. Abscissae: minutes of time along the orbit, invariant latitude Λ = arcos $(L^{-\frac{1}{2}})$; λ geographic latitude; LT local time; MLT geomagnetic local time; sza = solar zenith angle at the satellite.

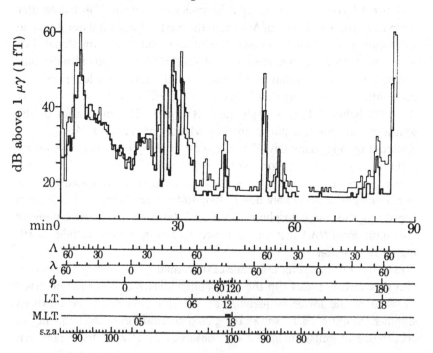

and L = 3 and propagating in the whistler mode to the magnetic conjugate in the southern hemisphere.

While a similar pattern emerges for emissions of the elf radiation, there is no evidence of enhanced emission of the elf radiation at 0.75 kHz over the industrialized regions. Pronounced elf emissions are observed on every satellite revolution, even under the quietest magnetic conditions, and cover a very large spatial extent. Assuming that the plasmasphere boundary is defined by the magnetic field line entering the atmosphere at the ionospheric trough elf zones must extend across the plasmapause without any marked discontinuity in intensity. However, when the ratio of mean to minimum signal is plotted against the invariant latitude it is found to increase sharply in passing from the low to the high-latitude side of the plasmapause. This suggests that the signal is of a steady hiss type within the plasmasphere, changing to chorus-type emission outside.

At 1.24 kHz the concentration over the industrial zones in the summer is already apparent.

Fig. 14.17. Global distribution of occurrence frequency of 3.2 kHz radiation for summer 1972 (days 117–213). Contours labelled 1, 2, - - - 10 represent occurrence frequencies 10, 20 - - - 100% - - - Contours of invariant longitude, △, X, O. Locations of Tokyo, New York and London.

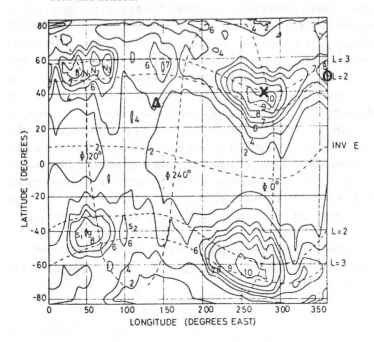

The occurrence distribution of the vlf radiation shows a strong seasonal dependence. Thus in northern winter there is little or no evidence of the maxima even over North America. This seems to be partly due to a lowered base of power-line resonance radiation and partly to greatly reduced wave amplication.

The Sheffield group using data from Ariel 3 and Ariel 4 have completed a systematic study of the diurnal as well as the seasonal and geographical variations of the elf and vlf noise. They also played a major part in developing theoretical understanding of the processes involved.

The launching of the ESA GEOS satellites has provided a rich source of further information on elf and vlf emissions in the magnetosphere. GEOS 1 did not attain the geostationary orbit for which it was intended but nevertheless provided much valuable data. However GEOS 2 was successfully launched into the desired orbit in March 1978. It could be transferred in longitude so that, in particular, the magnetic field line through the satellite reached the atmosphere in Northern Scandinavia. Under these conditions special interest attached to rocket- and ground-based observations in the auroral zone. The Sheffield group, together with the group from the University of Sussex under G. Martelli (see p. 306), has been associated with the wave experiments on GEOS 1 and 2 and is taking part in the analysis of the very extensive data obtained.

It remains to draw attention to some of the rocket experiments carried out by the Sheffield and Southampton groups in collaboration to study elf and vlf emissions using Petrel and Skylark rockets. On 10 November 1970 a Petrel rocket was launched from S. Uist with a payload which made possible observations of vlf noise in narrow-band channels from 2.2 to 5.4 kHz. At the time, chorus signals were being received on the ground. The vlf field strength recorded from the rocket exhibited a modulation with period equal to half the spin period of the rocket. This is consistent with propagation of the signals as single-plane waves along the geomagnetic field lines. Further analysis showed that the angle between the wave normal and the spin axis increased by about 10° in passing through a sporadic E layer located between 104 and 110 km altitude. In the same flight a burst of chorus signals of remarkably high frequency, 9.6 kHz, was recorded over a 7 s period.

Several other rocket flights with similar payloads were carried out subsequently, including a number during the High-Latitude Campaigns of 1973–74 and 1976–78.

4 Medium-frequency terrestrial radio noise

Already observations of terrestrial radio noise at medium frequencies had been made in the course of the cosmic radio noise experiment carried on Ariel 2 (see Chapter 5, p. 92). Ariel 3 carried a radio noise experiment specially directed towards the observation of thunderstorms from space vehicles (see Chapter 5, p. 100). The receiver frequencies were chosen in the standard bands at 5, 10 and 15 MHz where there were few authorized transmitters. Two frequencies were selected in each band on either side of the standard frequencies with sufficiently narrow bandwidth (1580 Hz) to avoid interference from the authorized transmitters. Observations made in a pass over Europe and Africa have been shown in Chapter 5, Figure 5.16. They show separate thunderstorm areas quite clearly but in general interference from transmitters proved to be more serious than expected, partly due to ionospheric effects and partly to the operation of a new standard transmitter, at the precise frequency of one receiver, which started up about the time of launch.

Ariel 4 carried a joint experiment[24] prepared by the Appleton Laboratory and the Nuffield Radio Astronomy Laboratories of the University of Manchester at Jodrell Bank. In these experiments a narrow-band receiver was used which could either sweep from 0.25 to 4.0 MHz every 16 s or operate at a fixed frequency of 2 MHz. The aim was to study the world-wide distribution of the noise bands and their correlation with the electron and proton fluxes observed in the 5 eV to 50 keV energy ranges by equipment carried on Ariel 4 (see Chapter 5, p. 103). Broadly speaking, two types of radio noise were observed.

The first are the normal noise bands in which the frequency is less than both the electron plasma frequency f_p and gyrofrequency f_B or is greater than both and less than $(f_p{}^2 + f_B{}^2)^{1/2}$. Figure 14.18 illustrates two examples, one in which $f_p > f_B$ and one in which $f_p < f_B$. In both the frequency regions concerned, the refractive index for either the ordinary or extraordinary ray is much greater than unity, suggesting that the source is Cerenkov radiation emitted by electrons of suprathermal energy. Evidence in support of this conclusion was obtained by P.A. Smith from the Appleton Laboratory who used data from Ariel 4 on the ambient electron concentration obtained from the Birmingham experiment and on electron fluxes in the range 244 eV to 10.8 keV measured by the University of Iowa experiment, to calculate the frequency spectrum to be expected and found quite good agreement with the direct observations. The noise band shows maximum intensities in the auroral zone, a result consistent with the proposed Cerenkov source.

A detailed study has been made of the distribution of the normal noise in invariant latitude and local time for the northern auroral zone. The two noise bands show significant differences in the diurnal distributions suggesting that they arise from precipitating particles of different energies, the lower frequency band with lower energy particles, 200 eV and below, and the higher with more energetic particles. This is consistent with data obtained from the University of Iowa experiment and with the detailed theory of Cerenkov radiation in a warm plasma.

The second type of noise is observed sporadically over the full frequency range covered, including frequencies which could not propagate, in the sweeping mode and is strongest over two centres, one in Eastern Europe and the other the Eastern Seaboard of the United States. The origin of this noise is uncertain though its location near the strong 14.8 kHz transmitter in Maine may be significant.

Fig. 14.18. Normal noise bands (*a*) $f_p > f_B$, $\Lambda = 77°$, geomagnetic local time = 23 h (*b*) $f_p < f_B$, $\Lambda = 74°$, geomagnetic local time = 19 h 50, observed from the Ariel 4 satellite.

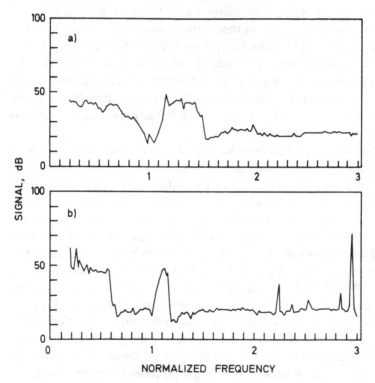

5 Particles and magnetic fields in the magnetosphere and beyond

The year 1968 was a special one for European scientists concerned with studies of the magnetosphere. No less than three ESRO satellites – ESRO II (Iris), ESRO I (Aurora) and HEOS 2 were successfully launched containing equipment specially designed to study particles and magnetic fields in the magnetosphere. These were followed by ESRO 1B (Boreas), HEOS II, GEOS 1 and GEOS 2, so there has been abundant scope for research work of this kind. Several British groups have played a very full part including particularly those at IC, the Appleton Laboratory and UCL (MSSL), while others have been concerned either with analysis of the superabundant data or in less ambitious though important experiments.

As an example of the latter, we refer to a number of experiments carried out from Petrel rockets in 1969–70, especially by the group at the University of Southampton. These were concerned with the study of particle acceleration processes during magnetospheric substorms. Two rockets were launched from S. Uist in October 1969, one during a period of strong chorus vlf emission and one in a quiet period at the same time of day. Bursts of energetic particles were observed in the former flight about one hundred or so times as intense as in the quiet period. Very much greater intensities still were observed in a third flight on 21 March 1970 during a strong magnetic disturbance. Several other flights were carried out successfully. Despite the immense data-gathering power of satellites there remains a need for rocket experiments, provided that launching can take place at short notice when a particularly interesting situation arises or when co-ordination with a satellite pass is required.

Changes in the flux of electrons with energies $> 30\,\text{MeV}$ during magnetospheric substorms were observed on many occasions by the group at the Appleton Laboratory using their equipment on ESRO 1 and 1B. Comparison with similar data from other satellites was used to help in disentangling variations in time and space. Periodicities of 1.2s were discerned in the data from ESRO I similar to those observed by the Southampton group from a Petrel launched into a vlf chorus event during the satellite pass.

The solar proton events of 18 November 1968 and 25 February 1969 were observed by both the groups from the Appleton Laboratory and from IC using proton-detection equipment in ESRO I, ESRO 1B and ESRO II. Both groups obtained evidence that protons with energies $\sim 100\,\text{MeV}$ reached the earth by direct entry through the magnetosphere at dawn and by entry through the magnetospheric tail.

During 25 February 1969 event, particle measurements were made from the ESRO II satellite by the IC group in collaboration with the Centre d'Etudes Nucléaires de Saclay, while particle and magnetic field measurements made outside the magnetosphere by the IC group from HEOS I were available. From the ESRO II data the flux of protons with energies > 27 MeV and > 90 MeV showed enhancements in the auroral zone region and in the central polar region relative to the remainder of the polar cap. These enhancements were appreciably less marked at energies of a few MeV. Possible interpretations were worked out using the HEOS I data and by trajectory computations within the magnetosphere.

The enhanced intensities of protons with energies between 100 and 300 MeV observed in both 25 February and 18 November events by the IC group were found to show a detailed structure. The results obtained by both the IC and Appleton Laboratory groups were consistent for protons of about 100 MeV energy.

The IC group also carried out measurements of particle fluxes from the satellite ISEE 3 which was launched into an orbit around the earth–sun libration point L1. Bursts of protons with energy > 100 keV in the magnetosphere were observed and a number of solar proton events studied.

Extensive observations of fluxes of suprathermal electrons and protons (0.5 to 500 eV) in the magnetosphere have been made by the UCL (MSSL) group from the ESA satellites GEOS 1 and GEOS 2 using the equipment referred to on p. 305. The fortuitous orbit for GEOS 1 proved to be quite a useful one as it permitted a full scan in local time of the low-latitude magnetosphere between 2.5 and 7 R_E where R_E is the radius of the earth. Careful allowance had to be made for the background due to photoelectrons released from the spacecraft. Large day-to-day movements of the boundary of the plasma were observed which were not connected with local time or geomagnetic activity. In the plasma trough beyond the plasmapause, the plasma is cold and particle densities of 1 to 10 cm^{-3} were observed though particles with energy up to 500 eV were sometimes detected. The particles in the plasma sheet on the other hand are so energetic that only the low-energy tails of the distributions were observed.

While the GEOS 1 spacecraft, with an apogee of 7R_E, normally remained within the magnetosphere, this region was so compressed during intense magnetic storms such as those of 21 September 1977 and 2 December 1977 that the orbit crossed into the magnetopause boundary layer. Under these conditions electron fluxes initially at energies as high as 500 eV and density 20 cm^{-3} were seen to cool to 20 eV with increasing penetration into the magnetopause. Protons were also observed with similar density and twice the mean energy and with a drift velocity of about $50\,km\,s^{-1}$ along the boundary.

GEOS 2 in geostationary orbit, is an observatory from which observations of the magnetosphere have been made synoptically and have already provided a very large volume of data from which, for example, transitions across the plasmapause, plasmasheet and magnetopause are clearly distinguished, and correlation sought with the data obtained by the other equipment on GEOS 2.

Turning now to magnetic measurements in the magnetosphere, the IC group has observed magnetic pulsations during times in which HEOS 1 was passing through the region. These were of periods ranging between 400 and 700 s, were of nearly sinusoidal wave form, mainly compressional (see Figure 14.19) and lasting 1–2 hours. Four such events were selected for detailed study. They are essentially micropulsations of the type observed at ground as of giant PC5 type, occurring quite frequently at a distance of 8 to 12 R_E, at local times between 16.00 and 20.00 but very rarely sunward of this zone. Both the polarization characteristics of the weak transverse component and the longitudinal extent agree well with the present theory of micropulsations. Other observations of magnetic pulsations have been made from the ISEE satellites.

The interplanetary magnetic field has been systematically studied by the IC group over the period December 1968 to August 1974 using the fluxgate magnetometers which they operated in HEOS I and II. The instruments operated very reliably over the entire lifetime, $7\frac{1}{2}$ years, of HEOS I and of HEOS II with an accuracy considered to be about $\pm 1\%$. Usually hourly

Fig. 14.19. Magnetic field component fluctuations, in gamma, showing micropulsations, observed by Hedgecock from HEOS 1.

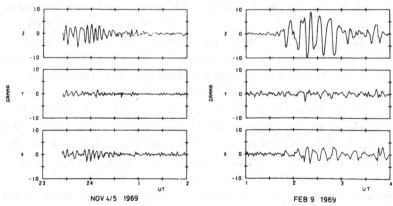

averages of the field components and vector magnitudes were calculated from about 100 instantaneous vector measurements sampled every 32–48 s. For the study of the interplanetary field, observations made inwards of the bow shock were rejected.

Owing to the solar rotation, the solar surface field extending from the solar corona is twisted into an Archimedian spiral. This was confirmed by early measurements from spacecraft which also showed that the field was divided into two or four sectors of approximately equal solar longitude corresponding to average radial field directions that were mainly either inward or outward along the spiral. The IC measurements showed that the field, over the period studied, had exhibited a two-sector structure at most times except for a few intervals lasting for a few solar rotations when a four-sector pattern was observed. Sometimes a sharp change of field occurs in passing through a sector boundary, as exemplified in Figure 14.20, and sometimes it is gradual or diffuse extending over 5–6 h.

Considerable interest attaches to the change of so-called dominant polarity[25] with heliographic latitude, taking advantage of the fact that during a year the earth moves through a range of $\pm 7.3°$ of heliographic latitude. The best fit to the rate of change of dominant polarity with latitude indicated that, on the average, the sector boundaries at 1 astronomical unit from the sun are inclined at about 12° to the solar equator. It was found that the rate of change reversed in sense during the 1970/71 period.

6 Cosmic rays

The problem of the source of modulation of cosmic rays in the neighbourhood of the earth was already a major concern at the time of the choice of payload for Ariel 1 and was one of the motivations for the inclusion of the cosmic ray payload from IC (see Chapter 5, p. 76). It was suggested at a quite early stage that the cosmic ray modulation arose from scattering by magnetic field fluctuations but it had not been possible to obtain a quantitative verification of this concept. However the measurements from HEOS I and II provided abundant data from which to study these fluctuations with a sufficiently high statistical accuracy. Theoretically, the cosmic ray modulation should correlate with that part of the power spectrum which is at a frequency in gyro resonance with the particles. That is, the spatial wavelength of the interplanetary fluctuations matches the helical path of the charged particles' orbit. It was surprising not to see this expected correlation in the HEOS magnetometer data, and the apparent correlation with the zero frequency limit is not well understood.

Between 1969 and 1972 the neutron cosmic-ray monitor at Deep River showed an increase in cosmic ray intensity of about 15%, not inconsistent with the data for HEOS I and II. However, the decrease in cosmic ray neutron monitor rate lags behind the increase in the power level of the interplanetary field fluctuations, measured from the earth, by 1–2 years suggesting that near-earth observations do not give a good measure of the average conditions at a given time in the solar modulation region.

The cosmic-ray experiment of Ariel 6 carried out by the group from the University of Bristol under P. Fowler was very successful, performing as planned over the entire life of the satellite from its launch in June 1979 until February 1982 when it was switched off. Very marked differences were

Fig. 14.20. Instantaneous vector measurements of the interplanetary field for three hours on either side of the sector boundary on 13 April 1969, derived by Hedgecock from data obtained from the magnetometer on HEOS 1. The polarity reversal occurs within a period of 5–10 min near noon.

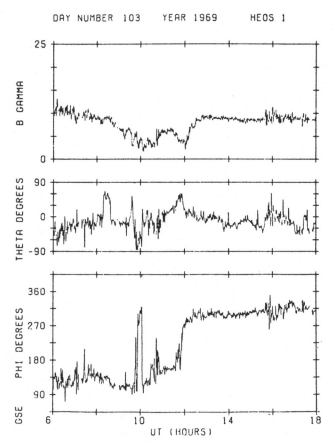

observed between the relative abundances of the elements with atomic number > 60 in the cosmic rays and in the interstellar medium. These differences are so large it must be assumed that the mixture of elements in the cosmic rays is different from that in the solar system or interstellar space.

Finally, we refer to the experiment flown by the group from Leeds University under P.L. Marsden, in the ESRO 2 satellite. It consisted of a Cerenkov scintillation electron detector to measure the flux and energy spectrum of cosmic-ray electrons and operated very satisfactorily for the first 3000 orbits between the launch of the satellite and the failure of the onboard tape-recorder. Analysis of the data showed that the primary electron-energy spectrum may be represented over the energy range 3 to 25 GeV by

$$j\ (E)\ dE = 265\ E^{-2.6}\ dE\ (\text{m}^2\ \text{s}\ \text{sr}\ \text{GeV})^{-1}$$

representing a higher intensity than in previous measurements.

15

The contribution from British space scientists to astronomy

British space scientists have devoted a great deal of attention to research in those branches of astronomy which became practicable only after the availability of space vehicles. We discuss here especially their major contributions to ultra-violet and X-ray astronomy both solar and cosmic. Most of the British work in γ-ray astronomy and in infra-red astronomy (outside the atmospheric window) has been carried out very effectively using balloons and so does not fall within the scope of the present account. The very recent successful launching and operation of the infra-red astronomy satellite IRAS points the way to a strong involvement of British space scientists in infra-red astronomy also.

We begin by discussing first solar ultra-violet and X-ray astronomy, then cosmic ultra-violet astronomy followed by cosmic X-ray astronomy and brief sections on γ-ray and infra-red astronomy.

1 Solar X-ray and ultra-violet astronomy

The first X-rays entering the atmosphere from an external source were observed by T.R. Burnight[1] in 1948 who exposed Schumann plates with thin beryllium filters, carried to an altitude of 96 km in an Aerobee rocket. On recovery and development, the plates were found to be blackened, showing that they had been exposed to radiation capable of penetrating the beryllium. The first quantitative measurements of X-rays, which clearly originated in the sun, were made by H. Friedman, H. Lichtman and E.T. Byram[2] a year later using a photon counter flown in a V2 rocket. The study of solar X-rays then developed rapidly. In 1960 Friedman[3] was able to take pinhole photographs of the sun in X-rays.

We have described in Chapter 3, p. 33, how solar X-ray studies were initiated in Britain by Boyd and his collaborators at University College

London, at an early stage in the Skylark rocket programme. From the start, E.A. Stewardson, Head of the physics department at Leicester University, who was an X-ray physicist, took a close interest. In 1960 K.A. Pounds, who had been a member of Boyd's group, joined the Leicester physics department and this was the beginning of a second major X-ray astronomy group in the UK. It was a real advantage that the Leicester group was involved in X-ray physics as well as in space experiments. From the outset there was close collaboration between the UCL and Leicester groups. Initially work on radiation detectors was carried out at Leicester, the electronic logic and payload technology at UCL, while both took part in the planning and data analysis.

The discovery of the high temperature of the solar corona was a great surprise at the time (1942). From its estimated electron temperature, close to 10^6 K, derived from the intensity ratios of spectrum lines in the visible emitted by highly ionized atoms, it seemed very likely that the corona would be a source of X-rays. The early rocket observations, including the pinhole camera pictures, showed that this is indeed the case but the emission is far from uniformly spread over the solar disc. Instead it is mainly concentrated above active (i.e. sunspot) regions which, by and large, correlate quite well with the regions of enhanced intensity of emission of 10.7 cm radio waves. From time to time a sudden outburst of energy is observed (see Chapter 3, p. 32). It appears that the structure of the solar corona is determined by the leakage of magnetic flux into the corona from the underlying chromosphere and photosphere. Great interest attaches to the determination of the electron temperature and concentration with good space, time and wavelength resolution so that diagnostic methods depending, for example, on comparison of intensities of spectral lines, or from measurements of absolute spectral intensities, may be applied. The high space and time resolution are required because many important solar phenomena such as flares are highly localized and develop very rapidly.

Much of this work required equipment to be flown on a stabilized rocket or on a satellite with controlled orientation. The proportional counter spectrometer in the payload of Ariel 1 recorded the total intensity of X-radiation from the solar disc in the wavelength range 5 to 12 Å. The occurrence of solar flare showed up clearly against this background (see Figure 5.9) but it could not be localized. The UCL and Leicester groups planned an extensive programme following Ariel 1. They proposed that X-ray cameras should be carried on each Skylark rocket as a standard auxiliary experiment. The development of the proportional counter spectrometer would continue as well as that of crystal X-ray spectrographs stable to a few minutes of arc. In anticipation of the availability of sun-stabilized Skylark rockets they planned to study the distribution of X-ray sources over the sun using a grazing incidence parabolic reflector with a suitable detector at the focus. This instrument depends on the fact that X-rays of a few keV energy are totally reflected from certain surfaces at angles < about 2°. A simple parabolic reflector will focus paraxial rays so it can be used as a light collector with a field of view defined by an aperture placed in the focal plane in front of an X-ray detector. A single reflector cannot form a high quality image but nevertheless with its use a field of view of 1' diameter may be obtained. This compares very favourably with what can be achieved with mechanical collimators. Further reference to the use of these reflectors is made on p. 367 in dealing with cosmic X-ray astronomy.

In addition, the UCL group planned to develop a small grating monochromator for observations of solar radiation in the ultra-violet, including specially the 304 Å resonance line of He II. It is to be expected that solar UV radiation will show much less variation in intensity over the solar disc so that useful observations could be obtained from unstabilized rockets. Nevertheless, it was planned to use stabilized rockets for these observations, when they became available. The UCL group also planned to monitor solar Lyα radiation using the nitric oxide ionization chamber flown unsuccessfully in Ariel 1.

Apart from its intrinsic interest the study of the UV spectrum is of special importance in the wavelength range below the Lyman limit of 912 Å. This is because absorption by interstellar hydrogen may be expected to block off these radiations from all but a few nearby stars.

Meanwhile in March 1961 the Controlled Thermonuclear Research Division at Culham had put forward a proposal, under the names of R. Wilson, D.B. Shenton and R.S. Pease, to the British National Committee for Space Research, concerning ultra-violet spectroscopy of the solar corona and the stars. We shall return to the latter below (see p. 353) and discuss only those concerning the sun, at this stage. The plan assumed the availability of at least the solar pointing Skylarks with an accuracy of stabilization of about 1° along the solar vector. It was proposed to develop a fine, servo-controlled optical alignment system to maintain the image of the sun tangentially to the slit of a normal incidence spectrograph to an accuracy of a few seconds of arc. In the first instance the wavelength range covered would be from 400 to 2000 Å while the 2000–3000 Å range would be covered in later flights. The proposed payload also included two grazing incidence spectrographs with appropriate metal films over their entrance slits to eliminate scattered light but allow transmission in wavelength bands from 400 Å to soft X-rays.

As we shall see, all of these programmes developed very rapidly and effectively. Furthermore the UK groups benefited very substantially through acceptance of their proposals to fly solar radiation experiments in US satellites, especially those of the OSO (Orbiting Solar Observatory) series and later in the Solar Maximum Mission.

Thus two proposals were accepted for launch in OSO D (later to become OSO 4). One, from UCL, was a grating monochromator to measure the intensity of the 303.8 Å HeII line while the second, jointly from UCL and Leicester, was designed to measure the total X-ray intensity in the 1.3 to 18 Å and 44 to 70 Å bands, using proportional counters developed from those in Ariel 1.

For OSO F (later to become OSO 5) UCL and Leicester proposed to study discrete sources of solar X-rays using a grazing incidence parabolic reflector and defining slits with proportional counters as detectors. This experiment was also accepted.

Finally, in this series of observatory satellites, a monochromator to measure absolutely the total flux of important solar lines – He I, He II, O, N and highly ionized Fe, designed and built by the UCL group was accepted for the payload of OSO G (later OSO 6).

Before long, preparations began to take part in the Solar Maximum Mission (SMM) planned by NASA to study solar flares during the maximum of the solar cycle in 1979–80. It was proposed to launch a satellite in October 1979 carrying seven experiments. Of these, two proposals were accepted which involved active participation by UK groups.

The first was a collaboration between UCL (MSSL) led by Culhane, an Appleton Laboratory group led by Gabriel, and the Lockheed Palo Alto Research Laboratory in the USA. It proposed to fly a set of seven plane crystal spectrometers with fine collimation ($\sim 10''$ of arc) scanning all atomic transitions important for flare and plasma diagnosis in the wavelength range 1.5 to 23 Å together with a set of eight curved ('bent') crystal spectrometers providing simultaneous fixed coverage of the wavelengths in certain important spectral intervals with very good time and wavelength resolution, at the cost of broader collimation ($6' \times 6'$ of arc). The combination of both instruments would make possible observations of flares with high spatial, time and wavelength resolution.

The second experiment was a hard X-ray imaging spectrometer proposed in collaboration between the group at Birmingham University under A.P. Willmore[*] and one at the Space Research Laboratory, Utrecht. This would, as its prime function, provide images of the active region around flares with spatial resolution of 8" of arc over a field of view of $3.5' \times 3.5'$ and a time resolution better than 0.5 s depending on the mode of operation. The photon energy/wavelength covered would be 3.5–30 keV (3.7–0.4 Å). Such hard X-rays should provide information on the basic flare mechanism. The detector would consist of an array of 576 point anode proportional counters. The Birmingham group would be responsible for the data-handling electronics, the flight power supply, the computer-controlled ground support equipment, development of the software and the X-ray test of the collimator.

These relatively very sophisticated experiments, typical of those carried in the SMM, show how the subject had progressed since the days of Ariel 1.

Results from the solar experiments in satellites, including the SMM, are discussed in p. 350).

1.1 Use of stabilized Skylarks

We have already given some account of the first payloads flown in Stage 1 stabilized Skylark rockets. In particular, the Culham group obtained beautiful ultra-violet spectra of the solar corona an example of which is reproduced in Figure 9.2. In later flights they continued to fly experiments to observe in the ultra-violet. In particular, in March 1968 they obtained for the first time solar UV spectra of both limb and disk. Such observations were aimed at deriving the temperature structure in the outer atmosphere of the sun by a method which did not depend on absolute photometry, knowledge of atomic collision cross-sections or the chemical composition of the layers. Following a second successful flight on a sun-stabilized Skylark (SL 606) on 17 April 1969 an analysis of the data obtained on this and the earlier flight showed the method to be feasible and already established that the results obtained from optically thin lines are consistent with a steep temperature change between 10^4 and 3×10^5 K on the transition region between chromosphere and corona. A further flight (SL 902) was carried out in August 1971. Analysis of the data confirmed the steep temperature rise, extended results to temperatures of order 10^6 K and provided an estimate of the height of the transition region above the visible limit as 1800 ± 80 km.

Data of this kind are very important for checking the validity of theories of coronal heating.

To extend this analysis to higher temperature regions requires the study of shorter wavelengths and hence the use of grazing incidence techniques. The ARU[5] group at Culham therefore included two grazing incidence spectrographs as well as the fine-pointing UV instrument in their solar payloads. In the flight of Skylark SL408 in March 1968, solar spectra were obtained by these instruments in the extreme ultra-violet (150–800 Å) and in the soft X-ray region (15–50 Å) which were the best quality available at that time. The rich spectra revealed, among other things, the presence of 'ghost' transitions apparently associated with the resonance lines of He-like ions such as N*VI*, O*VII* and Ne*IX*. The same group went on to identify these previously unobserved lines as forbidden intercombination transitions in the He-like ions and also 'satellite' lines associated with doubly excited states in the lower Li-like stage of ionization. Observations of these quickly became a very important diagnostic for solar studies (see below) and will also be important for cosmic X-ray astronomy.

An echelle spectrograph was also flown by the Culham group in stabilized Skylark SL 604 on 22 April 1969 and yielded the first high-resolution solar Fraunhofer spectrum in the range 2000 to 2200 Å. This was extended to wavelengths below 1900 Å in a second successful flight (SL 803) on 7 April 1970. Further interesting and important observations were made, in the latter flight, of emission line profiles of a number of lines. Those from S*iii*, S*iiii*, C*ii* and C*iv* all showed a large broadening which was certainly non-thermal. This clearly pointed to possession by the emitting ion of a large non-thermal kinetic energy of an order of magnitude greater than the thermal value. Results of analysis in these terms of the data from the flight and that from a further flight (SL 1102) on 20 March 1973, in which the spectral resolution was 0.025 Å, have been used as a basis for theoretical study of coronal heating.

Another important source of the basic data for such studies was first introduced through a collaboration between the Culham group and a group from Queen's University, Belfast, under D.J. Bradley, flying an interferometer for the first time. It was a Fabry–Perot etalon used in conjunction with an echelle spectrograph and the experiment was designed to produce high-resolution profiles of the Mg*ii* line at 2800 Å emitted from different regions of the solar disk and limb. Its first flight, in a stabilized Skylark (SL 606), took place successfully in December 1968 and was followed on 27 November 1969 by a second flight in which the optics were improved so that the wide variation of the line profiles from limb to disk and through active regions was observable in detail. For further experiments it became possible to use large, thin-envelope balloons which had been developed to reach altitudes to which the 2800 Å line could penetrate.

One of the last experiments to be flown on a stabilized Skylark launched from Woomera was a co-operative one between UCL (MSSL) and the ARU to measure the coronal helium/hydrogen abundance ratio. It used a grazing incidence telescope (see p. 341) in association with a grazing incidence spectrometer to record the intensities of selected wavelengths in the range 150–1335 Å so that measurements could be made of the resonant scattering from the corona of the Ly α line of HI at 1216 Å and HeII at 304 Å. The recovery parachute failed during the flight in May 1978 (SL 1305) but fortunately the essential data were recorded by telemetry. This experiment has been rebuilt in an enhanced payload to be carried on the second flight of Spacelab, planned for April 1985. It will systematically study hydrogen and helium abundances, temperatures and densities in coronal structures.

In parallel with the ultra-violet experiments an intensive programme of solar X-ray measurements were carried on stabilized Skylark rockets aimed towards implementation of the programmes outlined on p. 340, above. Both the Culham group[6] and the Leicester University group[7] obtained pinhole photographs of the sun in X-radiation. Figure 15.1 reproduces one such photograph obtained by the Leicester group from the second stabilized Skylark flight (SL 302) in the waveband 10–40 Å on 16 December 1964. These were the first true images of the solar corona because the earlier pinhole photographs were obtained from rockets whose roll about the solar vector was uncontrolled so that the image was smeared over a circular arc.

Again in an early stabilized flight (SL 304) launched on 5 May 1966, Evans, Pounds and Culhane[8] measured the absolute intensities of 28 identified and four unidentified solar X-ray emission lines in the wavelength band 11–22 Å using two slitless crystal spectrometers. From these results, which were among the first obtained in this spectral region, it was possible to derive information about temperatures both in the quiet corona and in active regions. As an example of later measurements also made with a crystal spectrometer, but with a collimator to limit the field of view so that only one active region was observed at a time, we show in Figure 15.2 spectra obtained over the wavelength band 14.7 to 173 Å for three different active regions by Brabben and Clencross[9] from UCL (MSSL) in a stabilized Skylark rocket flight on 10 December 1971. Further flights with increasingly sophisticated instrumentation followed.

1.2 *Solar eclipse observations*

The ARU and Imperial College London (IC) took part in collaboration with Harvard College Observatory and York University Toronto in an experiment in which an Aerobee rocket was launched from Wallops I. in the USA into the path of the total eclipse on 7 March 1970. Two slitless spectrographs covering the ranges 850 to 2150 Å and 200 to 3000 Å with absolute intensity calibration were carried and very high quality spectra were recorded in the 850 Å to 2150 Å range which are still unique. Many new spectral lines were observed for the very first time and their subsequent identification was important for establishing energy levels in a number of atomic species. The data are also of major value for studying the nature and structure of the chromosphere and corona.

Since the experiment by its very nature was a high risk operation, its great success was particularly pleasing. Some aspects seemed so fortunate that there was much post-mission discussion at the launch site. The weather was perfect and the rocket was launched on time (in a window measured in tens of seconds). Expected to have sufficient power only just to reach totality, the rocket over-performed to a statistically highly unlikely extent and carried

Fig. 15.1. X-ray photograph of the sun taken in the 10–40 Å band with a 2′ resolution, from the sun-stabilized Skylark SL 302, launched at Woomera on 16 December 1964.

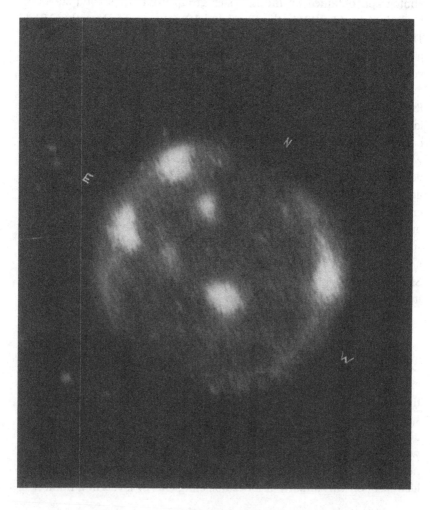

the payload exactly through the optimum trajectory for observation. The recovery parachute opened perfectly but, of course, the recovery vessel – a small coastal boat stationed at the expected descent point in the Atlantic – was completely out of touch, so much so that recovery would have normally been in doubt. However, stationed almost precisely at the actual descent point was the US Aircraft Carrier 'Guam' waiting for a US Naval payload also

Fig. 15.2. Spectra of three active regions on the sun. Lines of FeXVII at 15.01, 15.25 and 16.77 Å are marked as a,b,c respectively. The spectral feature at 16.01 Å is also indicated as d.

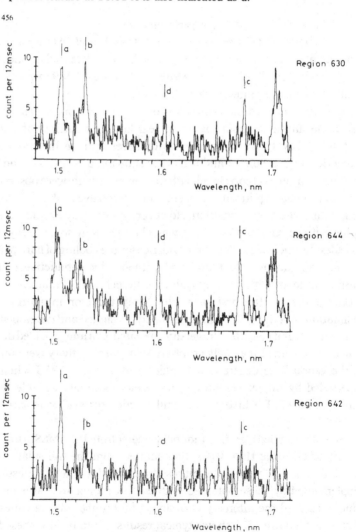

launched into the eclipse. Recovery was effected immediately and the payload returned to base by helicopter within an hour of launch.

A great deal of interesting results were derived from an extensive analysis of the flight. For example it was found that across an active region the electron density and pressure increase as the temperature increases, the hotter material existing in loops towards the centre of the region from which the main emission occurs. Also the intense coronal Lyman α was first observed, due to resonance scatter of the chromospheric emission by residual neutral hydrogen.

1.3 Results from satellite experiments

The three OSO satellites OSO 4, OSO 5 and OSO 6 carrying UK experiments among others in their payloads were all launched successfully on 18 October 1967, 22 January 1969 and 9 August 1969 respectively and in all cases the instruments operated well.

The proportional counters on OSO 4 observed X-ray emission from the full disc of the sun throughout the seven-year life of the satellite. It was found that the X-ray bursts associated with flares can be divided into two categories, one impulsive and the other showing a relatively gradual rise and fall. X-ray events correlated with microwave events were observed to be delayed by about 2 min with respect to the microwave peak, the delay being longer the softer the X-radiation. However, many X-ray events were found to have precursors and evidence was obtained that some X-ray activity precedes the microwave. The first evidence for the cooling of flare plasma by thermal conduction was found and a theory for conductive cooling of plasmas in loops developed by Culhane, Vesechy and Phillips.

OSO 5 carried the first parabolic X-ray reflector into orbit, giving collimation of perhaps $1\frac{1}{2}°$. This instrument, designed and developed jointly between UCL and Leicester University, operated well and provided data from which, for the first time, daily graphs of solar X-ray activity were published by the World Data Centre C at Boulder until January 1973 when it was superseded by higher resolution data which became available with the launch of OSO 7. The instruments, still functioning well, were turned off in July 1975.

OSO 4 also carried a grating monochrometer from UCL (MSSL) to measure the intensity of the HeII 304 Å line emitted from the full solar disk as a function of time. This instrument also functioned well and its results were supplemented by those obtained with the UK monochromator on OSO 6 which observed the intensity variations of a number of solar lines in the ultra-violet. Figure 15.3 shows typical results[10] obtained for these lines in

349

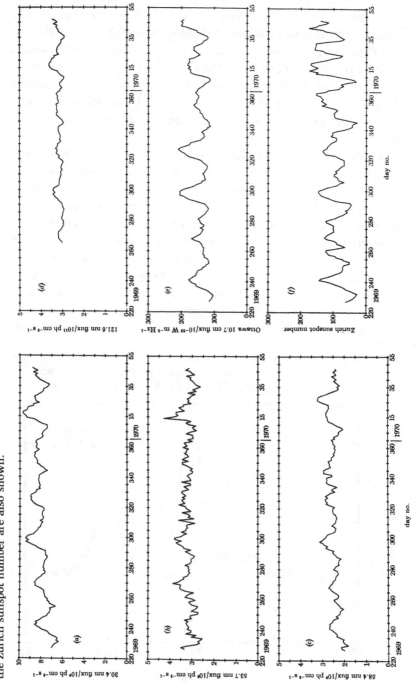

Fig. 15.3. Variation of the intensity of solar radiation of wavelength (a) 304 Å, (b) 537 Å, (c) 584 Å and (d) 1216 Å with time during 1969–70, observed with the ultra-violet spectrometer on OSO 6. For comparison the variations in the 10.7 cm flux and the Zurich sunspot number are also shown.

comparison with the observed flux of 10.7 cm microwave radiation and the Zurich sunspot number. The results for the 304 Å line agree quite well in general with those obtained with the instrument on OSO 4.[11] It will be seen that there are quite good correlations between all the variations shown.

The X-ray observations from OSO 4 and 5 were preliminary as it was necessary to improve the space, time and wavelength resolution in order to study individual flares in detail. Opportunity for UK scientists to carry out such observations came with the introduction of the Solar Maximum Mission.

The SMM was successfully launched in February 1980 and all on-board instruments operated well. In particular, the X-ray polychromator designed, constructed and operated by a collaboration between UCL (MSSL), the RAL and Lockheed Missiles and Space Co. Palo Alto Research Laboratory observed, with high spectral, spatial and temporal resolution in the wavelength range 0.15 to 2.5 nm, over 50 major flares and very many more minor ones. Unfortunately the fine pointing control of the satellite failed in November 1980 so that data could no longer be obtained with an instrument such as the polychromator with a restricted field of view.

The instrument was designed to operate in a wavelength range which should be most useful in providing diagnostic information for the determina-

Fig. 15.4. A typical Fe Kα high resolution spectrum from a solar flare on 5 July 1980 observed with the X-ray polychromator on SMM. The origin of the components F1, F2, F3 and F4 is not understood.

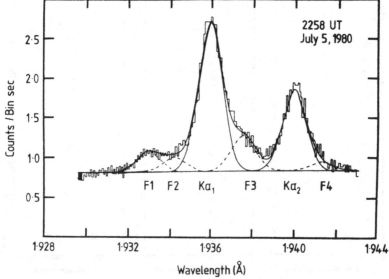

tion of electron temperatures and concentrations and other parameters in individual flares. As a bonus, it became clear from observations made in two flares that the Kα 1 and 2 lines of Fe$\textsc{ii}$ near 1.9 Å (see Figure 15.4) are mainly excited through inner shell photoionization of neutral Fe atoms in the photosphere by hard X-rays emitted from hot plasma of the flare. Prior to this work collision ionization by penetrating beams of non-thermal electrons was also thought to be possible. Using the observations of the intensity of the hard X-rays with energy $\geqslant 7.1$ keV made by the Hard X-ray Imaging Spectrometer developed, constructed and operated by the Space Research Institute, Utrecht and the University of Birmingham, a model for this mechanism has been derived and gives results in good agreement with observations. From this an ion abundance N(Fe)/N(H) of 5.5×10^{-5} is derived if all the flares studied are at a reasonable height above the photosphere. An estimate of the height of the flare may be made from the variation with heliocentric angle of the ratio of the Kα intensity to the intensity of the continuum. Again by determining this ratio as a function of time for these flares it has been deduced that in two cases the flare plasma is moving upwards with velocities of 25–30 km s^{-1}.

On the long wavelength side of the resonance lines of two electron ions such as Ca$\textsc{xi}$ and Fe$\textsc{xxv}$ 'satellite' lines appear which arise from the transitions.

$$1s^2nl \rightarrow 1s2nl.$$

The principal mechanism for production of the excited states was suggested by Gabriel and Jordan[12] to be through dielectronic recombination (inverse autoionization)

$$1s^2 + e \rightarrow 1s2pnl.$$

Theoretical evaluation of the wavelengths and intensities of the resulting 'satellite' lines was then carried out by Gabriel, Jordan and Paget,[13] Gabriel and Paget[14] and Gabriel.[15] The intensity relative to the resonance line of a particular 'satellite' increases with atomic number of the ion, being as low as 10^{-2} for O$\textsc{vii}$ but increasing to 10^{-1} for Si$\textsc{xiii}$ and to 0.5 for Fe$\textsc{xxv}$.

An alternative mode of production of the excited state of the 'satellite' line for $n = 2$ is by impact inner shell ionization of the three electron ions

$$1s^22s + e \rightarrow 1s\ 2s2p + e.$$

Gabriel and Paget also considered the theory of this process. In general, the dielectronic recombination is more important when the autoionization rate of the excited state is large and inner shell ionization when it is small. Thus the latter process becomes relatively more important for large atomic number though even for Fexxv it is considerably less effective than dielectronic recombination, assuming that the populations of the two and three electron ions are in thermodynamic equilibrium.

From the relative intensity of a satellite line arising from dielectronic recombination to that of the resonance line the electron temperature may be obtained. The corresponding relative intensity for inner shell excited satellites gives the relative abundance of three to two electron ions in the plasma. It can then be checked whether the radiating plasma is in equilibrium between ionization and recombination or not.

It is clear that observation of satellite lines in solar flare spectra is of considerable importance for diagnostic purposes. The polychromator in the SMM was chosen to operate a high resolution at wavelengths appropriate for observation of satellites associated with the resonance lines of relatively heavy two-electron ions.

Figure 15.5(*a*) shows a solar flare spectrum about the resonance line of

Fig. 15.5. Comparison of solar flare spectra of (*a*) calcium and (*b*) iron observed with the bent crystal spectrometer in the X-ray polychromator experiment computed using atomic theory. The notation for the lines is as follows:

$1s^2 2p - 1s2p^2$	$^2P^0 - {}^2P$ multiplets, a,b,c,d	$1s^2 2p - 1s2p2s$	$^2p^0 - {}^2S$	o,p
	$^2P^0 - {}^4P$ multiplets, e, f, g, h, i	$1s^2 2s - 1s2p2s$	$^2S - ({}^1P)^2P^0$	q,r
	$^2P^0 - {}^2D$ multiplets, j, k, l		$^2S - {}^4P^0$	u,v
	$^2P^0 - {}^2S$ multiplets, m, n	$1s^2 - 1s2s$	$^1S - {}^3S$	z
$1s^2 - 1s2p$	$^1S - {}^1P^0$ multiplets, w			
	$^1S - {}^3P^0$ multiplets, x, y			

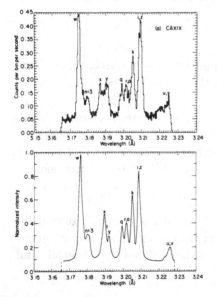
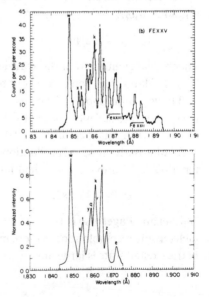

CaxIX obtained with the bent crystal spectrometer of the polychromator[16] compared with a calculated spectrum assuming an electron temperature of 12×10^6 K. The origins of the different lines are given in the Figure caption. Apart from dielectronic recombination leading to essentially three-electron spectra, the lowest forbidden transitions of the two-electron system must also be included. It is clear that the agreement between theory and experiment is very good.

For the satellite lines associated with the resonance line of FeXXV the agreement, is seen, by reference to Figure 15.5(*b*), to be less satisfactory at the longer wavelengths. This is due to the neglect of contributions from FexxIII ions in this figure. These occur also at shorter wavelengths and so, to some extent, confuse the interpretation.

As an example of the type of observation made with the hard X-ray imaging spectrometer (see p. 345) we show in Figure 15.6 intensity profiles in six energy bands of the emission associated with a solar flare on 10 April 1980.

The gradual change in the profiles as the X-ray energy increases is clearly seen in the rate at which the emission falls after a peak is reached. It is also noteworthy that there are fluctuations present before the flare onset which are more marked the softer the X-rays concerned. In Figure 15.6 the profiles arising from these different regions of the flare are shown for a soft and hard X-ray band in each case. The fluctuations are confined to the region C. It is found that the emission from region B, which shows a peak emission in relatively soft X-rays, is consistent with a thermal spectrum with a temperature of $25 \pm 5 \times 10^6$ K. On the other hand, for regions A and C which show the maximum count rates for hard X-rays, the best fit to the energy spectrum is a power law.

These examples have been selected to illustrate the high performance of the instruments and we do not attempt here to discuss further their relation to the physics of flares. But for the failure of the pointing mechanism much more would have been obtained as all the other intruments were also working very well. However, at the time of writing, SMM is scheduled for repair in orbit in April 1984 by astronauts using the Shuttle.

2 Cosmic ultra-violet astronomy

It has already been described in Chapter 6, p. 147, how from a very early stage the Solar and Stellar Astronomy Working Group of the British National Committee for Space Research was concerned with the design of experiments to observe ultra-violet radiation not only from the sun but also from cosmic sources. Included among these experiments was one to be flown in a Skylark without attitude control to make observations of stars in the Southern sky over a wavelength range between 1700 and 2000 Å. This experiment, a description of which has already been given in Chapter 9, p. 194, was designed and constructed by D.W.O. Heddle and his collaborators at UCL. It was launched from Woomera on 1 May 1961 and although two of the three attitude measuring devices, a magnetometer and a recoverable camera, failed, it was possible to reduce the data obtained so that ultra-violet radiation from 22 identified stars was detected and the ratios, in each case, of

Fig. 15.6. Intensity–time profiles of the flare of April 10 (*a*) for energy bands, as indicated, for the flare as a whole (*b*) for three regions of the flare indicated as A, B and C, the upper profiles being for the 3.5–8 keV energy band and the lower for the 16–30 keV band, for each region.

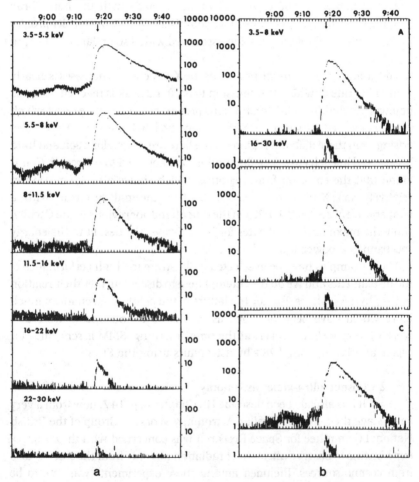

a

b

the flux at $1900\,\text{Å}$ to that at $5390\,\text{Å}$ determined. These were the first observations made of stars in the Southern sky at ultra-violet wavelengths. Further unstabilized rocket experiments were carried out both by the UCL group and a group at the Royal Observatory Edinburgh under H.E. Butler. The former group aimed at developing designs for high-spectral resolution of UV sources while the latter was concerned with broadband measurements of sky brightness in the ultra-violet.

2.1 LAS→UVAS→IUE

During this period also, as described in Chapter 6, p. 152, the Solar and Stellar Astronomy Working Group devoted much effort, in collaboration with RAE, to the design of a satellite with a payload consisting of an ultra-violet spectrometer operating in the wavelength range 1200 to $3000\,\text{Å}$ with a spectral resolution as high as $1\,\text{Å}$.

A very important development in March 1961 was the proposal put forward by the Controlled Thermonuclear Research Division of the UKAEA at Culham for a programme of ultra-violet spectroscopy of the solar corona and the stars. We have already described the major consequences of this new involvement of senior scientists, particularly R. Wilson, for UK solar ultra-violet astronomy. It had even greater impact on the UK programme for cosmic observations in the ultra-violet.

Quite close collaboration was soon worked out between the Culham group and the UCL group under Boyd and Boksenberg. By early 1963 they were already closely involved in design studies for the payload of the large astronomical satellite (LAS) of ESRO. The subsequent history of the LAS project has been described in Chapter 6, p. 152. Already by the end of 1966 it was quite clear that the LAS project, which was to be based on the UK design, would not proceed, for financial reasons, so that it appeared that the great amount of highly skilled scientific and technical effort put into the preparation of the payload design would be wasted. However, Wilson and his collaborators were not prepared to give up at this stage. He considered that, if the scientific and technical package was regarded as a single system with give and take between scientific and technical requirements, the cost could be reduced substantially. He had no difficulty in convincing his UK colleagues, so a revitalized design study was initiated under the auspices of the erstwhile Management Board which would have been set up within the UKAEA if the LAS project had gone ahead. A Project Study Group was set up with R. Wilson as Project Leader, E.G. Warnke and D.T. Boffin of the UKAEA as Study Manager and Study Secretary respectively. The other members were P.J. Barker (UKAEA) and N.J. Crawley, A. Boksenberg (UCL), G.W.

Brown and W.G. Hughes (RAE), W.M. Burton and W.G. Griffin (Astrophysics Research Unit, Culham), P.E.G. Cope and J.J. Werner (Elliott Bros, London Ltd), H.S. Mettam and F.A. Roberts (E.A. Space and Advanced Military Systems Ltd).

Armed with information about the technical side of the LAS plans obtained from ESRO, Wilson and his colleagues re-examined the scientific requirements in relation to their impact on the technology and produced a new overall design known as the Ultra-Violet Astronomy Satellite (UVAS). In this they had the assistance of a number of industrial firms in the UK.

The Study Group had the advantage of being able to draw on the extensive experience gained in the LAS design studies, especially of the complexities involved and the identification of problem areas. They also had the advantage of being able to carry out the design study as an integrated system design for the whole project. Also while image storage devices were known to be advantageous in principle during the LAS studies they had not been sufficiently developed for use in space vehicles.

In the final report on the UVAS proposal issued in 1968, the scientific aim was restricted to the prime objective of the LAS – high-resolution UV spectroscopy of bright stars. As in LAS the new proposals included a Cassegrain telescope but feeding an echelle spectrograph which is 10–20 times more powerful spectroscopically. This permitted a relaxation of tolerances in the pointing, mechanical and thermal satellite sub-system. It was also proposed by Boksenberg to use a television camera as detector because of the very large gain involved. This made it possible to reduce the telescope aperture from 80 cm to 45 cm and would compensate for loss of flux in the echelle spectrometer during the extra reflections involved. On the other hand, it would be necessary to develop image tubes capable of operation at ultra-violet wavelengths as well as being space qualified. The new optical system would not operate efficiently below 1200 Å so this was taken as the low wavelength limit of observations.

Further economy in design was achieved by eliminating stars of magnitude 8 and 9 from the observing programmes – LAS studies had shown how expensive is the observation of faint objects both in observing time and in the accuracy required of the coarse guidance systems. In this way it was reduced from 1′ to 10′.

Finally it was planned to observe no more than one object per orbit as compared with many objects for LAS, at least in its early life.

The UVAS proposal was submitted to ESRO in 1968 and received a very favourable report from the LPAC. Nevertheless it was not finally accepted, again on financial grounds. Despite this second disappointment, Wilson was not to be denied and wrote privately to L. Goldberg, the well-known US astronomer, who was then Chairman of the Space Science Board in the USA (see Chapter 4, p. 56), outlining the proposal. It was considered seriously by the Board who passed it on to the Goddard Space Flight Centre for detailed evaluation. The Director, J. Clarke, asked L. Meredith to set up a new assessment team and this reported very favourably. An invitation was sent to Wilson to visit Goddard to discuss the project and in 1968 he spent one week there in detailed discussions, particularly with Meredith and D. Kruger, who was later to be the first project manager, and with J. Mitchell the Director for Astrophysics and Physics at NASA headquarters. The American scientists made a very important suggestion that the satellite should be launched into a geosynchronous orbit so that it could be operating more or less in real time as an observatory for use by the astronomical community as a whole. It was also proposed that facilities should be incorporated which, through degradation of the spectral resolution when required to 6 Å instead of the usual 0.1 Å, would permit the observation of faint objects.

Wilson returned with the positive proposal that the UK should join with the USA in the new project which at that time was referred to as SAS D, forming part of the American small astronomical satellite programme. While a final decision on participation was not required until late 1971 a sum of £137,000 was required for participation in full design studies and this was voted by the SRC at the end of 1970.

It was agreed to invite ESRO to join the project in some form. After initial hesitation they finally agreed to do so by supplying the solar paddles and the European operation ground station. Under these circumstances NASA would provide the spacecraft, the optical and mechanical components of the scientific instrument and the US ground observatory and spacecraft control software, the UK Science Research Council in co-operation with UCL would provide the television cameras and baffles used to record the spectroscopic data, and also for acquisition.

The image processing software would be developed jointly by NASA and the Appleton Laboratory of the SRC. The satellite would be placed in geosynchronous orbit over the Atlantic Ocean and operated 16 hours a day from the US ground observatory at the Goddard Space Flight Centre for US sponsored observers, and 8 h a day for the UK and ESRO sponsored observers from a ground station at Villafranca in Spain. At least at first the observing time at the European station could be divided equally between the UK and ESRO and allocated by them independently.

The co-operation between the UK Science Research Council and UCL had to be, in fact, a very special one. While at UCL, Boksenberg, who became project scientist for the UK side of the programme, was a vital member of the project team, there was need to provide supporting manpower for him as well as very expensive and sophisticated research equipment. Just before it became clear that the project was likely to be accepted in all three agencies it was decreed by the senior administration at UCL that no further large grants could be accepted by Departments from research councils because of the load thrown on the College administration in processing and supervising the considerable expenditures involved. However, co-operation between J.F. Hosie, the Director of the Astronomy Space and Radio Division of the SRC and Chairman of the SAS D Project Committee of the SRC, J. Saxton, the Director of the Appleton Laboratory, and Massey at UCL overcame these difficulties. No large grants were ever made to UCL for the project but staff were seconded from the Appleton Laboratory and the necessary equipment loaned. The only problems were those of shortage of working space which could not be mitigated by any *ad hoc* arrangements. However, all concerned in the project were appreciative of its importance and were prepared to suffer overcrowding.

By the time the project was accepted as approved in the US Presidential Budget of 1973/4 its name had been changed to IUE, the International Ultra-violet Explorer. The SRC approved their participation on the terms proposed at about the same time as did ESRO.

The development of the detector system presented many difficult problems largely associated with the ultra-violet–visible conversion about which there was comparatively little technological experience. Between 1974 and 1975 the problem became acute because the original firm in the United States as prime contractor for the wavelength converter split off a major section which included those who had the experience and knowledge of a workable device. In this fission the know-how disappeared and for back-up ITT in the USA accepted a provisional contract which was in fact confirmed in 1976.

During the design period of UVAS, Elliott Bros. through their General Manager, were especially helpful and it was natural that they should play a major part in the development of the overall detection system. As by that time Elliott Bros. formed part of GEC UK it was the latter firm which was awarded the systems contract to deliver the full onboard hardware in full working order. However, because of the novelty of the UV-visible converter, it was clear that very skilful management of the UK side of the programme was essential and in May 1975, by agreement with Massey, within whose department Wilson was now Professor of Astronomy, Wilson was appointed project director of the UK side of the IUE, nominally on a half-time basis, with the understanding that P. Barker of the ARU, Appleton Lab. would be seconded to assist him on a full-time basis.

The SRC management team under Wilson took over the responsibility for the assessment and measurement of the properties of the converters delivered from ITT, for the bonding of the converters to the Videcon tube and for their delivery to the industrial firm (MSDS, a branch of GEC) in Portsmouth with a list of optimal operation parameters.

In July 1975 Wilson requested a further grant from the SRC for the IUE work of £658k which left the respective contributions from NASA, ESA and the SRC as £36m, £9.5m and £4m respectively. In view of the importance that the television camera would be ready in time the request was granted.

Through a personal intervention by the top management and introduction of three shifts a day, seven days a week working at MSDS it was still hoped in September 1976 to achieve the delivery of flight cameras to NASA by 30 November 1976 and in fact this date was met. After a further alarm in September 1977 that the launch date might be postponed from January till April 1978 owing to a possibly faulty main telescope mirror coating, which proved false, IUE was launched successfully on 12 January 1978. The onboard instruments functioned well from the start and are still operating well at the time of writing over five years later. This success must have been most gratifying to Wilson and his team who had finally brought to fruition a project on the same lines as the LAS.

The astronomical community soon realized the excellence of the new facilities for astronomical research made available from IUE and during the first three years it was used by astronomers from no less than 27 different countries. British astronomers have been very much involved. Up to October 1982, of the 475 scientific papers published in the main journals in which a refereeing system is used, on subjects arising from the use of IUE, as many as 27% have involved British authors as compared with the share (17%) of the total time allocated to the SRC.

As in solar physics, the progress of cosmic ultra-violet astronomy was helped very much by the existence of strong theoretical groups at UCL (under M.J. Seaton), at the ARD and at the Royal Greenwich Observatory.

Before discussing the most important results obtained by British scientists using IUE we must recall that while IUE was in the development phase, several of those much concerned in the latter work, including Wilson, Boksenberg, Barker and Warnke, were also vitally involved with the largest experiment, the sky scan over the wavelength band 1350–2550 Å, proposed jointly by ROE, Edinburgh and the Institut d'Astrophysique of the

University of Liège to be carried in the first three-axis-stabilized ESRO satellite TD-1 (see Chapter 6, p. 147). As described in Chapter 6, p. 149, the design, development and production of the equipment for this experiment was found, in 1968, to be far behind schedule largely because it was inadequately and inappropriately staffed. Wilson, Boksenberg and their colleagues assisted by support from the UKAEA were called in to rectify this position so that launch in early 1972 would be possible. Despite their involvement with the early planning of the IUE they performed an excellent rescue job, even improving greatly on the mode of operation of the spectrometer, so that TD-1 was successfully launched in March 1972. A description of the satellite and the instruments involved in the UV scanning experiment has already been given in Chapter 6, p. 150. An account is also given there of the early problems raised by the failure of the onboard tape recorder. During two six-monthly periods of operation some valuable scientific results were obtained and experience gained of value for the operation of IUE. We shall begin therefore with a brief account of these results and then proceed to discuss, rather selectively, the great range and scope of the research work carried out to date by British scientists using IUE.

2.2   Results obtained with the UV scan experiment on TD-1

Referring back to Chapter 6, p. 150, it will be recalled that the UV scan experiment as modified following Boksenberg's suggestion covered the wavelength band 1350–2550 Å in 60 channels with a grating spectrometer, and a single photometric channel with peak response at 2740 Å and bandwidth near 300 Å.

Apart from the sky surveys in this wavelength band, providing broadband ultra-violet spectra of many stars for the first time and making possible the determination of extinction as a function of wavelength in different directions within the galactic plane, one of the main scientific achievements was the first analysis of the chemical composition of Wolf-Rayet[18] stars.

The final catalogue published, the Catalogue of Stellar Ultra-Violet Fluxes, covered no less than 31,000 stars.

With such a wealth of data it is possible to study interstellar extinction in the ultra-violet in some detail. Such a study was carried out in a collaborative programme between UCL (Wilson) and ROE Edinburgh (Butler) by comparison of the spectra of reddened and unreddened stars of the same spectral type. Figure 15.7 shows the derived extinction curves for three directions in the galactic plane as indicated. The most striking features are the similarity of the curves and the strong maximum at 2200 Å which is almost certainly an absorption feature. In later work the analysis was extended to several hundred stars. These were divided into groups according to their galactic positions and mean extinction curves derived from each group. No significant difference was found between them. From the extinction curve, assuming the 2200 Å 'bump' to be due to absorption the distribution of interstellar dust within 2 kiloparsecs of the sun, causing the reddening was derived. It is concentrated strongly towards the galactic plane with a scale height of 110 pc.

These investigations were extended to the Magellanic clouds using data obtained from the scanning experiment.

The UCL group also studied the observations made on 9 Wolf-Rayet stars, in conjunction with visible spectra, to obtain information about their temperatures and chemical composition. By combining UV and visible data reliable temperatures were obtained and found to be $\sim 30,000$ K, somewhat lower than previously thought. The observed intensities of He, C and N lines in the ultra-violet and visible for four Wolf-Rayet stars were analysed theoretically to obtain abundance of these elements. For three WN stars the ratio of N/He was found to be higher than normal, C/He much lower, whereas for the one WC star investigated the C/He and N/He ratios were

Fig. 15.7. Mean ultra-violet extinction curves for three galactic regions.

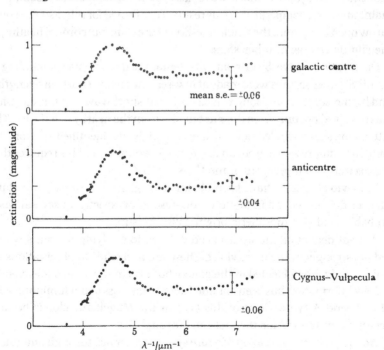

found to be close to normal. In all these stars, H/He was very low so that they are in the helium-burning phase with WN stars less evolved than WC stars.

2.3   Results from IUE[19]

In the previous section we have described results of great interest obtained with the relatively unsophisticated equipment aboard TD-1. What can we expect with the high-resolution spectrometry available on IUE? Hardly a single type of astronomical object from normal stars to quasars and clusters of galaxies is not open to study from IUE so that the background astronomical information gathered about ultra-violet-emitting objects is immense.

Low-resolution spectra[20] at short and long wavelengths have been observed for several near main sequence stars. In all cases emission was found of lines formed at temperatures up to about 10^5 K which must arise from the stellar corona. A detailed analysis[21] of the ultra-violet flux at low-spectral resolution and the emission line profiles at high resolution for αCmi (Procyon) made it possible to determine the coronal temperature more definitely as 3×10^5 K, rather lower than for the sun. However, it was found from energy balance considerations that either heating of the corona by acoustic waves or by shock-wave dissipation would be inadequate to maintain such a temperature. This result, if confirmed for other stars, would throw open once more the whole question of the source of coronal heating in the sun as well as in other stars.

Observations have been made of premain sequence stars including T Tau.[22] Six-h exposures were made at low-resolution and short wavelengths, and high resolution and long wavelengths. At short wavelengths, emission was observed not only from the star but also from the nearby nebulosity. The latter emission included several lines which were identified as from H_2 emitted by fluorescence from an excited vibrational state. This requires that the excitation temperature of the H_2 is 2000 K.

Ultra-violet spectra have been obtained for giant and super-giant stars. The continuous emission found in low-resolution spectra in several cases probably is of chromospheric origin as in the sun.

A great deal of attention has been devoted to studying hot stars which radiate strongly in the ultra-violet. These are accessible to observations by IUE not only when located in the galaxy but also in the Magellanic clouds.

New information has been obtained about mass loss in hot luminous stars of O, B and A types. Stars of this type in the Magellanic clouds became accessible in the UV for the first time through IUE.

An interesting observation of a rare system was made through analysis of spectral data on the eclipsing binary LB 3459 from which it was deduced that the temperatures of the two stars were respectively 65,000 K and 26,000 K, in which case they were both hot sub-dwarfs.

Much attention has been devoted to the study of Wolf-Rayet stars, large numbers of which have been observed in the galaxy and in the Magellanic clouds. Analysis of ultra-violet spectra from 15 galactic stars gave temperatures in the range 25–35,000 K. It appears from the high-resolution spectra that the nitrogen:carbon abundance ratio in WC stars is lower than at first thought (see, for example, p. 361)[23] From stellar wind transient velocities derived from high-resolution spectra, combined with data on the infra-red fluxes at 10 μm, accurate mass loss rates have been derived for 21 WR stars. These show a comparatively small spread (a factor of 4) and are much greater than could be attributed to radiation pressure alone. Comparison is made with corresponding mass loss rates of stars which have been thought to be precursors of the WR. This is in general 10 times smaller, showing that the mass loss rate in hot luminous stars is closely related to the stage reached by the star in stellar evolution rather than depending almost exclusively on the radiative luminosity.

The WR star HD 102163 is surrounded by a ring nebula. Sharp absorption lines have been observed in the spectra of the star which arise from a source moving with the expansion velocity of the ring and so must arise from absorption within it. This is the first observation of absorption lines associated with a nebula surrounding a hot star.[24] Analysis of the absorption spectra shows that the temperature of the absorbing gas in the ring is close to 60,000 K. This is probably a consequence of the interaction between the stellar wind from the star and the ambient interstellar material. Similar observations have been made for the ring nebula round WN5 HD 50896 and, for both cases, analysis of the data indicates that the nebula composition is close to that of the sun. This means that the rays are mainly composed of interstellar material swept up by the stellar wind rather than material ejected from the star.

The availability of ultra-violet spectra from IUE has considerably increased the scope for analysis of planetary nebulae, a subject initiated by Seaton some years ago. This analysis depends very much on the intensities of forbidden transitions between states of the same ground configuration of atoms and ions. Many of these are in the ultra-violet and were previously inaccessible to observation. As an example, Castor, Lutz and Seaton,[25] using IUE data, carried out a detailed analysis for the central star of NGC 6543, finding stellar wind velocities as high as 2100 km s^{-1} and mass loss rates greater than five times that which could be due to radiation pressure alone.

High-quality ultra-violet observations of X-ray binary systems, especially when taken together with X-ray measurements, help considerably in working out the details of the emitting system. Many observations of this kind have been made by IUE.

The effectiveness of IUE as an observatory which could be directed without delay to observe some special event, was demonstrated by the occurrence of Nova Cygni 1978. This nova was observed by IUE for 300 days from the fourth day after the outburst on 7 September 1978. The spectra obtained included the emission lines of HeII, C II, III and IV; NII, III, IV and V and OI, II, IV and V. A thorough analysis of these results was made by Seaton and his colleagues[26] using the best available data from atomic physics. 88 days after outburst they found an electron concentration of 8×10^{13} m^{-3} and electron temperatures ranging from 9500 K derived from the ionization balance for CIII, to 14,000 K from that for NV. The most interesting results, however, were the derived abundances of He, C, N and O shown in Table 15.1 in comparison with those for the sun. The abundances of all three of the heavier elements are much greater than in the sun, that of N being especially so. This is exactly what would be expected if the nova were produced by a runaway nuclear reaction, leading to ejection of a shell of material that can only occur if the abundances of C, N and O are much larger than for the sun. Moreover, if shell ejection occurs the enhancement of N will be greater than that of C and O.

The importance of observations on the ultra-violet for study of the interstellar medium depends on the fact that resonance lines of most elements fall in this wavelength region. Even extension of the range of observations from the atmospheric cut off at 2900 Å to 2470 Å made possible by flying the equipment to an altitude of 40 km in a 5.7×10^5 m^3 capacity balloon led to substantial progress as witnessed by the observations made in a collaboration between a UCL group and one from Queen's University Belfast.[27] With the large wavelength coverage available with IUE together with its high sensitivity the interstellar material may be studied in distant regions of our galaxy and also in other galaxies.

The presence of gas moving at high velocity arising from shock waves due

Table 15.1. *Abundance for Nova Cygni 1978 and for the sun*

Element	H	He	C	N	O
Nova Cygni	100	0.12	0.007	0.018	0.015
Sun	1.0	0.10	4.7×10^{-4}	9.8×10^{-5}	8.3×10^{-4}

to a supernova explosion or a stellar wind may sometimes be detected from observations of absorption spectra of background stars which show a high velocity component. Detailed studies, using IUE, of the Great Carina Nebula, where star formation is destroying a dense molecular cloud, have revealed[28] the highest velocities ($\simeq$350 km s^{-1}) and the highest degree of excitation, to date, for interstellar gas which has been detected in absorption.

As long ago as 1956[29] Spitzer had predicated that our galaxy must be surrounded by a hot halo or corona but no evidence of this was forthcoming until Ulrich *et al.*,[30] using the IUE, studied the absorption lines in the spectrum, of the brightest quasar 3C 273. They noticed that in the extensive absorption spectrum, which consist exclusively of galactic lines at zero–red shifts, the contribution from highly ionized species such as CIV was appreciably enhanced relative to that from the disc of the galaxy alone. After three years of observations with IUE of distant sources in directions making an appreciable angle with the galactic plane it was possible to carry out a thorough survey of absorption by the halo. From the relative abundances of CIV and SiIV derived from this survey the temperature of the hot gas in the halo was found to be 8×10^4 K. Information about the scale of the halo has been obtained from IUE observations of the absorption spectrum of luminous stars in the Magellanic clouds and of the supernova 1980 K in the external galaxy NGC 6946. Analysis of these observations indicates that the halo extends to at least 10–15 kpc in a direction normal to the galactic plane.

Many observations have been made of normal galaxies, particular attention being paid to the Larger Magellanic Cloud (LMC), which is rich in gas and dust and star formation is very active. Up to the present, most of the studies have concentrated on application to interstellar absorption analysis. However, evidence of a halo round the LMC as in our own galaxy has been obtained from absorption lines observed in high-resolution ultra-violet spectra of early type bright stars in the LMC.

A survey of the ultra-violet emission from the nuclei of spiral galaxies has been carried out to distinguish different spectral types.

Active galaxies have also been extensively observed using IUE. Thus the bright Seyfert galaxy WGC 4151 has been studied by several groups and this work is continuing. It appears that a black body at a temperature near 30,000 K probably exists in the galactic nucleus. No less than eight variable components of the emission from this galaxy were observed from IUE.

Observations have been made of the ultra-violet emission from three BL Lac objects (see p. 362). As in the visible, the spectrum, down to the short-wave ultra-violet cut-off, is found to consist of a continuum with no emission lines. From the intensity of this continuum ultra-violet photons are present to ionize any gas round the object and so giving rise to readily detectable Ly-α and Hβ lines. It seems that, in contrast to Seyfert galaxies and quasars, these objects do not possess an appreciable atmosphere.

Quasars have naturally been studied quite intensively. Figure 15.8 shows an early IUE spectrum obtained of the brightest quasar 3C273. By combining many spectra obtained on different occasions Boksenberg *et al.*[31] (1978) and Ulrich *et al.*[32] (1978) obtained data with a high signal to noise ratio. From the broad emission lines they derived an electron concentration of 1.6×10^8 cm^{-3} and mean electron temperature of 15,000 K.

3 Cosmic X-ray astronomy

From a very early stage British scientists were taking account in their space science planning of the possibility that X-rays emitted from cosmic sources might be observable. Thus at its meeting on the 2 October 1959 the Astronomy Sub-Committee of the British National Committee for Space Research 'noted that United Kingdom equipment is being developed for X-ray observation of the sun in the Scout Programme and it is likely that similar detectors could be developed for stellar sources. It had been suggested that radio souces might also emit X-rays and the first experiments might be devoted to observations of known radio sources, e.g. in our own galaxy. Since the object would be primarily to establish existence with an accuracy of 1°, it might be possible to produce a detector with a payload of 50 lb or less and a power consumption of 100 mW.'

Fig. 15.8. Short wave ultra-violet spectrum of the quasar 3C273 taken from IUE. The red shift of the spectrum lines due to the outward radial velocity of the quasar is clearly seen by comparison of the location of the Ly-α line of the quasar with that from the geocorona.

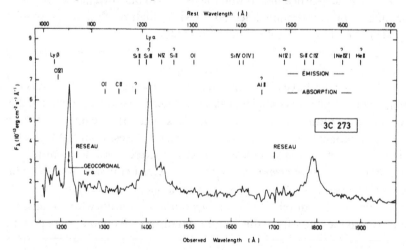

Undeterred by the absence of any evidence that observable cosmic X-ray sources existed, the UCL group under Boyd and Willmore proceeded with the design for a system using grazing incidence parabolic reflectors with proportional counters at the foci which could be included in a satellite to observe cosmic X-rays, if any. They also considered the design of detectors which could be used in Stage 3 stabilized Skylarks as announced in their research plans to the British National Committee at the end of January 1962. Somewhat later in that year Giacconi, Gursky, Paolini and Rossi announced their discovery of an astonishingly powerful X-ray source in the direction of the constellation Scorpio, known thenceforth as Scorpio X-1.

This discovery naturally encouraged a great growth in activity directed towards the detection and observation of cosmic sources. Boyd and Willmore proposed their grazing incidence reflector system for inclusion in the payload of the third US orbiting astronomical satellite OAO 3 as an auxiliary experiment to the main ultraviolet observing system to be carried. This proposal was accepted in 1963 but the satellite was not launched until 1972. Usually such a long delay would render the proposed experiments useless but by good fortune the launch took place at a most opportune time as we shall see.

Many cosmic X-ray observations from rockets followed the initial discovery by Giacconi and his colleagues. These included experiments using Skylark rockets both from the UCL group under Boyd and Willmore and the Leicester group under Pounds which lost no time in extending its solar X-ray studies to cosmic sources. However, despite this intense activity the total observing time world-wide from rockets only amounted to a few hours.

The situation changed dramatically with the launch of the first satellite devoted to observation of cosmic X-rays. This was the US satellite Uhuru launched from the Italian San Marco platform off the coast of Kenya in December 1970. Not only were a great number of new sources observed but it was now possible to check their variability, so expanding very much the scope of the subject. A further major discovery was that of extended sources of emission, particularly from clusters of galaxies.

With good fortune in launch timing, X-ray astronomers from the UK were in an excellent position to take advantage of these developments. OAO 3, henceforth known as Copernicus, including the X-ray grazing incidence reflector system, was launched in 1972. The equipment worked very well and provided the UK X-ray community, especially those at UCL, with a unique and powerful facility for observing X-ray sources, galactic and extragalactic, with high space and time resolution. Initially, it had been arranged that the Copernicus satellite would be at the disposal of the X-ray observers for 10% of the time but, in 1973, the Project Scientist for the main experiment, L. Spitzer, agreed to increase this to 20%. Special arrangements had to be made so that this time was used effectively. This involved assistance from staff at the Appleton Laboratory stationed at the Goddard Space Flight Centre as well as the maintenance there of scientists from the MSSL. A very large number and variety of observations were made over a period of years.

In October 1974, X-ray astronomers in the UK had the range and scope of their observations even further extended by the launch of the Ariel 5 satellite, an account of which has already been given in Chapter 5, p. 103. Contributions to the payload of this exclusively X-ray satellite came not only from groups at MSSL (R.L.F. Boyd and L. Culhane), Birmingham (A.P. Willmore) and Leicester (K.A. Pounds) but also from Imperial College London (H. Elliot and J.J. Quenby) who were concerned particularly with the observation of high-energy cosmic rays (quantum energy > 26 keV). It is interesting to note that Boyd, Pounds, Willmore and Elliot had all been involved in a major way in Ariel 1.

The Ariel 5 experiments all worked well for five years until the satellite re-entered the atmosphere on 14 March 1980. During much of this time the facilities available to UK X-ray astronomers were unsurpassed in the world and full advantage was taken of this strong position to make many new discoveries, some of which we shall be referring to below, and to obtain systematic data necessary for the ultimate understanding of the physics underlying the wide variety of phenomena observed.

Some months before the demise of Ariel 5 the final satellite of the Ariel series, Ariel 6, containing further equipment for observations in cosmic X-ray astronomy was launched successfully and continued until February 1982 to provide an important facility for UK astronomers even though, as explained in Chapter 5, there were problems of data recovery arising from powerful ground-based transmissions.

Before proceeding to discuss some of the most important results, obtained from the unprecedented resources for observation which became available to the UK we shall describe briefly some of the features of the Copernicus satellite and of the X-ray equipment aboard. An account has already been given of the experiments on Ariels 5 and 6 in Chapter 5.

The satellite weighed 2200 kg and the main body (see Figure 15.9) was an octagonal structure 3 metres in length and 2.03 metres across. The 1.22 m diameter centre section housed the principal experiment, the ultra-violet spectrometer for high-resolution studies of stellar spectra. The secondary, X-ray, experiment was contained in the upper two outer bays shown in Figure 15.9. A drawing of the instrument is shown in Figure

Fig. 15.9. A structural model of the Copernicus satellite.

15.10. It was about 91 cm long and consisted of the three reflecting mirror systems, a collimated proportional counter and a star tracker to measure any gross misalignment of the instrument with respect to the axis of the spacecraft. Great care was necessary to keep the background count rate to a tolerable level which in the event prevented useful observations being made over the waveband 20–100 Å but good data were obtained from 1–20 Å. The fields of view of the mirrors could be varied by changing the aperture at each focus. For 1–3Å, the field of view of 3° was fixed by the collimators, but for 6–18 Å the field of view could be chosen to be 2′, 6′ or 10′ and for 3–9 Å 1′, 2′, or 10′, the large aperture being generally used for studying time variations in the intensity emitted by a particular source.

3.1   Results from rocket experiments

From 1970 to the conclusion of the Skylark programme in 1978, 27 Skylarks were launched with X-ray astronomy payloads. In addition, a number of UK payloads have been flown in other rockets in collaboration with groups in the USA and in West Germany.

The first Skylark launched with cosmic X-ray equipment aboard was SL

Fig. 15.10. The Cosmic X-ray telescope used in the Copernicus satellite.

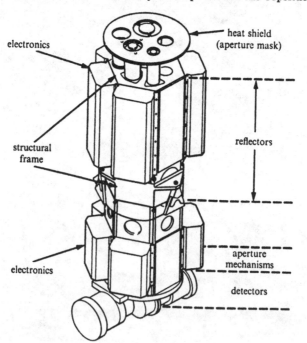

118 which contained equipment designed by Leicester University to scan the Southern Sky for X-ray sources. The detector was a proportional counter of 300 cm² collecting area and data were analysed at five energy channels ranging from 0.9–4.7 keV. SL 118, which was unstabilized, was launched on 10 April 1967 and several new sources were discovered. For two following flights, the Leicester group provided proportional counters with much greater collecting area (3000 cm²). Figure 15.11 reproduces a photograph of the nose cone of one of these Skylarks (SL 723) open to show the counter array, the largest to be flown up to that time. Both the launching of this unstabilized Skylark on 12 June 1968 and a similar one on 1 April 1969 were successful. The first measurements were made of the spectrum of Virgo XR-1 source and several new sources were observed.

In an early stabilised Skylark flight, SL 901, launched on 19 March 1970 the Leicester group introduced an X-ray sensor system which successfully locked onto the Scorpio X-1 source so that the pointing accuracy in roll was refined to better than 10'. After this was achieved, a large LiF crystal spectrometer was switched on to carry out a high wavelength resolution study between 6.3 and 6.8 keV to search for evidence of the resonance radiation from Fexxv. No such evidence was found.

The UCL (MSSL) group flew two large (1000 cm²) proportional counters in a successful experiment on a stabilized Skylark (SL 905) on 11 November 1970 to survey the centre of the galaxy, the large and small Magellanic clouds and several pulsars. This flight was related to three others, SL 971, 1021 and 973 on 14 July 1970, 14 October 1970 and 7 October 1971 in which a rotation collimator was introduced together with the large proportional counter array so that a large area of the sky could be surveyed with good space and wavelength resolution.

The power of lunar occultation methods to determine source positions with high precision was demonstrated when the Leicester and MSSL groups observed the occultation of the source GX 3 + 1 on two different occasions using equipment flown on Skylarks 1002 and 974 respectively, the launch dates being 27 September and 24 October 1971. As a result the source was located to ± 0.5″, still the most precise location of any cosmic X-ray source. A careful search carried out at the Royal Greenwich Observatory recorded a 16th magnitude star which is located within this error box. A further lunar occultation – that of the source GX 2 + 5 – was observed by the MSSL group on 30 January 1973 using Skylark 1205.

As described in Chapter 9, p. 208, the first successful Stage 5 (star pointing) stabilized Skylark SL 1011 was flown in 1973. This carried equipment designed by the Leicester group which used for the first time a

Fig. 15.11. The nose cone of Skylark SL 723 open to show the counter array.

variable spacing modulation collimator, capable of providing the spatial structure of extended sources.

On 4 February 1974, the MSSL group flew a large thin window proportional counter on Skylark SL 1203. This was the first flight of a low-energy X-ray detector. It included a highly sophisticated control system which maintained gas density in the detector within a range of ±0.15%. The flight established Gould's belt as an X-ray absorbing region and also discovered an absorbing ridge in Hydra.

In October 1974 a joint experiment was carried out by the Leicester group with groups from the Max Planck Institute at Garchung and the University of Tübingen, in West Germany. This took advantage of a lunar occultation to investigate the spatial distribution of X-ray emission from the Crab Nebula. By using a scintillation counter system as well as a proportional counter array, X-rays in the energy range 15–150 keV were observed. It was found, from the flight of Skylark SL 1304, that the emission of these hard X-rays is centred about 1–15″ NNW of the Crab radio pulsar.

The MSSL group made the first high-resolution detection of an X-ray line from a cosmic source using a Bragg crystal spectrometer flown in Skylark SL 1012 on 4 October 1974. This was the OvII Ly-α line at 19 Å in the spectrum of the supernova remnant Puppis A. A comparison of this result with a later observation by Ariel 5 allowed a new determination of the distance to this source.

A sophisticated imaging system was flown in two NASA Astrobee F rockets by a collaboration between the Leicester group and a group at MIT in the USA, the first on 27 July 1977 and the second on 8 March 1978. These were the first ever flights of imaging X-ray telescopes for non-solar studies. The equipment consisted of a position-sensitive proportional counter in the focal plane of an imaging X-ray telescope with a focal length of 114 cm and a field of view of 2.5°. The telescope aspect angle was given to 2″ accuracy from information provided from a 16 mm camera and a star tracker. Two-dimensional images were obtained for three supernova remnants, the Cygnus Loop, (see Figure 15.12), Puppis A and IC 443. It was found that, for all three, there is only weak correlation between the spatial distributions of X-ray, optical and radio emission.

3.2 Observations from Copernicus[33]

At the time of launching of Copernicus, galactic and extragalactic sources had been observed and limited observations of time variability had been made. Thus, four transient galactic sources had been observed in which the intensity rose from an immeasurably small value to a maximum

Fig. 15.12. X-ray image of the Cygnus loop supernova remnant at 0.15–1 keV superimposed on a Palomar Sky Survey print. The point response function of the telescope is shown at the bottom left-hand corner.

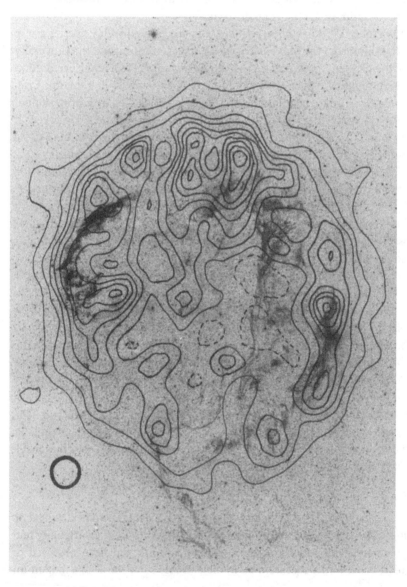

in a matter of days and then decayed back to a very low intensity once again on a similar time scale. The so-called X-ray bursters from which sudden bursts of emission occur for periods of 1 to 10 s at recurrent intervals of minutes to hours had not yet been discovered.

Special importance attaches to observation of time variability as the period of variations imposes an upper limit on the linear dimensions of the source. Thus if τ is the period, it cannot be less than the time taken for light to travel across the source. So if d is the maximum linear dimension of the source, $d < c\tau$. The first observations made located the source GX2 + 5 with higher accuracy than hitherto, an observation of special value for the rocket observation of lunar occultations of this source planned for January 1973 (see p. 371).

The second observation was a study of the Cygnus X-3 source from the region of which the intensity of the radio emission had increased over 100 times. No corresponding increase in X-ray emission was found. However, it was discovered that Cygnus X-3 varied in intensity with a period close to 5 h as may be seen from the data reproduced in Figure 15.13 obtained from the 1-3 Å detector. The period was actually measured as 4.792 ± 0.001 h. The form of the light curve (the variation of X-ray intensity with time) differs

Fig. 15.13. X-ray emission in 1–3 Å and 3–9 Å wavelength bands from Cygnus X-3. The spectral hardness is the ratio of the 3–9 Å counting rate to the 1–3 Å.

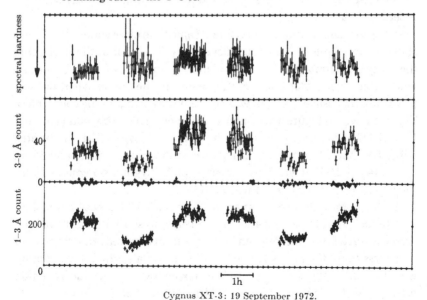

Cygnus XT-3: 19 September 1972.

from the characteristics of other sources showing periodic variation. That for Cygnus X-3 is smooth and the minimum flux is near zero whereas, for example, the Hercules X-1 source shows sharp transitions from high to low with no zero flux at the minima.

Special interest attaches to the Cygnus X-I source because it is the most likely example of a binary in which the compact object is a black hole.

While Cygnus X-I was observed at an early stage, it was not until 1971 that its position was determined to 1' of arc. A radio search of the region within the error box led to the discovery of a radio source which appeared suddenly at the end of March 1971. At nearly the same time, a sudden change of X-ray flux from Cygnus X-I occurred suggesting in fact that it is coincident with the radio source. The position of this latter source could be determined to high accuracy (< 1″ of arc) and within this very small error box a spectroscopic binary with a 5.6-day period, called HDE 226868 was known, from optical observations, to exist. It was therefore reasonable to suppose that Cygnus X-I, the transient radio source and spectroscopic binary were coincident. The special interest arose from the fact that, from optical analysis of the binary system, the mass of the normal star is greater than $10\,M_s$, where M_s is the mass of the sun, and that of the companion is greater than $8\,M_s$. If this companion is indeed compact then, according to theory, it must be a black hole in order to possess so great a mass. Assuming the theory to be valid, then it remains to establish directly the association of Cygnus X-I with the binary system and to verify that the X-ray source is indeed compact.

Strong evidence of a direct association came from observations by Mason, Hawkins and Sanford from Copernicus. They showed that the strength of the X-ray emission in the 1–3 Å band from Cygnus X-I dropped by a small amount at a particular phase of the motion in the binary when the star producing the visible emission was between the earth and the companion. This is shown on Figure 15.14 and 15. Similar results were obtained from OSO 7 by Li and Clark. There seems little doubt that Cygnus X-I is associated with HDE 226868. Further confirmation came from observations made from Ariel 5 (see p. 104) in which a modulation of the Cygnus X-1 emission with the binary period of 5.6 days was found.

Finally, from continued observation of Cygnus X-I it has been found that, from time to time, bursts of X-radiation of period as short as 1 ms have been detected. Such fast variations can only arise from a source extending over a range less than 300 km. A region of this scale on a normal star radiating with this intensity is a very unlikely possibility because the radiation pressure in the emitting range would greatly exceed the inward gravi-

Fig. 15.14. A typical example of the X-ray emission signals from
Cyg X-1 observed by Mason, Hawkins and Sandford with the grazing
incidence telescope on Copernicus. A sudden reduction in the signal
intensity followed by a quick recovery occurs between 11 and 13 h
after the record begins. This is associated with a sudden temporary
increase in hardness of the radiation as shown by the upper record
which gives the ratio of the intensities in the 1–3 Å and 3–9 Å X-ray
wavelength bands.

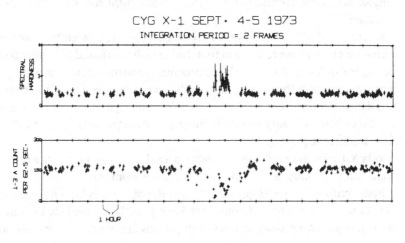

Fig. 15.15. Variation of the radial velocity with the phase of the
motion observed for the spectroscopic binary HDE 226868 associated
with Cyg X-I. The points refer to measurements by different observers.
The bars ʜʜ show the phases at which X-ray absorption events such
as shown in Fig. 15.14 occurred.

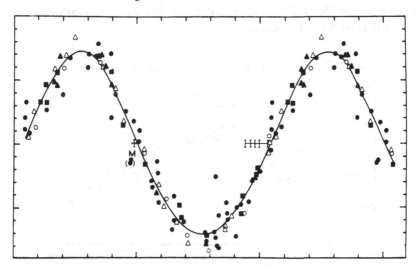

tational force. The probability that the compact object in Cygnus X-I is a black hole is therefore very high.

Observations were made on many other variable sources including a search for time variations, on a variety of time scales, in the emission from the Crab nebula. A feature of the work was the number of occasions in which collaboration was arranged with astronomers working in other wavelength ranges, including ground-based optical, radio, infra-red and ultra-violet wavebands.

Scorpius X-I, the first X-ray source discovered, was carefully observed with much better time resolution than hitherto. It was found to exhibit both flaring and quiesent states, the latter remaining constant to within a few per cent over the past three years. Four other sources were found to show a variability pattern very similar to Sco X-I but Cyg X-2, which is also associated with a binary source involving a blue star, was found to show quite a different pattern.

Vela X-I was observed over a 10-day period covering its 9.8-day binary cycle and variability was found on time scales from minutes to days.

Radio pulsars vary with periods of the order of seconds but observations from Copernicus, confirmed from Ariel 5 (see p. 104) revealed the existence of regular pulsating X-ray sources with periods of several minutes, the so-called slow rotators.

As an example, Figure 15.16 shows observations made of the brightening and fading of the transient source A 1118–61, initially referred to as Cen X-mas because in 1974 it reached a peak intensity on Christmas day. The period is seen to be 6¾ min. It seems likely that these slow rotators are neutron stars like radio pulsars but occurring in binary systems in association with normal stars.

Fig. 15.16. The flaring and fading of Cen X-mas.

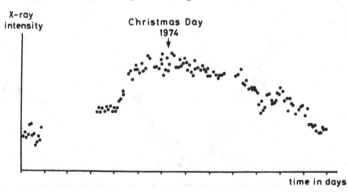

A good example of co-operation between observers at different wavelengths was the programme designed to search for X-rays associated with the radio and optical flaring of the red dwarf star UV Ceta in October 1974. Radio observations were made from Jodrell Bank and optical from Catania Observatory, Italy. Two flares at optical and radio wavelengths were observed on 19 October. In each case X-ray observations from Copernicus began between the time of peak optical and radio flaring but in neither case was any X-ray emission detected.

Another interesting example of observation of a transient X-ray source was made on Aquila X-I, which shows intense X-ray flaring on time scales of hundreds of days. The manoeuvrability of Copernicus made it possible to take quick looks at Aquila and during one of these it was found to be flaring. Many observations were taken over the following two months, including simultaneous X-ray and ground-based optical photometry and spectrometry. The pattern followed by the X-ray flaring was found to be different from usual and at one stage during the decline of intensity there was a sudden large increase.

Many observations were made of the emission from supernova remnants to determine, with much higher resolution than hitherto, the distribution of intensity over the emitting region. For Puppis A the peak of the X-ray emission was found to be displaced relative to that of peak radio intensity as may be seen from Figure 15.17. On the other hand for Cassiopeia there is quite close agreement between the intensity distributions – there are two peaks of X-ray emission which are coincident with peaks of radio emission.

With good fortune, two lunar occultations of the Crab nebula occurred from which quite good data on the X-ray source structure were obtained. The results confirmed earlier observations that the peak X-ray intensity is offset from the radio pulsar (see p. 373).

Many observations were made of extragalactic sources. Of these the most exciting were those in 1975 of Centaurus A, a strong radio source, which was found to have increased its X-ray emission more than four-fold since its first observation from Uhuru, 18 months before. This was the first variable extragalactic X-ray source to be discovered.

A search of earlier radio data showed evidence of variability of the radio emission over the same period. The behaviour of Cen A was monitored every three or four weeks and it gradually settled back towards its original level of intensity as observed by Uhuru.

Operations from Copernicus were terminated at the end of 1980 after $7\frac{1}{2}$ years of very successful X-ray observations, from which it has only been possible to select a few examples for discussion here.

3.3   *Observations from Ariel 5*[34]

The experiments on Ariel 5 were designed to carry out a wide variety of different observations of X-ray sources. A rapid sky survey could be performed by the experiment from the University of Leicester (K.A. Pounds) which could locate sources with an accuracy $\sim 12'$ and over an energy spectrum from 0.9 to 18 keV. For survey purposes separate measurements were made dividing the detectors into two pairs, the so-called high (2.4–19.8 keV) and low (1.2–5.8 keV) energy fluxes. The time resolution of these detectors was about 2 min. Figure 15.18 illustrates results obtained in a 74-orbit exposure of the galactic plane, towards the galactic centre, obtained during the first year of operation. The increased intensity towards the galactic centre is very clear.

Fig. 15.17. Contours of soft X-ray emission (solid lines) superposed on the radio map, for Puppis A.

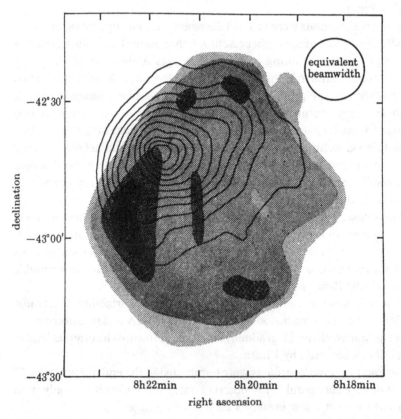

381

Fig. 15.18. 74-orbit exposure of the galactic plane obtained with the Ariel 5 Sky Survey instrument. The scan is in two parts centred respectively on the Galactic Centre and anti-Centre regions.

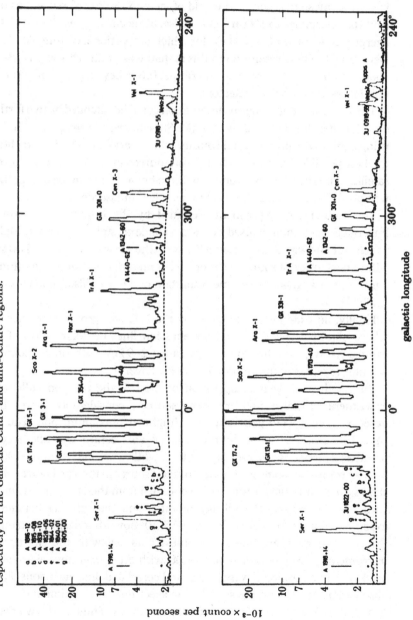

The most accurate determination of the locations of sources could be made from the experiment designed by UCL (MSSL) and the Space Research Department of the University of Birmingham,[35] which was aligned along the spin axis of the system and used a rotation-modulation collimator. When a known X-ray source was in the field of view to use as reference, position could be determined to 1' of arc. Otherwise, the accuracy was about 10'. For energies in the range 1.9–18 keV, for which proportional counter detectors were used, the time resolution in this system was $\frac{1}{2}$ min. This was raised to about 1 min for the low-energy range 0.6–6 keV for which electron multipliers were used as detectors.

The spectral energy range of observation could be extended by two further experiments. One designed by UCL (MSSL) covered the range 1.4–30 keV using a multiwire proportional counter with a beryllium window to obtain the best possible spectral resolution for sources in directions close to the satellite axis. The time resolution was 1 min but a 'pulsar' mode of operation was possible in which it was as low as a few milliseconds.

The high-energy experiment, designed by the Cosmic ray group at Imperial College London, used a caesium iodide crystal and photomultiplier as detector and was able to observe X-rays in the energy range 26–1200 keV with an accuracy of source location $\sim 2°$ and a normal time resolution of 4 min. A pulsar mode with time resolution of a few milliseconds was also available.

Finally, an experiment, designed by the Leicester group, was included whose objective was to search for emission lines or polarization of the emitted X-rays, covering the energy range 2–9 keV with time resolution ~ 1 min together with a pulsar mode as above.

During the first year of operation of Ariel 5, the rotation collimator experiment observed 8–9 transient sources, considerably more than expected. These generally very bright at maximum, one, Tau XT (A 0535 + 26), very close to the galactic anti centre was the second brightest source observed up to that time while a second, A 0620 − 00 in Monoceros was for some weeks the strongest source in the sky. Figure 15.19 shows the light curve for this source obtained from the observations made with the sky survey, the all sky monitor and the rotating collimator experiments, all in Ariel 5, while Figure 15.20 shows the energy spectrum of the emission near maximum brightness obtained with the sky survey instrument. Association of A 0620 − 00 with an optical nova was established through positional coincidence and correlated time variations. From observations made with the scintillation spectrometer experiment on Ariel 5, a phase difference was found between variations of high and low energy

Fig. 15.19. The light curve for A 0620–00 using data obtained from instruments on Ariel 5 - - - sky survey (Elvis *et al.* 1975) —— All Sky Monitor · rotation modulation collimator.

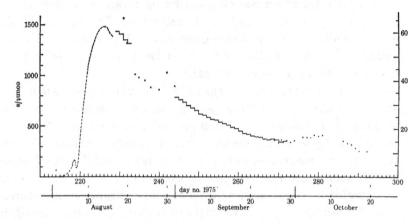

Fig. 15.20. The energy spectrum of A 0620–00 obtained near maximum brightness.

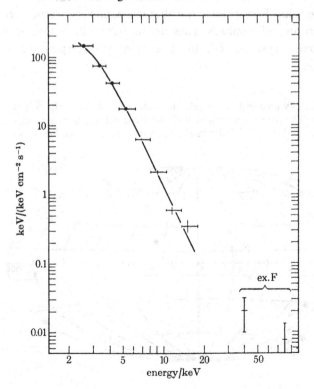

flux – the intensity of X-rays with energy 30 keV was still increasing even when the flux at lower energies had decreased to one-fifth of its maximum value. Similar behaviour was observed during a transient flaring stage of Cygnus X-I in May 1975, the peak and subsequent decline of which was also observed with the rotation collimator instrument. The transient Tau XT, the second brightest source of soft and medium energy X-rays, proved to be the strongest hard X-ray source observed.

The transient sources observed in 1974–75 and lasting weeks or months were all located close to the galactic plane. Later, several much briefer transients were found and the final picture produced by the Sky Survey Instrument (Figure 15.21) shows 35 transients with a majority of the fast type. From the observations made by the rotation modulation collimator and sky survey instruments a great amount of information was obtained about the location, intensity, time variation and spectra of transient sources. It was found from observations with the former instrument that, among the first transients observed, one was modulated with a period of 6.75 min and a second of 1.7 min. These were the first slow rotators observed. They have already been referred to on p. 375 in connection with observations from Copernicus where it has been pointed out that the periods concerned are much longer than for radio pulsars. They are probably characteristic of neutron stars in binary systems as distinct from single stars responsible for the radio pulsations.

Fig. 15.21. Sky distribution in galactic coordinates of the Ariel 5 X-ray transients. The squares represent the long-lived events, lasting weeks to months, and the circles the brief or fast transients visible for hours or less.

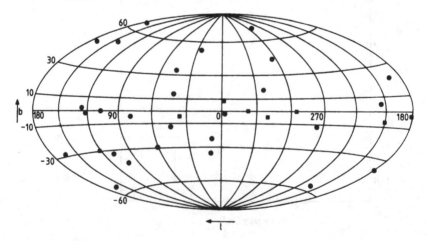

X-ray spectra of the Cas A and Tycho supernova remnants were obtained with multi-wire proportional counter. These showed emission lines of Fexxv and allowed estimates of both the ion abundance and the temperature of the shock-heated gas.

In 1976, X-ray bursters were discovered and were soon observed by the rotation collimator instrument in Ariel 5. In fact on every occasion in 1976 when the instrument was pointed close to the galactic centre X-ray bursts were seen. For at least two of the sources a steady background of X-ray emission was observed from an accurate position within the less well-defined error box deduced from the bursts. The observations certainly show a marked concentration of the bursts towards low galactic latitudes. High time resolution studies were made of the very strong burster MXB 1930-335 which showed that the bursts, each a few seconds long, came in pulse trains, the pulses within each train showing some periodicity typically of period 15 to 20 s. Combined operation of the multi-wire proportional counter and the high-energy scintillation spectrometer showed that the hardest X-ray emission is delayed relative to that at lower energies by some seconds. The spectrum of this source integrated over a burst was found to be relatively hard and to remain remarkably constant despite the strong variability of the source. The high energy spectrometer looked for gamma bursts in a region of sky viewed by the collimated proportional counter. Some high energy bursts were seen in conjunction with energetic proportional counter events and these could only have come from one or two sources, defined by the proportional counter field of view rather than the wider, 11°, scintillator field of view. This suggested that X-ray bursts can sometimes extend to γ-ray bursts energies. In fact γ-ray bursts might represent the high energy tails of X-ray bursters.

Very great progress was made with the instruments on Ariel 5 in observing extragalactic X-ray sources. From the results obtained by the sky survey instruments, 50% at least of these sources were found to be associated either with active (mainly Seyfert) galaxies or with clusters of galaxies. X-rays of variable intensity were observed from several of the former indicating that the emission arises from an extremely small region within the galactic nucleus.

Using the rotation-modulation collimator and multiwire proportional counter experiments, detailed studies of the emission from a number of extragalactic sources were carried out, including the quasar 3C 273, the strongest extragalactic source, M 87, the Seyfert Galaxy NC-C 4151 and the four bright clusters of galaxies, Virgo, Perseus, Coma and Centaurus. Figure 15.22 shows the spectra of the latter four observed with the multi-wire instrument. A feature is seen in the spectrum of the Perseus cluster near 7 keV which can be identified as arising from emission lines, the K lines of Fe xxv and Fe xxvi. Similar, weaker, features were also identified in the spectra of the Centaurus cluster and also the source 0627–544. The existence of these spectrum lines supports the view that the X-rays arise from a hot plasma, at temperatures of the order 5×10^7 K. Two X-ray spectra of

Fig. 15.22. Observed X-ray spectra of (*a*) Coma cluster of galaxies (*b*) Virgo cluster (*c*) Perseus cluster (*d*) Centaurus cluster.

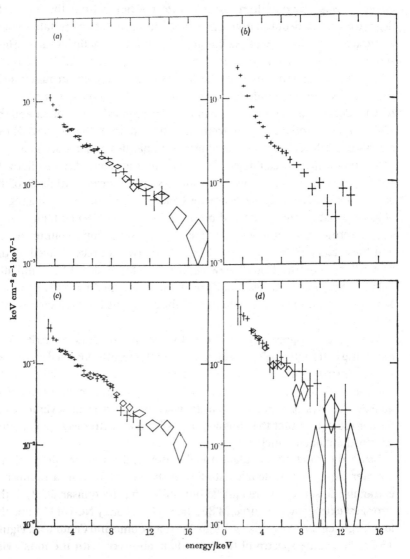

NGC 415 were obtained a year apart. The second showed a substantial increase in the quantity of absorbing material surrounding the active nucleus of the galaxy. In particular, an X-ray absorption edge due to iron atoms was detected for the first time.

With the wealth of data gathered from Ariel 5 in particular it became possible to search effectively for correlations between X-ray features and those observed optically. From the sky survey instrument X-ray luminosities for many cluster sources were obtained and found to correlate with the richness of the cluster in B–M type galaxies. On the other hand there is anticorrelation with the spiral galaxy content. This is consistent with the X-ray emission originating in hot intergalactic gas. If so, the possibility is opened of obtaining information about the composition of this gas, important for any theory of the evolution of galaxies. A further correlation found from observations made by the multi-wire instrument is one between the X-ray temperature and the velocity dispersion in the cluster observed optically.

In 1977 the observations of a cyclotron line at 64 keV in the X-ray emission from the binary pulsar Hex I was announced by Trumper. No time was lost in detecting this line with the scintillation spectrometer on Ariel 5 and a careful measurement of its intensity gave a value of 2.1×10^{-3} photons cm^{-2} s^{-1} in good agreement with that observed by Trumper. According to relativistic quantum mechanics the cyclotron line was emitted in a magnetic field of 6.9×10^8 T (6.9×10^{12} g). Although a careful search was made, no corresponding line was found in the emission from another binary pulsar G X 301-2.

Several catalogues of the locations, spectra and identifications (if any) of sources were published from the abundant data obtained from the Ariel 5 instruments. Thus the 2A catalogue, obtained from sky survey observations included information on 150 sources of which 40 were new. Most of the extragalactic sources were included. A further catalogue confined to sources in latitudes < 10° included about 150 sources, mostly galactic. Figure 15.23 shows a sky map of 287 X-ray sources observed by the Ariel 5 sky survey. A final 3 Å catalogue also published by the Leicester group, included 250 sources, galactic and extragalactic. 38 of these were transients.

3.4 Ariel 6

The X-ray astronomy results obtained with Ariel 6 have been somewhat meagre for the reasons given in Chapter 5 but there is still the prospect that with proper processing more can be extracted from the extensive data obtained.

The Leicester group participated in the NASA HEAO-B (Einstein) X-ray observatory satellite programme through collaboration with the Smithsonian Astrophysical Laboratory in designing a high-resolution image tube for this project and assisting in its production. The Einstein satellite was launched in November 1978 and was extremely successful. Having early access to the data obtained enabled the Leicester group to publish several papers including an X-ray map of the galactic centre, the detection of an X-ray jet in the quasar 3C273 (see Figure 15.8) and a study of several supernova remnants.

4   Gamma ray astronomy

Research in gamma ray astronomy has been carried out by groups at Imperial College London and at Southampton University but most of their work has involved the use of balloons rather than space vehicles and so will not be discussed here. However the Southampton group flew an acoustic spark chamber for detecting gamma ray photons in the NASA satellite OGO 5. The angular resolution was about 1° for 200 MeV photons and a sensitivity for point sources of about 10^{-6} photons cm^{-2} s^{-1}. An isotropic flux greater than about 10^{-6} photons cm^{-2} s^{-1} sr^{-1} could be detected. The results obtained confirmed that the gamma ray emission is greatest from the region of the galactic plane.

While pursuing further investigations using balloons, the Southampton

Fig. 15.23. The Ariel 5 Map of the X-ray sky at 2–15 keV shown on Altoff equal area projection in galactic coordinates. Each source is represented by a symbol of diameter roughly proportional to the log of the mean source flux.

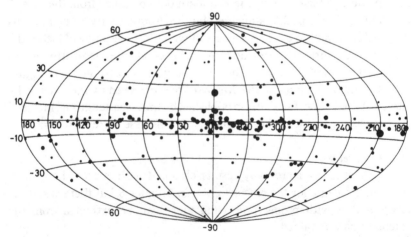

group co-operated with the Universities of Leiden and Milan, the Centre for Nuclear Research (Saclay) and Max Planck Institute for Extraterrestrial Research (Munich) in planning a large spark chamber experiment for gamma ray astronomy to be flown on an ESRO satellite known as COS-B. At first this proposal was strongly criticized by members of the Space Policy and Grants Committee in the UK as representing too expensive use of a large space project when they felt that the same results could be attained with much less expense using balloons. They did not feel able to recommend financial support for the Southampton group to take part in the project. Because of this criticism ESRO arranged for a special meeting to take place at Noordwijk on 1 June 1970 to discuss the scientific and other aspects of the COS B project which was strongly supported by continental scientists. Massey was invited to take the Chair at this meeting, which went far towards convincing the UK scientists of the scientific value of the project, so that it received their support as part of the ESRO programme. The SPGC did not, however, reverse its decision about a grant to Southampton so COS-B went ahead without any direct UK participation. It proved to be a great success.

5 Infra-red astronomy

As with gamma-ray astronomy, until very recently UK groups working in infra-red astronomy at wavelengths which cannot penetrate to ground through the atmospheric windows, used balloons to transport their observing equipment to the necessary levels. A number of groups have taken part in such work, especially at UCL (Jennings), IC (Ring and Joseph), Queen Mary College (Bastin, Clegg and Martin) and the Royal Observatory Edinburgh. The first opportunity to take part in infra-red observations from satellites came from UK participation with US and Dutch scientists in the IRAS satellite programme. This is the first satellite to be able to undertake a thorough survey of cosmic infra-red sources (it will be the Uhuru of infra-red astronomy). The UK is responsible for data recovery and analysis, operated by the Rutherford Appleton Laboratory, and for certain aspects of the software. At the time of writing (1983) IRAS has been successfully launched and the instruments aboard are operating very satisfactorily but it is too early to discuss scientific results.

16

Concluding remarks

In this account we have attempted to cover the development of space science in the UK over a period a little short of 30 years from the initial steps taken in 1953 to the present time (1983). The concluding date has been arbitrarily chosen – it does not represent any special stage reached. UK space scientists are concerned in a wide variety of projects covering many scientific disciplines. In cosmic X-ray astronomy they are much involved in the ESA EXOSAT satellite to be launched shortly, as well as with bilateral collaborations with West Germany and with Japan. IUE continues to operate very successfully as an ultra-violet observatory satellite. In optical astronomy UK scientists are planning to make maximum possible use of the space telescope and will certainly be concerned in the design and development of instrumentation for the ESA astrometry satellite Hipparcos. The IRAS satellite already pouring out new information about cosmic infra-red sources will provide data for UK scientists to work with in infra-red astronomy. In addition UK X-ray astronomy experiments will be flown on Space Lab 1 and Space Lab 2.

A number of UK groups will be involved in instrumentation for GIOTTO, the ESA probe to fly by Halley's Comet in 1986 and for the SPM which will be the first to make observations of the heliosphere well out of the ecliptic plane. This vehicle, being developed by ESA, will also be launched in 1986. Further scope for studies of the heliosphere is being provided by the AMPTE programme towards which UCL (MSSL) and RAL are contributing not only instrumentation but one of the three satellites involved.

The ability of a physics group in a university not only to design, develop and construct instruments for operation in space vehicles but also to engineer the satellite itself brings to mind some of the doubts raised at an early stage about the possibility of establishing an effective space science

390

programme. It was, for example, considered by many that it would not be possible to design and construct instruments which would operate for long periods of time without maintenance. It was even more strongly felt that even if advanced technological institutes might achieve this it would certainly be beyond the capability of university groups. And yet the UK programme was based on university physics departments which proved fully able to meet all requirements. University participation on the scale which developed was also thought to be unlikely because of the organization of university research. It meant for example that staff engaged in research in space science would have to spend considerable periods of time abroad while research students, hoping to obtain results for incorporation in a PhD thesis within three years, might have their hopes dashed by a rocket failure or the postponement of a development over which they had no control. In many ways these problems arise in all branches of big science – there are perhaps a wider range of possibilities in space science than in particle physics. Soon after the establishment of NIRNS in 1958 it was accepted by the University Grants Committee that departments engaged in research in some aspects of big science would be allowed a higher staff/student ratio than other departments. Given the will, arrangements could be worked out to allow for greater absence of staff involved in research in big science, even in a department such as the physics department at UCL which was actively involved not only in space science but also particle physics. Besides leave for staff there was also the problem that a research student would normally undertake his research work under supervision at the University to which he was attached. At UCL these problems were avoided by designating certain areas such as CERN, Woomera, etc. as extensions of the College for these purposes. In these respects the Provost of University College during the growth of these activities, Sir Ifor Evans, later Lord Evans, played a most valuable part through his flexible and sympathetic approach. His contribution to the effectiveness of UCL in space science[1] was a major one. Other universities solved the problem in various ways so that there was a rapid spread of research activity in space science throughout university physics departments and this has persisted (see Appendix F) Instrumental development in the subject makes such demands on scientific skill and originality that research students did not have to rely always on flight data to obtain PhD material.

From the academic point of view it is surely an advantage that, in a significant fraction of university departments, research is carried out in subjects which both scientifically and technically are in the forefront even though they may be expensive and require use of external facilities. It should also be of great importance nationally in that effective progress in big science calls not only for development and use of the most advanced technologies in the world today but also for great skill in scientific management, systems engineering, data retrieval and analysis, international negotiations and finance. Participation in such work at university post-graduate level is clearly of high educational value and should also be an excellent preparation for industrial employment. Unfortunately there is inadequate appreciation of the advantages that may be gained by closer contact between those undertaking pure research in big science and those concerned with applications – the technologies are much the same and if anything are more advanced in pure science.

The need for close association between those concerned with pure and applied space research is especially clear in remote sensing, the potential of which is now beginning to be realized. One of the major problems here is the overwhelming amount of data which has to be processed. Much experience in these problems both on the hardware and software side has been gained by space scientists and this needs to be tapped. Furthermore the design of the most suitable sensors for different purposes is an area to which space scientists could make important contributions.

Even apart from the direct contributions which can be made by pure space scientists it is well to remember how important nationally during the Second World War was the contribution made by University scientists whose main interests both before and after were in pure science. The presence of a pool of experts in such a demanding subject as space science, no matter how pure, is a very substantial national asset. No one could doubt that this pool exists today in Britain but it is important to preserve it and to make better use of the unique expertise which exists within it.

APPENDIX A1

National space science programme of unstabilized Skylark rockets

Results are classified as:
s   success
ps partial success
f   failure
vf rocket vehicle failure
sf system failure (e.g. telemetry)

	Launch date	Vehicle reference	Experiment	Group	Altitude attained (km)	Result
1957	23 Jul.	SL 03[b]	Flight test two photometers	Bel	84	sf
	13 Nov.	SL 04[b]	Wind structure by grenade method	UCL	123	s
			Wind profile, falling dipoles[a] tracked by radar	IC		s
			Electron concentration by rf probe	Bir		f
1958	2 Apr.	SL 06[b]	Flight test a dummy probe	UCL	42	s
	17 Apr.	SL 07	Wind structure by grenade method	UCL	146	s
			Wind profile by falling dipoles	IC		s
			Electron concentration	Bir		ps
	20 May	SL 05[b]	Flight test two airglow photometers	Bel	149	ps
	18 Jun.	SL 18	Repeat of SL 07		3	vf
	19 Jun.	SL 09	Grenades	UCL	144	f
			Falling dipoles	IC		f
			Electron concentration	Bir		s
	19 Jun.	SL 08[b]	Grenades	UCL	154	f
			Falling dipoles	IC		f
			Electron concentration	Bir		f
	3 Dec.	SL 11	Temperature and wind by observation of ejected sodium vapour glow	Bel	126	s
1959	4 Mar.	SL 21	Grenades	UCL	30	vf
			Falling dipoles	IC		
			Sodium vapour	Bel		
	8 Jul.	SL 14	Electron and ion concentration by Langmuir probes	UCL	93	ps
			Solar X-ray detectors	UCL/Lei		ps
	19 Aug.	SL 17	Langmuir probes	UCL	144	s

	Launch date	Vehicle reference	Experiment	Group	Altitude attained (km)	Result
1959	17 Sep.	SL 12	Ion mass spectrometer	Bir	132	f
			Solar Lyman α detectors	UCL		ps
			Solar X-ray detectors	UCL		s
	24 Sept.	SL 15	Ion mass spectrometer	Bir	158	s
			Sporadic E probe	UCL		s
	30 Nov.	SL 16	Sodium vapour	Bel	151	s
	30 Nov.	SL 60	Grenades	UCL	163	s
			Falling dipoles	IC		f
	1 Dec.	SL 38	Grenades	UCL	159	ps
			Falling dipoles	IC		f
			Dielectric probe	Bir		s
1960	21 Apr.	SL 62	Grenades	UCL	141	ps
			Falling dipoles	IC		f
	16 Jun.	SL 10	Sodium vapour	Bel	129	f
			Micrometeorite detector	Bel		f
	10 Aug.	SL 61	Grenades	UCL	171	ps
			Falling dipoles	IC		ps
	26 Aug.	SL 13	Sodium vapour	Bel	109	s
			Micrometeorite detector			ps
	17 Nov.	SL 67	Grenades	UCL	247	s
			Falling dipoles	IC		s
			Sodium vapour	Bel		f
	24 Nov.	SL 49	Propagation of CW radio to measure ionospheric properties	RAE/ UCW	160	s
			Soft solar X-ray detector	Lei		s
			Positive ion probe	UCL		s
	7 Dec.	SL 50	CW radio experiment	RAE/ UCW	35	vf
			Soft solar X-ray detector	Lei		
1961	13 Feb.	SL 36	Grenades	UCL	146	s
			Falling dipoles	IC		s
	6 Mar.	SL 63	Grenades	UCL	231	s
			Grenade glows	UCL		ps
			Falling dipoles	IC		s
			Sodium vapour	Bel		s
	5 Apr.	SL 64	Grenades	UCL	158	s
			Grenade glows	UCL		s
			Falling dipoles	IC		s
	1 May	SL 43	Stellar UV detection	UCL	155	s
	4 Jul.	SL 65	Grenades	UCL	70	s
			Grenade glows	UCL		f
			Falling dipoles	IC		s
	1 Aug.	SL 34	UV camera	UCL	124	f
	27 Sep.	SL 40	Solar X-ray photos and spectra	UCL/Lei	152	s
			Proton magnetometer	IC		s
			Sporadic E probe	UCL		s
	24 Oct.	SL 37	Solar X-ray photos	UCL/Lei	142	s
			Micrometeorite detection	JB		ps
			Sporadic E probe	UCL		s
	8 Nov.	SL 35	Ozone distribution	MO	152	s
			Micrometeorite detection	JB		f
			Sporadic E probe	UCL		s

	Launch date	Vehicle reference	Experiment	Group	Altitude attained (km)	Result
1961	24 Nov.	SL 83	Solar X-ray photos	UCL/Lei	–	vf
			Sporadic E	UCL		
			CW radio experiment	RAE/UCW		
	6 Dec.	SL 42	RF electron probe	Bir	226	s
			Solar X-ray photos	UCL/Lei		s
			Sporadic E probe	UCL		s
1962	5 Mar.	SL 66	Grenades	UCL	233	s
			Grenade glows	UCL		s
			Falling dipoles	IC		ps
			Sodium vapour	Bel		f
	6 Mar.	SL 160	Grenades	UCL	173	s
			Grenade glows.	UCL		s
			Falling dipoles	IC		ps
	6 Mar.	SL 161	Ditto	Ditto	140	s
			Ditto	Ditto		s
			Ditto	Ditto		s
	6 Apr.	SL 163	Ditto	Ditto	238	s
			Ditto	Ditto		s
			Ditto	Ditto		s
	14 Apr.	SL 162	Ditto	Ditto	–	vf
	20 Jun.	SL 45	Solar X-ray photos	UCL/Lei	216	f
			Solar Lyman α detectors	UCL		ps
			Grenades	UCL		ps
			Eject and observe falling sphere	UCL		f
			Positive ion probe	UCL		s
	5 Jul.	SL 164	Grenades	UCL	111	s
			Falling dipoles	IC		s
	15 Aug.	SL 109	Electron density by effects on lf radio waves	Shf/RRS	111	s
			Lyman α detector	UCL		s
			Positive ion probe	UCL		s
	11 Sep.	SL 44	Electron probe	UCL	232	s
			Ion probe	UCL		s
			Langmuir probe	UCL		s
	20 Sep.	SL 108	Electron density by effects on lf radio waves	Shf/RRS		s
	15 Oct.	SL 82	Deployment of a long wire aerial	Cam	154	f
			Measurement of electron density	Cam		f
			Sporadic E probe	UCL		s
	13 Nov.	SL 114	Solar X-ray counter spectrometer	UCL/Lei	206	f
			Micrometeorite detectors	JB		f
			Ozone measurements	MO		ps
	28 Nov.	SL 165	Grenades	UCL	230	s
			Grenade glows	UCL		s
			Falling dipoles	IC		s
	4 Dec.	SL 166	Grenades	UCL	105	s
			Grenade glows	UCL		s
			Falling dipoles	IC		s

	Launch date	Vehicle reference	Experiment	Group	Altitude attained (km)	Result
1963	1 Mar. D	SL 84	RF probe	Bir	197	s
			CW propagation	UCW/ RAE		ps
			X-ray counters	Lei		s
			X-ray cameras	Lei		s
			Lyman-α telescope	UCL		s
			Sporadic-E spike probe	UCL		f
	11 Mar. D	SL 85	RF probe	Bir	211	s
			CW propagation	UCW/ RAE		s
			X-ray counters	Lei		s
			X-ray cameras	Lei		s
			Lyman-α telescope	UCL		ps
			Sporadic-E spike probe	UCL		s
	17 Apr. T	SL 167	Grenades	UCL	235	s
			Grenades (glow cloud)	UCL		ps
			Falling sphere	UCL		s
			Sodium vapour	Bel		f
	25 May T	SL 115	Ozone distribution	MO	207	s
			Micrometeorites	JB		s
			X-ray cameras	Lei		s
	29 May D	SL 127	LF propagation	RRS	207	s
			Resonance probe	RRS		s
			Sporadic-E	UCL		f
			X-ray cameras	Lei		s
	18 Jun D	SL 126	LF propagation	RRS	208	s
			Resonance probe	RRS		ps
			Sporadic-E	UCL		s
			X-ray cameras	Lei		s
	20 Jun. D	SL 46	Langmuir probe	UCL	232	ps
			Electron temperature probe	UCL		s
			Lyman-α chambers	UCL		ps
			Sporadic-E grid probe	UCL		s
			X-ray cameras	Lei		s
	3 Oct. N	SL 81A	Long wire aerial impedance	Cam	146	ps
			Sporadic-E	UCL		s
	16 Oct. T	SL 168	Grenades	UCL	186	s
			Grenades (glow cloud)	UCL		ps
			Sodium vapour	Bel		f
			Falling sphere	UCL		f
	15 Oct. N	SL 169	Grenades	UCL	136	s
			Grenades (glow cloud)	UCL		s
			Window	IC		s
	16 Oct. N	SL 170	Grenades	UCL	130	s
			Grenades (glow cloud)	UCL		s
			Window	IC		ps
	15 Oct. T	SL 171	Grenades	UCL	206	s
			Grenades (glow cloud)	UCL		ps
			Sodium vapour	Bel		f
			Falling sphere	UCL		f

	Launch date	Vehicle reference	Experiment	Group	Altitude attained (km)	Result
1963	19 Nov. D	SL 103	RF probe	Bir	–	
			CW propagation	UCW/ RAE		vf
			X-ray counters	Lei		
			X-ray cameras	Lei		
			Lyman-α telescope	UCL		
			Sporadic-E spike probe	UCL		
	2 Dec. D	SL 104	RF probe	Bir	–	
			CW propagation	UCW/ RAE		vf
			X-ray counters	Lei		
			X-ray cameras	Lei		
			Lyman-α telescope	UCL		
			Sporadic-E spike probe	UCL		
1964	10 Mar. D	SL 129	Galactic γ-rays (spark chamber)	Sth	175	s
			Geomagnetic field (magnetometer)	IC		s
			Solar X-rays (non-imaging cameras)	Lei		s
			Sporadic-E (probes)	UCL		f
	12 Mar. D	SL 128	Galactic γ-rays (spark chamber)	Sth	176	ps
			Geomagnetic field (magnetometer)	IC		s
			Solar X-rays (non-imaging cameras)	Lei		s
			Sporadic-E (probes)	UCL		f
	11 Apr. D	SL 136	Ozone distribution (broad-band detector)	MO	202	s
	20 Aug. D	SL 121	Electron density (pulse propagation)	UCW/ RAE	192	ps
			Electron density (cw propagation)	UCW/ RAE		ps
			Solar X-rays (pin-hole camera)	Lei		s
	1 Sep. D	SL 137	Ozone distribution (broad-band detector)	MO	211	s
	17 Sep. D	SL 120	Electron density (pulse propagation)	UCW/ RAE	192	s
			Electron density (cw propagation)	UCW/ RAE		s
			Solar X-rays (pin-hole camera)	Lei		f
	24 Sep. D	SL 133	Plasma environment (Q meter)	RSRS	180	s
			Plasma frequency (resonance probes)	RSRS		s
			Solar X-rays (non-imaging cameras)	Lei		f
	29 Sep. D	SL 132	Plasma environment (Q meter)	RSRS	175	s
			Plasma frequency (resonance probes)	RSRS		s
			Solar X-rays (non-imaging cameras)	Lei		ps

	Launch date	Vehicle reference	Experiment	Group	Altitude attained (km)	Result
1964	27 Oct. N	SL 47	Sporadic-E (probes)	UCL	146	ps
			Non-solar X-ray background (telescope)	UCL/Lei		f
			Stellar UV emission (camera)	UCL		f
1965	17 Mar. N	SL 140	Sporadic-E (positive ion probes)	UCL [1]	19	vf
			Aerial impedance (probe)	Shf		vf
			Stellar UV radiation	ROE		vf
	25 Mar. D	SL 141	Sporadic-E (positive ion probes)	UCL [1]	174	s
			Aerial impedance (probe)	Shf		s
			Stellar UV radiation	ROE		ps
	29 Apr. T	SL 464	Wind profile (falling dipoles)	IC	0	vf
			Wind velocities (lithium vapour)	Bel		vf
			Wind structure (grenades)	UCL [2]		vf
	N	SL 363	Wind profile (falling dipoles)	IC	137	ps
			Wind structure (grenades)	UCL [2]		s
	N	SL 364	Wind profile (falling dipoles)	IC	132	s
			Wind structure (grenades)	UCL [2]		s
	N	SL 461	Wind profile (falling dipoles)	IC	U	ps
			Wind structure (grenades)	UCL [2]		s
	30 Apr. N	SL 462	Wind profile (falling dipoles)	IC	140	s
			wind structure (grenades)	UCL [2]		s
	N	SL 463	Wind profile (falling dipoles)	IC	135	s
			Wind structure (grenades)	UCL [2]		s
	T	SL 362	Air density (falling sphere)	UCL [1]	175	f
			Wind structure (grenades)	UCL [2]		sf
	11 May T	SL 361	Wind velocities (lithium vapour)	Bel	181	s
			Air density (falling sphere)	UCL [1]		s
	14 May N	SL 139	Electron density (lf signal receiver)	RSRS	187	s
			Electron density (current probe)	RSRS		
	19 May D	SL 138	Electron density (lf signal receiver)	RSRS	179	s
			Electron density (current probe)	RSRS		s
			Solar X-rays (non-imaging camera)	Lei		s
	1 Jul. D	SL 106	Solar X-rays (non-imaging camera)	Lei	130	s
			Electron and ion concentration (Langmuir and positive ion probes)	UCL [1]		s
			Electron temperature (probe)	UCL [1]		s
			Effect of rocket on local environment	UCL [1]		s
			Effect of rocket motion and photoemission on probe currents	UCL [1]		s
	14 Jul. N	SL 48	Stellar UV spectra (photometers)	UCL [1]	182	s
			Attitude measurement (Moon detectors/horizon detectors/fluxgate magnetometers)	UCL [1]		s

	Launch date	Vehicle reference	Experiment	Group	Altitude attained (km)	Result
1965	29 Jul. D	SL 105	Solar X-rays (non-imaging camera)	Lei	0	vf
			electron and ion concentration (Langmuir and positive ion probes)	UCL [1]		vf
			Electron temperature (probe)	UCL [1]		vf
			Effect of rocket on local environment	UCL [1]		vf
			Effect of rocket motion and photo-emission on probe currents	UCL [1]		vf
	18 Oct.	SL 39	Night airglow (photometer)	Bel	205	s
	19 Nov. D	SL 130	Wind velocities (lithium vapour)	Bel	138	f
			Electron density and temperature (rf probe)	Bir		s
			Solar X-rays (non-imaging camera)	Lei		
			Sporadic-E field discontinuities (proton precession magnetometer)	IC		s
			Sporadic-E ionization (positive ion probes)	UCL [1]		s
	13 Dec.	SL 131	Wind velocities (lithium vapour)	Bel	130	s
			Electron density and temperature (rf probe)	Bir		s
			Solar X-rays (non-imaging camera)	Lei		
			Sporadic-E field discontinuities (proton precession magnetometer)	IC		s
			Sporadic-E ionization (positive ion probes)	UCL [1]		s
1966	31 May	SL 422	Detection of plasma waves in rocket wake (probes)	MSSL	161	s
			Aerial impedance (dipole and parallel-plate probes)	Shf		s
	2 Jun.	SL 327	Ozone concentration (solar UV absorption detector)	MO	176	s
			Solar intensity 1800–2400 Å (photo-cells)	MO		s
			Electron density profile (cw propagation)	UCW		s
	9 Jun.	SL 328	Ozone concentration (solar UV absorption detector)	MO	184	s
			Solar intensity 1800–2400 Å (photo-cells)	MO		s
			Electron density profile (cw propagation)	UCW		s
	23 Jun.	SL 421	Detection of plasma waves in rocket wake (probes)	MSSL	170	s
			Aerial impedance (dipole and parallel-plate probes)	Shf		s

	Launch date	Vehicle reference	Experiment	Group	Altitude attained (km)	Result
1966	25 Oct.	SL 321	Electron density (39 Mc/s dielectric probe)	Bir	211	s
			Electron temperature (double Langmuir probe)	Bir		s
	25 Oct.	SL 322	Electron density (39 Mc/s dielectric probe)	Bir	119	f
			Electron temperature (double Langmuir probe)	Bir		f
	27 Oct.	SL 324S	Electron density (39 Mc/s dielectric probe)	Bir	173	s
			Electron temperature (double Langmuir probe)	Bir		s
	27 Oct.	SL 323	Electron density (39 Mc/s dielectric probe)	Bir	212	s
			Electron temperature (double Langmuir probe)	Bir		s
	27 Oct.	SL 324	Electron density (39 Mc/s dielectric probe)	Bir	–	f
			Electron temperature (double Langmuir probe)	Bir	–	f
1967	4 Apr.	SL 426	Electron density profile (cw propagation)	UCW	220	f
			Electron density profile (hf pulse propagation)	UCW		s
	10 Apr.	SL 118	Altitude and intensity profiles of 8645 and 7619 Å molecular oxygen (photometers)	Bel	167	s
			Attitude (airglow horizon detector)	Bel		s
			Rocket-induced ionosphere disturbance (3.6 MHz ground equipment)	Bel		s
			X-ray survey of southern sky (proportional counter spectrometer)	Lei		s
			Attitude (infra-red horizon detector)	Lei		ps
	12 Apr.	SL 119	Altitude and intensity profiles of 8645 and 7619 Å molecular oxygen (photometers)	Bel	160	s
			Attitude (airglow horizon detector)	Bel		s
			Rocket-induced ionosphere disturbance (3.6 MHz ground equipment)	Bel		s
			X-ray survey of southern sky (proportional counter spectrometer)	Lei		s
			Attitude (infra-red horizon detector)	Lei		ps
	21 Apr.	SL 425	Electron density profile (cw propagation)	UCW	218	f
			Electron density profile (hf pulse propagation)	UCW		s

	Launch date	Vehicle reference	Experiment	Group	Altitude attained (km)	Result
1967	24 Aug.	SL 521	Lyman-α and Hα night sky brightness (ionization chamber, photometer)	Oxf	178	ps
	29 Aug.	SL 522	Lyman-α and Hα night sky brightness (ionization chamber, photometer)	Oxf	124	ps
	14 Nov.	SL 423	E-region current system magnetic field (rubidium magnetometer)	IC	178	s
			Small electron density variations (positively biased probe)	RSRS		s
			Atmospheric absorption of solar Lyman-α radiation	RSRS		s
	17 Nov.	SL 424	E-region current system magnetic field (rubidium magnetometer)	IC	181	s
			Small electron density variations (positively biased probe)	RSRS		s
			Atmospheric absorption of solar Lyman-α radiation	RSRS		s
1968	31 May	SL 761	Wind velocity and atmospheric density (grenades and TMA dispenser)	UCL	238	s
	31 May	SL 762	Wind velocity and atmospheric density (grenades and TMA dispenser)	UCL	242	s
	12 Jun.	SL 723	Cosmic X-ray survey (proportional counters)	Lei	184	s
	23 Jul.	SL 523	Electron density profile and layer heights (cw and pulse propagation)	UCW	258	ps
	5 Aug.	SL 524	Electron density profile and layer heights (cw and pulse propagation)	UCW	235	s
	6 Dec.	SL 725	Positive ion density in Es layer (probes)	MSSL	146	s
			Magnetic field in Es layer (Rb magnetometer)	RSRS		s
			Electron temperature (probe)	Bir		s
1969	23 Jan.	SL 726	Positive ion density in Es layer (probes)	MSSL	149	s
			Magnetic field in Es layer (Rb magnetometer)	RSRS		s
			Electron temperature (probe)	Bir		s
	1 Apr.	SL 724	Cosmic X-ray survey (proportional counters)	Lei	192	s
	15 Jul.	SL 722	Electron density (lf impedance probe)	Shf	216	s
			electron density profile and layer heights (pulse propagation)	UCW		ps
			Electron density profile (cw propagation)	UCW		ps

	Launch date	Vehicle reference	Experiment	Group	Altitude attained (km)	Result
1969	17 Jul.	SL 729	Day-time airglow (photometer)	Bel	244	s
			Electron temperature (probe)	MSSL		s
			Electron density profile (cw propagation)	UCW		s
	25 Jul.	SL 721	Electron density (lf impedance probe)	Shf	213	s
			Electron density and layer heights (pulse propagation)	UCW		s
			Electron density profile (cw propagation)	UCW		s
	30 Jul.	SL 730	Day-time airglow (photometer)	Bel	262	ps
			Electron temperature (probe)	MSSL		ps
			Electron density profile (cw propagation)	UCW		s
	16 Oct.	SL 861	Wind and temperature structure of atmosphere (grenades and TMA dispenser)	UCL	318	s
			Electric fields (barium ion cloud dispenser)	UCL		s
	17 Oct.	SL 862	Wind and temperature structure of atmosphere (grenades and TMA dispenser)	UCL	318	s
			Electric fields (barium ion cloud dispenser)	UCL		s
	22 Oct.	SL 821	Galactic X-ray spectra (proportional counters)	MSSL	216	vf
			Solar X-ray spectra (calibration of OSO 5 detectors)	MSSL		vf
			Monitoring space potential (probes)	MSSL		vf
1970	12 Mar.	SL 921	Position and energy spectra of X-ray sources (rotating modulation collimator and proportional counters)	MSSL	247	f
	16 Apr.	SL 728	Molecular oxygen and ozone concentrations (spectrometer and filter sensors)	MO	210	s
			Spectra and intensities – celestial X-ray sources (proportional counters)	UA		s
	10 Jul.	SL 727	Ionospheric currents (rubidium magnetometer)	RSRS	203	s
			Wind profile (positive ion plate probes)	MSSL		
			Ion concentration profile (wire probes)	MSSL		
			Celestial X-ray sources – spectra and intensities (gas-filled proportional counters)	UA		
	14 Jul.	SL 971	Positions and energy spectra of X-ray sources (rotating modulation collimator)	MSSL	223	s

	Launch date	Vehicle reference	Experiment	Group	Altitude attained (km)	Result
1970	8 Oct.	SL 972	Cosmic X-ray source measurements (proportional counters)	Lei	232	s
	14 Oct.	SL 1021	Position and energy spectra of X-ray sources (rotating modulation collimator and proportional counters)	MSSL	270	s
1971	2 Mar.	SL 922	Ionospheric current system F- and E-layer dynamic coupling of the ionosphere and neutral atmosphere	UCL/ MSSL/ RSRS/ Bir	223	s
			Neutral and ionospheric structure in E region			s
			Comparison of positive ion probe and chemical release technique of wind measurement			
			Comparison of field probe and ion cloud techniques of electron field measurement			s
						s
			Instrumentation Rubidium magnetometer TMA dispenser Barium gas generators Aluminium grenades DC electric field probe Langmuir probe Positive ion density plate probes			
	7 Oct.	SL 973	Position of celestial X-ray sources (rotating modulation collimator)	MSSL	230	s
	24 Oct.	SL 974	Lunar occultation of cosmic X-ray source GX3 + 1	MSSL	245	s
1972	16 Mar.	SL 1023	Comprehensive investigation of midday Sq current structure, including profiles of ionospheric neutral wind, electron density, magnetic field and electric field (electron density and electric field probes, positive ion probes, rubidium magnetometer and lithium/sodium burner)	UCL/ MSSL/ Bir/WRE	259	s
	18 Oct.	SL 1022	Resonance scattering of sunlight by upper atmosphere (photometers)	Bel	143	s
			Ion composition of lower ionosphere (cryogenic quadrupole mass spectrometer)	Shf		f

	Launch date	Vehicle reference	Experiment	Group	Altitude attained (km)	Result
British national programme (*Skylark* rockets launched from Andøya)
1973	16 Oct.	SL 1121	Particle acceleration processes during negative phase of substorm (electron analyser, particle detectors and electric field probe)	AL/Bir/ MSSL/ Sth	266	s
	30 Oct.	SL 1122	Pre-break-up phase of a steady auroral arc (plasma wave probe, proton and electron particle detectors, electron analysers and Langmuir probe)	AL/ MSSL/ Shf/Sth/ Sx	230	s
	16 Nov.	SL 1123	Influence of particle precipitation on winds in the upper ionosphere (electric field probe, cylindrical probe, electron density and temperature probes, particle detectors, TMA dispenser and vector magnetometer)	AL/Bir/ MSSL/ UCL	215	s
	5 Dec.	SL 1124	Auroral particles and fields (vlf sensors and broad-band receiver, dipole antenna, geiger counters, electrostatic analyser, electron density and temperature probes)	Bir/Shf	222	s
1974	10 Oct.	SL 1471	Electronic programme unit (receivers and timers)	AL	270	s
British national programme (*Skylark* rockets launched from Andøya)
	1 Nov.	SL 1221	Study of particle precipitation in an auroral arc (channel multipliers and electrostatic analysers)	AL/KTH	225	s
1976	21 May	SL 1271	Proving vehicle	BAC/ RAE	254	s
			Neutral winds (lithium trail)	UCL		
British national programme (*Skylark* rockets launched from Andøya)
	21 Nov.	SL 1422	Measurement of supra-thermal electrons and ions in an auroral arc (low energy electron analysers/Langmuir probe)	MSSL/ AL/Sx/ Sth	715	s
	11 Dec.	SL 1425	Measurement of structure and response of the thermosphere and ionosphere during a strong geomagnetic disturbance (sodium trail/dye laser tracking, three-axis probes, vector rubidium vapour magnetometer, sensors for electrons, ions, plasmas and electric field (DC and AC))	UCL/AL GSFC/ UCL NDRE MSSL	695	s

	Launch date	Vehicle reference	Experiment	Group	Altitude attained (km)	Result
			British national programme (*Skylark* rockets launched from Andøya)			
1977	13 Oct.	SL 1423	VLF waves in pulsating aurora (receivers and Langmuir probes)	AL/ MSSL/ Shf/UCL	787	s
			Suprathermal ions and electrons in range 5–500 eV (boom-mounted electrostatic analysers, fixed energy detectors and ion flow detectors)			
			Electrons and positive ions in range 0.5–30 keV (channel multipliers with electrostatic analysers)			
			Neutral winds by laser tracking (sodium thermite canister)			
	17 Nov.	SL 1421	Suprathermal ions and electrons in range 5–500 eV and thermal plasma (boom-mounted electrostatic analysers, fixed energy detectors, ion flow detectors, ion mass spectrometer and Langmuir probe)	AL/ GSFC/ MSSL/Sx	718	s
			Electrons and positive ions in range 0.5–30 keV (channel multipliers with electrostatic analysers)			
			British national programme (*Sklylark* rocket launched from Andøya)			
1978	10 Nov.	SL 1424	Vlf waves in chorus event (receivers and Langmuir probes)	AL/ MSSL/ NDRE/ Shf/UCL		s
			Neutral winds (TMA trail)			

[a] also known as 'window' or chaff'
[b] primarily for vehicle proving tests

N signifies a night-time launch
T signifies a twilight launch
D signifies a daytime launch.

UCL[1] – Space Research Group, R.L.F. Boyd
UCL[2] – Space Research Group, G.V. Groves

MSSL, the Mullard Space Science Laboratory of UCL, previously the Space Research Group, led by R.L.F. Boyd.

APPENDIX A2

Programme of attitude controlled Skylark experiments

The three types of Control System are identified as Stages 1, 3 and 5:
Stage 1   Low accuracy sun pointing
Stage 3   Higher accuracy sun pointing or moon pointing
Stage 5   High accuracy pointing to a selected star

	Launch date	Vehicle reference	Experiment	Group	Altitude attained (km)	Result	Control stage
	British national programme (rockets launched from Woomera)						
1964	11 Aug D	SL 301	Attitude control (sun sensors)	RAE	145	s	1
			Solar image control (± 5″ of arc)	Cul		s	
			Solar corona (normal incidence spectrograph)	Cul		ps	
			Solar corona (grazing incidence spectrograph)	Cul		f	
			Solar UV (pin-hole camera)	Cul		f	
			Solar X-rays (pin-hole) camera)	Lei		s	
			Solar X-rays (crystal spectrograph)	Lei		s	
	17Dec. D	SL 302	Attitude control (sun sensors)	RAE	167	s	1
			Solar corona (normal incidence spectrograph)	Cul		s	
			Solar corona (grazing incidence spectrograph)	Cul		s	
			Solar UV emission (pin-hole camera)	Cul		s	
			Solar X-rays (pin-hole camera)	Lei		s	
			Solar X-rays (crystal spectrograph)	Lei		s	
1965	9 Apr. D	SL303	Attitude control (sun sensors)	RAE	161	s	1
			Solar corona (normal incidence spectrograph)	Cul		s	
			Solar corona (grating pin-hole camera)	Cul		s	
			Solar UV (pin-hole camera)	Cul		s	
			Solar X-rays (pin-hole camera)	Lei		s	
			Solar X-rays (crystal spectrograph)	Lei		s	

	Launch date	Vehicle reference	Experiment	Group	Altitude attained (km)	Result	Control stage
1965	20 Oct.	SL 306*	Solar X-rays (pin-hole camera)	Lei	213	s	1
			Solar spectra (spectrohelio-graph)	Cul		s	
			Solar image in extreme UV (pin-hole camera)			s	
1966	2 Feb.	SL 307	Attitude control unit (stage 1)	RAE	216	s	1
			Solar X-rays (pin-hole camera)	Lei		s	
			Solar spectra (spectroheliograph)	Cul		ps	
			Solar image in extreme UV (pin-hole camera)	Cul		ps	
			Solar image in short wavelength radiation (pin-hole camera)	Cul		s	
	5 May	SL 304	Solar X-rays (plain crystal spectrometer)	Lei	203	s	1
			Solar X-rays (reflector-counter spectrometer)	Lei		s	
			Solar Lyman-α (scanning spectroheliograph)	UCL		s	
			Silicon solar cell performance	RAE		s	
	8 Dec.	SL 405	Solar corona spectrum from 500 to 3000 Å (normal incidence spectrograph	Cul	–	vf	1
			Solar corona spectrum from 10 to 500 Å (grazing incidence spectrograph)	Cul		vf	
			Solar image in short wavelength radiation (normal pin-hole camera)	Cul		vf	
			Solar radiation flux above atmosphere at 1425 to 1475 Å (*Ariel 3* type sensor)	MO		vf	
			Molecular oxygen concentration (solar radiation flux attenuation by atmosphere)	MO		vf	
			X-ray solar spectrum (pin-hole camera)	Lei		vf	
1967	14 Mar.	SL 406	Solar spectrum 10–500 Å (grazing incidence spectrograph)	Cul	200	ps	1
			Spectrum of solar corona 500–3000 Å (normal incidence spectrograph)	Cul		s	
			Solar image in extreme UV (pin-hole camera)	Cul		f	
			Solar disk X-ray spectrum (pin-hole camera)	Lei		s	
			Solar radiation flux 1425–1475 Å (*Ariel 3*-type sensor)	MO		s	
			Molecular oxygen concentration (attenuation of radiation)	MO		s	

	Launch date	Vehicle reference	Experiment	Group	Altitude attained (km)	Result	Control stage
1967	8 Aug.	SL 305	Solar X-ray spatial resolution 8–18 Å (scanning spectroheliograph)	MSSL	182	s	1
			Solar Lyman-α (ionization chamber)	MSSL			
			Solar soft X-ray high resolution spectra (counter crystal spectrometer)	Lei		s	
	1 Nov.	SL 407	Solar spectrum between 500 and 1540 Å (normal-incidence spectrograph)	Cul	206	s	1
			Solar spectrum between 12 and 400 Å (grazing-incidence spectrograph)	Cul		f	
			Solar X-ray image (pin-hole camera)	Lei		s	
			Solar radiation flux (*Ariel 3*-type sensor)	MO		s	
			Molecular oxygen concentration (attenuation of radiation)	MO		s	
1968	20 Mar.	SL 408	Solar spectrum between 140 and 500 Å (grazing-incidence spectrograph)	Cul	217	s	1
			Solar spectrum between 770 and 2810 Å (normal-incidence spectrograph)	Cul		s	
			Solar image in extreme UV (grating pin-hole camera)	Cul		f	
			Solar X-ray image (pin-hole camera)	Lei		s	
			Solar radiation flux (detector)	MO		s	
			Molecular oxygen concentration (attenuation of radiation)	MO		s	
	8 Jul.	SL 403	Cosmic X-ray emission (proportional counters scanning M-87 galaxy)	Lei	177	s	3
			Cosmic X-ray image (pin-hole camera stabilized on M-87 galaxy)	Lei		s	
	29 Aug.	SL 501	Intensity of solar emission lines (scanning spectrophotometer)	MSSL	210	ps	1
			Extreme UV solar spectrum (grazing-incidence spectrograph)	MSSL		ps	
			Electron temperature (probe)	MSSL		ps	
			Positive ion density (probe)	MSSL		ps	
			Electron density (probe)	Bir		ps	
	4 Dec.	SL 601	Solar Mg ii doublet profile (Fabry–Perot interferometer)	Cul/Bel	182	s	1
			Solar image (imaging camera)	Cul/Bel		s	

	Launch date	Vehicle reference	Experiment	Group	Altitude attained (km)	Result	Control stage
1969	3 Apr.	SL 502	Intensity of solar emission lines (scanning spectrophotometer)	MSSL	273	s	1
			Extreme UV spectrum (grazing-incidence spectrograph)	MSSL		s	
			Electron temperature (probe)	MSSL		s	
			Positive ion density (probe)	MSSL		s	
			Electron density (probe)	Bir		s	
	17 Apr.	SL 606	Solar UV spectra (grazing-incidence spectrographs)	Cul	202	sf	1
			Solar UV spectrum (normal-incidence spectrograph)	Cul		sf	
			Solar extreme UV image (normal- and grating-pin-hole camera)	Cul		sf	
			Molecular oxygen concentration (ionization chambers)	MO		s	
	22 Apr.	SL 604	Solar UV spectrum (échelle grating spectrograph)	Cul	180	s	1
	14 May	SL 404	Solar UV spectrum (grazing-incidence spectrograph)	Cul	178	vf	3
			Solar extreme UV image (grating-pin-hole camera)	Cul		vf	
			Solar soft X-ray image (pin-hole camera)	Lei		vf	
	21 Aug.	SL 605	Solar X-ray emission (high resolution crystal spectrometer)	Lei	197	f	1
			Solar X-ray image (pin-hole camera)	Lei		s	
	18 Nov.	SL 602	Cosmic X-ray sources (Fresnel shadowgraph camera)	MSSL	206	vf	1
	20 Nov.	SL 801	Solar UV spectrum (normal- and grazing-incidence spectrographs)	Cul	274	ps	1
			Solar UV intensity (ionization chambers)	MO		vf	
			Molecular oxygen concentration (ionization chambers)	MO		vf	
			Solar extreme UV image (pin-hole cameras)	Cul		vf	
	27 Nov.	SL 603	Solar Mg ii doublet profile (Fabry–Perot interferometer)	Cul/Bel	180	s	1
			Solar extreme UV image (pin-hole camera)	Cul/Bel		s	
1970	19 Mar.	SL 901	X-ray line emission – SCO XR-1 (Bragg spectrometer)	Lei	208	s	1
	20 Mar.	SL 401	UV stellar photography (objective prism Schmidt camera)	ROE	181	s	3
			In-flight calibration of camera (UV stellar scanning photometer)			s	

	Launch date	Vehicle reference	Experiment	Group	Altitude attained (km)	Result	Control stage
1970	24 Mar.	SL 802	Survey of discrete sources of X-radiation in the 1–10 KeV energy range (conventional and modulation collimated detectors)	Lei	179	sf	1
	7 Apr.	SL 803	Solar UV spectrum in range 1000 Å to 2100 Å (échelle diffraction grating spectrograph)	ARU	204	s	1
	16 Jul.	SL 811	Stellar spectroscopy (900 Å to 2300 Å) (high resolution objective grating spectrographs)	ARU/ MSSL	164	ps	5
	11 Nov.	SL 905	Study of X-ray sources in the energy range 0.5–12 KeV (proportional counter and probe measurements)	MSSL	183	ps	1
	20 Nov.	SL 904	Measurement of variation in X-ray background; origin of non-X-ray pulses (collimated proportional detectors)	Lei	209	s	1
	25 Nov.	SL 804	High resolution spectroscopy of the solar corona at 5 Å–25 Å (crystal spectrometer, field of view collimator and pin-hole camera)	Lei	182	ps	1
1971	29 Jan.	SL 1001	Small spatial fluctuations in the isotropic X-ray background (proportional counter)	MSSL	203	s	1
			Isotropic spectrum from 0.25 to 15.0 keV (photomultipliers)	MSSL		s	
	10 May	SL 812	X-ray survey (modulation collimator)	Lei	219	f	5
	5 Aug.	SL 902	Solar spectrography (normal- and grazing-incidence spectrographs)	Cul	272	ps	1
	27 Sep.	SL 1002	Lunar occultation of cosmic X-ray source GX3 + 1	Lei	222	s	1
	30 Nov.	SL 1101	High resolution solar corona X-ray spectra (crystal spectrometer) Solar X-ray photography Solar white light photography	Lei	193	s	1
	10 Dec.	SL 903	Solar X-ray spectrometry of small areas of solar disk (collimated Bragg crystal spectrometer)	MSSL	198	s	1
1972	17 Mar.	SL 1009	Solar line profiles in range 100–210 nm (échelle diffraction grating spectrograph)	RSRS	237	ps	1

Launch date	Vehicle reference	Experiment	Group	Altitude attained (km)	Result	Control stage
1972 27 Mar.	SL 1081	Proving rocket (zenith pointing attitude control system and earth resources photographic equipment)	RAE	297	s	1
23 Oct.	SL 402	Soft X-rays from small Magellanic clouds (large area collimated counter)	MSSL	210	f	3
		Lunar UV albedo and night-time ozone concentration	MO		s	
26 Oct.	SL 1003	Solar X-rays (crystal spectrometer and rotating collimator)	MSSL	194	s	1
		Solar X-ray photography	MSSL		s	
7 Dec.	SL 1202	Lunar occultation of cosmic X-ray source GX5-1.	Lei	248	f	1
11 Dec.	SL 1005	Resonance scattering from metal atoms and ions (photo-multiplier tubes with filters)	Bel	195	ps	1
		Abundance and distribution of positive ions in the lower ionosphere (cryogenic quadrupole mass spectrometer)	Shf		ps	
		Molecular oxygen concentration and study of use of ion-chambers (ion chambers)	MO		s	
1973 30 Jan.	SL 1205	Lunar occultation of GX2 + 5 (X-ray detector)	MSSL	234	ps	1
28 Feb.	SL 1010	Soft X-ray survey over energy band 0.2–2.0 keV (parabolic reflector telescope, 1.0–10 keV proportional counters)	Lei	199	ps	1
14 Mar.	SL 1004	Grazing-incidence limb/disk experiment to determine temperature/height structure of solar atmosphere (grazing-incidence spectrographs with imaging optics)	AL	226	s	1
20 Mar.	SL 1102	High spectral resolution study of solar emission line profiles from 140 to 200 nm (echelle spectrograph)	AL	225	s	1
17 Apr.	SL 1011	Accurate spacing of X-ray sources in Centaurus and Circinus regions (variable spacing modulation collimator telescope)	Lei	214	s	5
16 Jun.	SL 1111	High resolution spectroscopy of bright stars γ Vel (WC8 + 08) and ξ Pup (05f) in region 91.2 to 230 nm ($d\lambda = 0.03$ nm) (objective grating spectrographs)	AL/ UCL	220	s	5

	Launch date	Vehicle reference	Experiment	Group	Altitude attained (km)	Result	Control stage
1973	26 Nov.	SL 1206	Solar corona spectroscopy, X-ray photography, solar white light photography (scanning Bragg crystal spectrometers with proportional detectors)	Lei	238	s	1
1974	4 Feb.	SL 1203	Diffuse X-ray background near Radio Loops I & IV. EUV survey of Southern Galactic Hemisphere (proportional counters. EUV telescopes)	MSSL/ UCB	200	s	1
	24 Apr.	SL 1207	Ionospheric currents (rubidium vapour magnetometer)	UCL	177	s	1
			Daytime electric fields (barium ion cloud)	UCL/ MPI		s	
			Daytime wind profile (lithium thermite trail)	UCL/ WRE		s	
			Thermospheric temperature and winds using airglow Doppler shifts (ruggedised Fabry–Perot interferometer)	UCL		s	
	18 Jun.	SL 1104	High time resolution measurements of CenX-3 X-ray emissions (proportional counters and memory)	MSSL/ UCB	234	s	1
	4 Oct.	SL 1012	Crystal spectrometry of Puppis-A in search of line emission (proportional counters)	MSSL	196	s	5
	28 Nov.	SL 911	High resolution ultra-violet stellar spectroscopy (echelle spectrograph)	AL/ UCL	203	f	5

British national programme (*Skylark* rocket launched from El Arenosillo, Spain)

	Launch date	Vehicle reference	Experiment	Group	Altitude attained (km)	Result	Control stage
	7 Oct.	SL 1304	Lunar occultation of Crab Nebula (proportional counter, scintillation counter)	Lei/ MPI/ Tb	190	ps	1

British national programme (*Skylark* rockets launched from Woomera)

	Launch date	Vehicle reference	Experiment	Group	Altitude attained (km)	Result	Control stage
1975	25 Feb.	SL 1301	Measurement of solar atmospheric structure (grazing-incidence telescope/ spectrometer)	AL	282	s	1
	24 Jun.	SL 1105	Soft X-ray mapping of Vela-Puppis supernova remnants (one-dimensional focusing telescope)	Lei	180	s	1
	24 Nov.	SL 1112	Soft X-ray spectrum of Crab Nebula to search for inter-stellar oxygen edge (mirror and cooled Si (Li) detector/ titanium gas detector)	Lei/ Saclay	251	s	5
1976	30 Jan.	SL 1302	X-ray emission, quiet coronal structure (collimated X-ray detector)	MSSL/ Lock-heed	278	f	1

	Launch date	Vehicle reference	Experiment	Group	Altitude attained (km)	Result	Control stage
1976	12 May	SL 1115	Mapping Puppis A supernova remnant in 0.25–2 keV energy range (imaging X-ray telescope)	Bir/ MSSL/ GSFC	231	sf	5
	11 Jun.	SL 1212	Measurement of X-ray line emission from the Cygnus Loop (scanning X-ray crystal spectrometer)	MSSL	> 200	s	5
	17 Jun.	SL 1501	Imaging of the galactic centre X-ray source (pseudo-random mask telescope)	Bir	258	s	1
	4 Nov.	SL 1306	High time resolution study of Cygnus X-1 (large area proportional counter)	Lei	191	f	1
	3 Dec.	SL 1114	High resolution ultra-violet stellar spectroscopy (Cassegrain telescope and echelle spectrograph)	ARD	263	f	5
	British national programme (*Skylark* rocket launched from El Arenosillo)						
	17 Jul.	SL 1402	X-ray emission from Cas-A (X-ray crystal spectrometer)	MSSL/ MPG	256	f	1
	British national programme (*Skylark* rocket launched from Woomera)						
1977	28 Apr.	SL 1115B	Observation of Puppis-A in X-ray wavelengths (position-sensitive detector)	MSSL	–	sf	5
	British national programme (*Skylark* rocket launched from Woomera)						
1978	12 May	SL 1305	H/He abundance ratio in solar corona (UV spectrometer)	AL/ MSSL		s	1

APPENDIX B

British experiments and British/Commonwealth experiments in Petrel, Fulmar and Skua rockets

Nomenclature

Vehicle designation suffix indicating launch range site:

A: Andøya, Norway
E: El Arenosillo, Spain
G: Greenland
H: Hebrides Range, South Uist
K: Kiruna, Sweden

S: Sonmiani, Pakistan
T: Thumba, India
/C: indicates part of the Commonwealth Co-operative Programme

Vehicle designation prefix indicating type of rocket:

P: Petrel
MOP: Petrel in the Meteorological Office programme
F: Fulmar

C: Centaure in the Commonwealth programme
Su, SK: Skua

	Launch date	Vehicle reference	Experiment	Group	Apogee (km)	Result
1968	3 Feb.	P 5H	Electron density variations (positively biased probe)	RSRS	140	s
			Electron flux and energy spectrum (Geiger counters)	RSRS		s
	29 Feb.	P 6H	Electron density variations (positively biased probe)	RSRS	U[a]	sf
			Electron flux and energy spectrum (Geiger counters)	RSRS		sf
	6 Mar.	P 7H	Electron temperature in C, D and E regions	Bir	U	ps
			Fluxes of charged particles (scintillation counters)	Sth		f
	25 Mar.	P 8H	Electron temperature in C, D and E regions	Bir	135	s
			Fluxes of charged particles (scintillation counters)	Sth		s
	29 May	P 17H	Electron density (Langmuir probe)	Shf	U	sf
	1 Jun.	P 18H	Electron density (Langmuir probe)	Shf	131	ps
	1 Aug.	P 13H	Electron density (Faraday rotation, differential hf absorption)	RSRS	145	sf
			Electron temperature (Langmuir probe)	RSRS		sf
			Lyman-α radiation (detector)	RSRS		sf
			Solar X-rays (proportional counter)	RSRS		sf
			Electron flux and energy spectrum (Geiger counters)	RSRS		sf

414

	Launch date	Vehicle reference	Experiment	Group	Apogee (km)	Result
1968	11 Nov.	P 15H	Electron temperature in C, D and E regions	Bir	142	ps
			Sporadic-E formation	Sth		s
	30 Nov.	P 19H	Electron and ion measurements in D region	RSRS	U	s
			Radio propagation effects	RSRS		s
	5 Dec.	P 22H	Electron and ion measurements in D region	RSRS	144	ps
			Radio propagation effects	RSRS		ps
	9 Dec.	P 9H	Winter absorption anomaly and stratospheric warming study	UCW	118	ps
1969	3 Feb.	P 24K	Electron flux and energy spectrum	RSRS	157	sf
	11 Feb.	P 23K	Electron flux and energy spectrum	RSRS	157	ps
	14 Feb.	P 27K	Small-scale structure of lower ionosphere	Shf	149	s
	15 Feb.	P 28K	Small-scale structure of lower ionosphere	Shf	153	s
	17 Mar.	P 25K	Electric field gradient in lower ionosphere	Bir	148	sf
	17 Mar.	P 29K	Flux and energy spectrum of charged particles	Sth	151	s
	18 Mar.	P 11K	Electric Field in E region	Sth/Ssx	170	s
	29 Mar.	P 26K	Small-scale structure of lower ionosphere	Shf	150	s
	18 Jun.	P 34H	Electron density and collision frequency	RSRS	137	sf
			Solar-Lyman-α and X-ray fluxes	RSRS		sf
	19 Jun.	P 35H	Electron density and collision frequency	RSRS	136	s
			Solar Lyman-α and X-ray fluxes	RSRS		s
	28 Jul.	P 40H	Fine-scale structure of sporadic E	Shf	121	ps
	2 Aug.	P 12H	Fine-scale structure of sporadic E	Shf	126	ps
	15 Sep.	P 58H	Electron and ion measurements in D region	RSRS	141	sf
			Radio propagation effects	RSRS		sf
	30 Sep.	P 46H	Night ionization in D and E regions	RSRS	U	s
	1 Oct.	P 51H	Night ionization in D and E regions	RSRS	U	s
	1 Oct.	P 52H	Night ionization in D and E regions	RSRS	U	sf
	10 Oct.	P 20H	Sporadic-E formation	Sth/Bir	130	s
	11 Oct.	P 38H	Electric fields	Sth/Ssx	U	s
	14 Oct.	P 14H	Electric and positive ion density variations in D and E region	MSSL	U	s
			Electron temperature	MSSL		s
	17 Oct.	P 16H	Sporadic-E formation	Sth/Bir	U	ps
1970	21 Feb.	P 53H	Electron density and temperature (design assessment for *Ariel 4*)	Bir	148	s
	21 Apr.	P 36H	Measurement of charged particles	Sth	139	s

	Launch date	Vehicle reference	Experiment	Group	Apogee (km)	Result
1970	27 Jun.	P 39H	Small scale structure during sporadic-E	Shf	126	s
	8 Sep.	P 30H	Electron and positive ion densities in D and E regions	MSSL	147	s
	8 Sep.	P 32H	Electron and positive ion densities in D and E regions	MSSL	149	s
	11 Sep.	P 31H	Electron and positive ion densities in D and E regions	MSSL	U	s
	18 Sep.	P 10H	Winter anomaly and stratospheric warmings	UCW	144	s
	12 Nov.	P 21H	Winter anomaly and stratospheric warmings	UCW	85	s
	16 Nov.	P 37H	High latitude charged particles	Sth	U	s
	19 Nov.	P 47H	VLF electromagnetic fields in the ionosphere	Sth/Shf	129	s
	2 Dec.	P 81H	D-region electron and ion studies	RSRS	38	sf
1971	25 Jun.	P 48H	Electron density in D and E regions (Faraday rotation)	UCW	U	s
	26 Jun.	P 120H	Electron density and collisional frequency	RSRS	138	s
			Solar Lyman-α and X-ray intensity	RSRS		s
	28 Jun.	P 109H	Electron density and collisional frequency	RSRS	143	s
			Solar Lyman-α and X-ray intensity	RSRS		s
	14 Jul.	P 79H	Flux, pitch angle and energy of charged particles (electrostatic analyser)	Sth	143	s
	14 Jul.	P 93H	Flux, pitch angle and energy of charged particles (electrostatic analyser)	Sth	142	s
	13 Sep.	P 74H	Night ionization in the D and E regions	RSRS	18	sf
	22 Sep.	P121H	Electron density and collisional frequency	RSRS	137	s
			Solar Lyman-α and X-ray intensity	RSRS		s
	24 Sep.	P 122H	Electron density and collisional frequency	RSRS	125	s
			Solar Lyman-α and X-ray intensity	RSRS		s
	28 Sep.	P 49H	Winter anomaly and stratospheric warming		142	s
	1 Dec.	P 80H	D-region electron and ion measurements	RSRS	U	s
	3 Dec.	P 82H	D-region electron and ion measurements	RSRS	U	s
	6 Dec.	P 83H	D-region electron and ion measurements	RSRS	U	sf
	7 Dec.	P 84H	D-region electron and ion measurements	RSRS	U	ps
	10 Dec.	P 119H	Electron temperature (twin temperature probe, Langmuir probes)	Bir	U	s

	Launch date	Vehicle reference	Experiment	Group	Apogee (km)	Result
1971	12 Mar.	P 69K	1–100 keV auroral electrons (electron density probe)	RSRS	150	ps
	12 Mar.	P 92K	1–150 keV auroral electrons (electron density probe)	RSRS	60	sf
	6 Apr.	P 68K	Small-scale structure of ionization in lower ionosphere	Shf	134	s
	14 Apr.	P 61K	Charged particles at sub-auroral and auroral latitudes	Sth	154	s
	15 Apr.	P 66K	Neutral wind velocity, density and temperature profiles	UCL	131	s
			Electric field measurement	UCL		s
			Electron density during sub-storm	UCL		s
	16 Apr.	P 62K	VLF study in ionosphere	Sth/Shf	138	sf
			Electron density measurements	Sth/Shf		sf
	21 Apr.	P 64K	Electric fields in lower ionosphere	Bir	143	s
	21 Apr.	P 67K	Neutral wind velocity, density and temperature profiles	UCL	128	s
			Electric field measurement	UCL		s
			Electron density during sub-storm	UCL		s
	22 Apr.	P 63K	Electric field strength	Ssx/Sth	164	ps
	25 Apr.	P 54K	Electric field strength	Ssx/Sth	162	s
	2 May	P 65K	Electric fields in lower ionosphere	Bir	146	s

Commonwealth Collaborative programme (*Petrel* rockets launched from Sonmiani)

	Launch date	Vehicle reference	Experiment	Group	Apogee (km)	Result
1971	17 Nov.	P 123S/C	Winds in lower thermosphere (lithium trail release)	UCL	130	s
	25 Nov.	P 124S/C	Winds in lower thermosphere (lithium trail release)	UCL	U	s
1972	15 Jan.	P 85H	Electron density and collisional frequency; solar Lyman-α and X-ray intensity (radio receivers, probes, X-ray and Lyman-α sensors	RSRS	147	sf
	22 Jan.	P 107H	Electron density and collisional frequency; solar Lyman-α and X-ray intensity (radio receivers, probes, X-ray and Lyman-α sensors)	RSRS	142	ps
	17 Feb.	P 111H	N(h) (Faraday rotation and differential absorption)	UCW	134	s
	18 Feb.	P 117H	Multi-probe investigation of E region (Langmuir probe and rf electron density probes)	Bir	U	sf
	29 Aug.	P 118H	Multi-probe investigation of E region (Langmuir probe and rf electron density probes)	Bir	U	sf
	29 Aug.	P 130H	Solar Lyman-α and X-rays	RSRS	135	s
	30 Aug.	P 75H	Electron density, scattered Lyman-α, at night (radio receivers, probe and Lyman-α sensors)	RSRS	118	s

	Launch date	Vehicle reference	Experiment	Group	Apogee (km)	Result
1972	31 Aug.	P 108H	Electron density, scattered Lyman-α, at night (radio receivers, probe and Lyman-α sensors)	RSRS	146	s
	31 Aug.	P 76H	Electron density, scattered Lyman-α, at night (radio receivers, probe and Lyman-α sensors)	RSRS	26	sf
	5 Oct.	P 127A	Precipitating charged particles; electron density and temperature (Geiger counters, channeltrons; temperature probes)	Sth	U	ps
			Study of break-up stages of auroral substorm (high resolution electron temperature and density Langmuir probes)	Bir		ps
	7 Oct.	P 42A	Precipitating charged particles; electron density and temperature (Geiger counters, channeltrons; temperature probes)	Sth	U	s
			Study of break-up stages of auroral substorm (high resolution electron temperature and density Langmuir probes)	Bir		ps
	10 Oct.	P 55A	VLF radio hiss during an auroral substorm (loop aerial – seven narrow-band channels, and Langmuir probe)	Sth/Shf	U	ps
	21 Oct.	P 126A	Precipitating charged particles; electron density and temperature (Geiger counters, channeltrons; temperature probes)	Sth	117	ps
			Study of break-up stages of auroral substorm (high resolution electron temperature and density Langmuir probes)	Bir		ps
	23 Oct.	P 125A	Measurement of suprathermal electrons during an auroral event (electrostatic analyser)	MSSL	U	sf
	3 Nov.	P 129A	Polar chorus emissions (three orthogonal loop aerials and Langmuir probe)	Sth/Shf	U	s
	8 Nov.	P 41A	Precipitating charged particles; electron density and temperature (Geiger counters, channeltrons; temperature probes)	Sth	U	ps

	Launch date	Vehicle reference	Experiment	Group	Apogee (km)	Result
1972	8 Nov.		Study of break-up stages of auroral substorm (high resolution electron temperature and density Langmuir probes)	Bir		s

Commonwealth collaborative programme (*Petrel* rockets launched from Thumba)

	Launch date	Vehicle reference	Experiment	Group	Apogee (km)	Result
1972	4 Jan.	P 78T/C	Electric field strength and neutral wind measurements (barium and strontium thermite cloud releases)	Ssx/Sth/ PRL	130	s
	6 Jan.	P 95T/C	Electric field strength and neutral wind measurements (barium and strontium thermite cloud releases)	Ssx/Sth/ PRL	169	s
	7 Jan.	P 77T/C	Electric field strength and neutral wind measurements (barium and strontium thermite cloud releases)	Ssx/Sth/ PRL	170	s
	14 Jan.	P 89T/C	Precipitating charged particles; electron density in the ionosphere (geiger counter, channeltrons; electron density probe)	Sth/PRL	128	s
	18 Jan.	P 88T/C	Precipitating charged particles; electron density in the ionosphere (geiger counter, channeltrons; electron density probe)	Sth/PRL	23	sf
	2 Feb.	P 110T/C	Structure of the equatorial electrojet (magnetometer)	RSRS	134	s
	5 Feb.	P 59T/C	Structure of the equatorial electrojet (magnetometer)	RSRS	U	s
	13 Feb.	P 60T/C	Structure of the equatorial electrojet (magnetometer)	RSRS	132	s
1973	19 Feb.	P 116H	Atomic oxygen and electron density (silver-strip sensors and Faraday receivers)	UCW	124	ps
	22 Feb.	P 114H	Electron density and Lyman-α (Faraday receivers and Lyman-α chamber)	UCW	126	s
	23 Feb.	P 113H	Electron density and particle count during auroral event (Faraday receivers and ionization chambers)	UCW/ Sth	U	s
	20 Mar.	P 43H	Correlation between incidence of dawn chorus, particle precipitation and changes in D and E regions (particle detectors, electron density and temperature experiment)	Bir/Sth	129	s

	Launch date	Vehicle reference	Experiment	Group	Apogee (km)	Result
1973	31 Mar.	P 44H	Correlation between incidence of dawn chorus, particle precipitation and changes in D and E regions (particle detectors, electron density and temperature experiment)	Bir/Sth	126	s
	3 Apr.	P 131H	D-region electron density, collision frequency, solar Lyman-α and X-rays (radio receivers, Lyman-α and X-rays sensors)	AL	131	s
		P 133H	D-region and electron density, collision frequency, solar Lyman-α and X-rays (radio receivers, Lyman-α and X-ray sensors)	AL	133	s
		P 105H	D-region electron density, collision frequency, solar Lyman-α and X-rays (radio receivers, Lyman-α and X-ray sensors)	AL	137	s
		P 132H	D-region electron density, collision frequency, solar Lyman-α and X-rays (radio receivers, Lyman-α and X-ray sensors)	AL	140	s
		P 134H	D-region electron density, collision frequency, solar Lyman-α and X-rays (radio receivers, Lyman-α and X-ray sensors)	AL	143	s
		P 104H	D-region electron density, collision frequency, solar Lyman-α and X-rays (radio receivers, Lyman-α and X-ray sensors)	AL	136	s
		P 135H	D-region electron density, collision frequency, solar Lyman-α and X-rays (radio receivers, Lyman-α and X-ray sensors)	AL	134	sf
	19 May	P 56H	Diffuse X-ray background (scintillation telescope)	Bri	137	f
	29 Jun.	P 162H	Atomic oxygen, Lyman-α and electron density (silver-strip sensors, Faraday receivers and Lyman-α chamber)	UCW	U	ps
	23 Jul.	P 45H	Diffuse X-ray background (scintillation telescope)	Bri	143	s
	27 Jul.	P 163H	Atomic oxygen, Lyman-α and electron density (silver-strip sensors, Faraday receivers and Lyman-α chamber)	UCW	135	s

	Launch date	Vehicle reference	Experiment	Group	Apogee (km)	Result
1973	3 Sep.	P 97H	Photometry of nightglow features (photometers)	Bel	135	s
		P 98H	Photometry of nightglow features (photometers)	Bel	143	s
	6 Sep.	P 165H	Atomic oxygen, Lyman-α and electron density (silver-strip sensors, Faraday receivers and Lyman-α chamber)	UCW/ AL/EMIE	131	ps
	10 Sep.	P 102H	Neutral wind density and temperature (TMA dispenser)	UCL	170	s
		P 103H	Neutral wind density and temperature (TMA dispenser)	UCL	168	s
	15 Sep.	P 106H	elf/vlf. electron density and temperature (elf/vlf receiver and Langmuir probe)	Sth/Shf	124	s
	14 Oct.	P 100A	Neutral wind velocity profile (sodium trail tracked by lidar)	UCL/AL	166	s
	16 Oct.	P 141A	Particle acceleration processes in the aurora (channeltrons and Geiger counters)	Sth	148	s
	17 Oct.	P 101A	Neutral wind velocity profile (sodium trail tracked by lidar)	UCL/AL	136	sf
	30 Oct.	P 147A	Particle acceleration processes in the aurora (channeltrons and Geiger counters)	Sth	70	sf
		P 142A	Auroral electrons and protons (channel multipliers and Geiger counters)	AL	131	s
		P 143A	Auroral electrons and protons (channel multipliers and Geiger counters)	AL	134	s
	16 Nov.	P 150A	Electric fields	Bir	144	s
		P 145A	Neutral wind velocity profile (sodium trail tracked by lidar)	UCL/AL	175	s
	17 Nov.	P 149A	Particle interaction with upper atmosphere during recovery phase of substorm (electric field probe, cylindrical probe and electron density and temperature experiment)	Bir	137	f
		P 146A	Neutral wind velocity profile (sodium trail tracked by lidar)	UCL/AL	175	s
		P 148A	Particle interaction with upper atmosphere during recovery phase of substorm (rf electron density probe, double electron probe and particle detectors)	Bir/Sth	132	s
	2 Dec.	P 181A	Electric fields (barium thermite)	Ssx/Sth	U	s
		P 96A	Electric fields (barium thermite)	Ssx/Sth	U	s
	6 Dec.	P 71A	elf/vlf, electron density and temperature (elf/vlf receiver and Langmuir probe)	Sth/Shf	U	s

	Launch date	Vehicle reference	Experiment	Group	Apogee (km)	Result
1974	1 Apr.	P 178H	Electron density and atomic oxygen (receivers, probes and silver film resistors)	UCW/AL	134	s
		P 174H	Measurement of concentration of atomic oxygen (UV resonance lamp, UV detectors and Langmuir probe)	AL/UCW	142	s
	3 Apr.	P 173H	Measurement of concentration of atomic oxygen (UV resonance lamp, UV detectors and Langmuir probe)	AL/UCW	132	ps
		P 177H	Electron density and atomic oxygen (receivers, probes and silver film resistors)	UCW/AL	133	s
	30 Apr.	P 72H	Quiet ionospheric conditions (quadruple radio frequency [massenfilter])	Shf	126	ps
	3 Sep.	P 99H	Measurement of low energy charged particle fluxes near the plasma pause (electrostatic analysers, associated channeltron detectors)	Sth	137	s
	6 Sep.	P 94H	Measurement of low energy charged particle fluxes near the plasma pause (aspect magnetometers, Geiger counters)	Sth	173	s
	11 Sep.	P 90H	Mass spectrometry in D and E regions (short path length mass spectrometer)	MSSL	135	sf
	24 Sep.	P 50H	Airglow emissions (optical monochromator)	Bel	133	s
	27 Sep.	P 57H	Airglow emissions (optical monochromator)	Bel	136	s
	28Sep.	P 161H	Electron density (receivers and probe)	UCW	123	s
	29 Nov.	P 175H	Measurement of concentration of atomic oxygen (UV resonance lamp, UV detectors and Langmuir probe)	AL/UCW	135	ps
		P 179H	Electron density and atomic oxygen (receivers, probes and silver film resistors)	UCW/ AL/Bel/ Hd	132	s
	2 Dec.	P 91H	Mass spectrometry in D and E regions (short path length mass spectrometer)	MSSL	130	s
	7 Dec.	P 183H	Electron density and atomic oxygen (receivers, probes and silver film resistors)	UCW/ AL/Bel/ Hd	132	s
	7 Dec.	P 176H	Measurement of concentration of atomic oxygen (UV resonance lamp, UV detectors and Langmuir probe)	AL/UCW	137	s

	Launch date	Vehicle reference	Experiment	Group	Apogee (km)	Result
		P 128H	Nightglow volume emission rates (photometers)	Bel	135	s
	17 Dec.	P 86G	Twilight/daytime neutral wind observations in the Polar Cusp region (lithium thermite release, ground-based observations by scanning photometers)	UCL	175	s
		P 87G	Twilight/daytime neutral wind observations in the Polar Cusp region (lithium thermite release, ground-based observations by scanning photometers)	UCL	175	a
	18 Dec.	P 166G	Twilight/daytime neutral wind observations in the Polar Cusp region (lithium thermite release, ground-based observations by scanning photometers)	UCL	175	s
		P 167G	Twilight/daytime neutral wind observations in the Polar Cusp region (lithium thermite release, ground-based observations by scanning photometers)	UCL	175	s

Commonwealth collaborative programme (*Petrel* and *Centaure* rockets launched from Thumba, India)

	Launch date	Vehicle reference	Experiment	Group	Apogee (km)	Result
1975	9 Feb.	P 154T	Twilight neutral wind measurements (TMA trail release)	UCL	165	s
		P 158T	Electric field measurements (long boom probe)	UCL/ PRL/ ISRO	140	s
		P 155T	Daytime neutral wind measurements (lithium thermite trail)	UCL	165	s
		P 159T	Electric field and plasma measurements (long boom probe)	UCL/ PRL/ ISRO	130	f
		C (1)	Magnetic fields (vapour magnetometer)	UCL/ PRL/ ISRO	U	ps
	19 Feb.	C (4)	Magnetic field measurements ionospheric currents (rubidium vapour scalar magnetometer)	UCL/ PRL/ ISRO	U	s
		P 156T	Daytime neutral winds (lithium thermite trail)	UCL	165	s
		C (2)	Magnetic fields (vapour magnetometer)	UCL/ PRL/ ISRO	U	s
		P 157T	Twilight neutral wind and temperature measurements (TMA trail)	UCL	165	s

	Launch date	Vehicle reference	Experiment	Group	Apogee (km)	Result
1975	15 Mar.	P 70T	Precipitating particles (channeltron–electron multipliers)	Sth	U	s
		P 184T	Precipitating particles (channeltron–electron multipliers)	Sth	U	s
	19 Mar.	P 185T	Precipitating particles (channeltron–electron multipliers)	Sth	107	s
	3 May	P 164H	Electron density (receivers, spin-sensors and probe)	UCW	112	s
	8 Sep.	P 136H	Atomic oxygen concentration and electron density (UV lamp and detector/Langmuir probe)	AL	141	s
	9 Sep.	P 153H	Electron density (receivers, vhf reference, magnetometer and probe)	UCW	129	s
	9 Sep.	P 168H	Airglow emission measurements (photometer)	Bel	130	s
	28 Nov.	P 152H	Electron density and Lyman-α (receivers spin-sensors, ionization chamber and probe)	UCW	129	s
	28 Nov.	P 151H	Atomic oxygen concentration and electron density (UV lamp and detector/Langmuir probe)	AL	140	s
1976	6 May	P 115H	Electron concentration by Faraday rotation (radio receivers, spin sensors analysed by digital electronics probe)	UCW/ MSSL	136	s
	6 May	P 186H	D-region ion mass (short path length rf mass spectrometer)	MSSL	150	s
	4 Jun.	P 144H	Measurement of vlf emissions (three-axis magnetic field sensors with filtered channels) Electron density (Langmuir probe)	Shf/Sth	133	s
	4 Jun.	P 187H	Measurement of DC electric field and vlf emissions (single axis electric field sensor with wideband telemetry) Electron density (Langmuir probe)	Shf/Sth	144	ps
	5 Jul.	P 188H	Atomic oxygen measurement (Langmuir probe and resonance lamp (UV $\lambda = 130$ nm)	AL/UCW	140	sf
1977	7 Feb.	P 192H	Night-time atomic oxygen (resonance lamp)	AL	146	s
	7 Feb.	MOP1	Ozone concentration from 45 to 90 km (UV lunar photometer) Night-time Lyman-α flux and electron density (Lyman-α chambers and Langmuir probes)	AL/MO UCW	127	ps

	Launch date	Vehicle reference	Experiment	Group	Apogee (km)	Result
1977	11 Feb.	P 189H	Day-time atomic oxygen (resonance lamp)	AL	142	s
	27 Jun.	P 190H	Electron density by Faraday rotation and Langmuir probe, and molecular oxygen by Lyman-α extinction (radio receivers, Langmuir probe, ionization chambers with attitude sensor)	UCW	124	s
	27 Jun.	P 196H	Day-time atomic oxygen (resonance lamp)	AL	130	s
	16 Oct.	P 207A	Neutral winds by laser tracking (sodium thermite canister)	AL/UCL	207	s
	Fulmar rockets launched from Andøya					
	16 Oct.	F3	Neutral winds and temperatures by electric field energetic particles (twin Fabry–Perot long-boom probe channeltrons) Electrons in range 0.5–30 keV (channel multipliers with electrostatic analysers)	AL/UCL	260	s
	17 Nov.	F1	Electrons and positive ions in range 0.5–30 keV (channel multipliers with electrostatic analysers)	AL/Sx	722	s
	15 Dec.	F4	VLF waves in chorus event (receivers and Langmuir probe)	Shf	255	ps
1978	2 Feb.	P 199H	Night-time atomic oxygen (resonance lamp and Langmuir probe)	AL/MPI/ UCW	140	s
	2 Feb.	P 203H	Ion composition (cryogenically cooled mass spectrometer)	AL/MPI	112	s
	2 Feb.	P 112H	Electron density and solar Lyman-α (Faraday rotation and Langmuir probe)	AL/UCW	135	s
	6 Feb.	P 200H	UV airglow (photo-multiplier with filters and absorption cells, Langmuir probe)	AL	136	s
	6 Feb.	P 202H	Electron density and night-time Lyman-α (Faraday rotation and Langmuir probe)	AL/UCW	125	sf
	7 Feb.	P 204H	Ion composition (cryogenically cooled mass spectrometer)	AL/MPI	108	s
	7 Feb.	P 198H	Atomic oxygen (resonance lamp and Langmuir probe)	AL	139	s
	13 Feb.	MOP 2	Night-time ozone concentration and temperatures above 65 km. (UV photometers and hot-wire gauge)	MO/		ps
	13 Feb.	MOP 4	Day-time ozone concentration and electron density (UV photometers and Faraday rotation)	MO/ UCW		sf

	Launch date	Vehicle reference	Experiment	Group	Apogee (km)	Result
1978	30 Jul.	P 205K	Atomic oxygen concentration (resonance lamp and Langmuir probe with PCM telemetry)	AL/SSC	149	s
	13 Aug.	P 206K	Atomic oxygen (resonance lamp and Langmuir probe with PCM telemetry)	AL/SSC	150	s
1979	25 Jan.	P 216K	Particle detectors for electron spectra and angular distribution for auroral pulsations	AL	183	s
	27 Jan.	P 215K	Ditto	AL	179	s
	28 Mar.	P 208A	VLF wavefield and Langmuir probe for electron density	Shf	169	sf
	20 Jul.	P 209H	Atomic oxygen by resonance fluorescence and absorption	AL/UCW	149	ps·
	20 Jul.	P 210H	Ditto	AL/UCW	145	s
	20 Jul.	P 211H	Ditto	AL/UCW	145	s
	24 Sep.	P 217E	Electrostatic and electromagnetic effects of impulsive injection of plasma into ionosphere	Sx	30	f
	3 Oct.	P 218E	Ditto	Sx	145	s
1980	26 Jul.	P 223K	VLF propagation in earth-ionosphere wave guide, in the ionosphere and magnetosphere	Shf	168	sf
	11 Nov.	P 221K	Atomic oxygen by resonance fluorescence and absorption. Langmuir probe, magnetometers and temperature measurements	AL/UCW	170	s
	15 Nov.	P 220K	Ditto	AL/UCW		sf
	30 Nov.	P 222K	Ditto	AL/UCW		s
1981	7 Feb.	P 219K	Ditto	AL/UCW		s
	1 Dec.	P 224K	Measurement of neutral winds and temperatures by chemical release trail–lithium/sodium	UCL		s
	9 Dec.	P 225K	Ditto	UCL	190	f
	16 Dec.	P 226K	Ditto	UCL		s
Experiments in *Skua* rockets						
1967	8 Jul.	Su 1H	D-region pressures	UCW	73	s
	9 Jul.	Su 2H	D-region pressures	UCW	81	s
1968	21 Jun.	Su 3H	Positive ion densities	MSSL	0	f
	24 Jun.	Su 4H	Positive ion densities	MSSL	100	s
	2 Jul.	Su 5H	Positive ion densities	MSSL	U	ps
	8 Jul.	Su 6H	Positive ion densities	MSSL	100	s
	23 Nov.	Su 7H	Positive ion densities	MSSL	U	ps
	5 Dec.	Su 7aH	Positive ion densities	MSSL	U	ps
	29 Mar.	Su 8H	Electron density profiles (ionospheric absorption)	UCW	88	ps
	7 Aug.	Su 9H	Electron density profiles (ionospheric absorption)	UCW	94	s
	8 Nov.	Su 10H	Electron density profiles (ionospheric absorption)	UCW	U	ps

	Launch date	Vehicle reference	Experiment	Group	Apogee (km)	Result
	16 Nov.	Su 11H	Electron density profiles (ionospheric absorption)	UCW	U	ps
	5 Dec.	Su 12H	Electron density profiles (ionospheric absorption)	UCW	U	s
1969	12 Dec.	Su 24H	D-region measurements during stratospheric warming	UCW	79	ps
	16 Dec.	Su 25H	D-region measurements during stratospheric warming	UCW	89	s
1970	9 Jan.	Su 26H	D-region electron density and absorption	UCW	82	s
	12 Jan.	Su 22H	D-region air densities	UCW	85	s
	11 Apr.	Su 27H	D-region electron density and absorption	UCW	U	s
	11 Apr.	Su 28H	D-region electron density and absorption	UCW	U	s
	14 Apr.	Su 16H	Ion densities in lower F region	MSSL	92	s
	14 Apr.	Su 17H	Ion densities in lower D region	MSSL	92	s
	17 Apr.	Su 18H	Ion densities in lower D region	MSSL	95	s
	24 Jun.	Su 23H	D-region air densities	UCW	U	s
	1 Jul.	Su 29H	Ionosphere – neutral atmosphere relationships, particularly during periods of stratospheric warming	UCW	U	s

Commonwealth collaborative rocket programme (*Skua* rockets launched from Thumba)

	Launch date	Vehicle reference	Experiment	Group	Apogee (km)	Result
1970	18 Feb.	Su 13T/C	Ion densities in lower D region	MSSL	92	s
	15 Apr.	Su 15T/C	Ion densities in lower D region	MSSL	93	telemetry failure
	16 Apr.	Su 14T/C	Ion densities in lower D region	MSSL	91	s
1971	1 Jul.	SK 39H	Ionosphere/neutral atmosphere relationships	UCW	U	s
	14 Jul.	SK 36H	Electron density in D region	UCW	U	s
	30 Jul.	SK 37H	Electron density in D region	UCW	U	ps
	15 Sep.	SK 38H	Electron density in D region	UCW	82	s
	14 Oct.	SK 41H	Electron density in D region	UCW	91	s
	19 Nov.	SK 42H	Electron density in D region	UCW	U	s

British national programme (*Skua* rockets launched from Kiruna)

	Launch date	Vehicle reference	Experiment	Group	Apogee (km)	Result
	8 Jan.	SK 83/1	Wind and temperature (sonde)	UCL	T[b]	f
	15 Jan.	SK 83/2	Wind and temperature (sonde)	UCL	T	s
	17 Jan.	SK 83/3	Wind and temperature (sonde)	UCL	T	f
	19 Jan.	SK 83/4	Wind and temperature (sonde)	UCL	T	s
	21 Jan.	SK 83/5	Wind and temperature (sonde)	UCL	T	s
	23 Jan.	SK 83/6	Wind and temperature (sonde)	UCL	T	ps
1971	25 Jan.	SK 83/7	Wind and temperature (sonde)	UCL	T	s
	27 Jan.	SK 83/8	Wind and temperature (sonde)	UCL	T	ps
	29 Jan.	SK 83/9	Wind and temperature (sonde)	UCL	T	s
	31 Jan.	SK 83/10	Wind and temperature (sonde)	UCL	T	s
	3 Feb.	SK 83/11	Wind and temperature (sonde)	UCL	T	s
	8 Feb.	SK 83/12	Wind and temperature (sonde)	UCL	T	ps
	9 Feb.	SK 83/13	Wind and temperature (sonde)	UCL	T	s
	11 Feb.	SK 83/14	Wind and temperature (sonde)	UCL	T	s
	12 Feb.	SK 83/15	Wind and temperature (sonde)	UCL	T	ps

	Launch date	Vehicle reference	Experiment	Group	Apogee (km)	Result
1972	15 Jan.	SK 43H	N(h) (Faraday rotation and differential absorption)	UCW	87	s
	19 Apr.	SK 45H	N(h) (Faraday rotation and differential absorption)	UCW	87	s
	17 May	SK 44H	N(h) (Faraday rotation and differential absorption)	UCW	93	s
	12 Oct.	SK 19A	D-region measurements (barium-lined shaped charge)	Ssx	U	s
	23 Oct.	SK 20A	D-region measurements (barium-lined shaped charge)	Ssx	U	s
	29 Oct.	SK 21A	D-region measurements (barium-lined shaped charge)	Ssx	U	s
			Commonwealth collaborative programme (*Skua* rockets launched from Thumba)			
	5 Jan.	SK 30T/C	Electric fields and neutral winds (shaped charge test rounds)	Ssx/PRL	1.7	sf
	18 Jan.	SK 31T/C	D-region measurements (barium-lined shaped charge release)	Ssx	40	sf
	31 Jan.	SK 32T/C	D-region measurements (barium-lined shaped charge release)	Ssx	80	s

[a] U indicates altitude not available

[b] T indicates that the rocket was tracked but the apogee is not available.

The extensive Meteorological Office programme using Skua rockets is not included.

APPENDIX C

Experiments in the Ariel series of satellites and in the British-launched Prospero satellite

	Launch date	Satellite reference	Experiment	Group
1962	26 Apr.	Ariel 1	Radio frequency plasma probe to measure electron density	Bir
			Langmuir probes to measure electron density and temperature	UCL
			Positive ion mass spectrometer	UCL
			Solar X-ray spectrometer	UCL/Lei
			Solar Lyman $-\alpha$ radiation detector	UCL
			Solar aspect angle detector	UCL
			Cosmic ray detector	IC
1964	27 Mar.	Ariel 2	Galactic radio noise received on long wire aerial	Cam
			Spectrometer and broad-band detection of ozone distribution	MO
			Detectors sensitive to penetration by micrometeorites	Man
1967	5 May	Ariel 3	Electron density and temperature of topside ionosphere by plasma probe	Bir
			Radio reception of galactic noise over 2 to 5 mHz band	Man
			Radio reception of very low frequency radiation	Shf
			Vertical distribution of molecular oxygen by ion chamber detector	MO
			Radio reception of terrestrial radio noise	RSRS
1971	11 Dec.	Ariel 4	Improved electron density and temperature probes	Bir
			Improved radio reception at very low frequencies	Shf
			Lightning flash impulse counter	Shf/RSRS
			Measurement of ionospheric radio noise in the mHz range	Man/ RSRS
			Measurement of energy spectra of electrons and protons	Univ. of Iowa

	Launch date	Satellite reference	Experiment	Group
1971	28 Oct.	Prospero	Determination of micrometeoroid flux by 'ion splash' impact detector	Bir
			Various technological experiments	RAE
1974	15 Oct.	Ariel 5	Measurement of source positions and X-ray sky survey in energy range 0.3 to 30 keV	UCL
			Spectra of discrete sources in the 2 to 30 keV energy range	UCL
			X-ray sky survey in energy range 1.5 to 20 keV	Lei
			Measurement of polarization of X-rays in energy range 1.5 to 8 keV	Lei
			Study of high energy X-ray sources up to 2.0 meV	IC
			All-sky monitor in the energy range 3 to 6 keV	GSFC
1979	2 Jun.	Ariel 6	Detection of heavy cosmic ray primary particles	Bri
			Observation of rapid time fluctuations in X-ray sources	Lei
			Observation of very soft X-ray emissions	UCL/Bir

APPENDIX D

British experiments in ESRO sounding rockets and in ESRO/ESA scientific satellites

NOMENCLATURE

Rockets: In each case the ESRO vehicle designation is given. The prefix indicates the type of rocket.

S, SK: Skylark   A, AE: Arcas (USA)
C, CE: Centaure (French)   P: Petrel

The altitude attained, or Apogee, is given where this is known.

U: indicates that the Apogee is not available.
T: ·indicates that the rocket was tracked but no record of altitude is available.

Launch site: This is indicated for each group of rockets.

A: Andøya   S: Sardinia
K: Kiruna   W: Woomera
E: Euboea, Greece

British experiments in ESRO sounding rockets

Site	Launch date	Vehicle reference	Experiment	Group	Altitude attained (km)	Result
1965 S	31 Mar.	S 03	Electron temperature probe	UCL	U	vf
	3 Apr.	S 03	Repeat	UCL	U	s
	11 Aug.	S 05	Stellar UV emission photometer spectrophotometer	ROE	215	sf ps
	30 Sep.	S 04	Grenade experiment	UCL	U	s
			Wind profile by trimethylaluminium release	UCL		s
	2 Oct.	S 04	Repeat	UCL	U	s
			Repeat	UCL		s
	14 Dec.	C 06	Solar X-rays by broad-band spectrometer	Lei/UCL	U	f
			Air density	Lei/UCL		f
1966 A	5 Feb.	CE 016	Auroral ionization (positive ion probes)	UCL	>90	ps
			Auroral electron densities (impedance probes)	Shf		f
	11 Feb.	CE 015	Auroral ionization (positive ion probes)	UCL	U	ps
			Auroral electron densities (impedance probes)	Shf		f
	22 Feb.	CE 013	Auroral ionization structure (positive ion probes)	UCL	111	ps
	24 Feb.	CE 014	Auroral ionization structure (positive ion probes)	UCL	118	ps

Site	Launch date	Vehicle reference	Experiment	Group	Altitude attained (km)	Result
1966	9 Mar.	CE 012	Positive and negative ion concentrations (ion energy spectrometer)	UCL	95	ps
			Electron energy in the aurora (channel multiplier, electrostatic energy analyser)	RSRS		ps
	10 Mar.	CE 011	Positive and negative ion concentrations (ion energy spectrometer)	UCL	115	ps
			Electron energy in the aurora (channel multiplier, electrostatic energy analyser)	RSRS		ps
E	15 May	CE 005	Solar X-ray flux (proportional counters)[a]	Lei	U	ps
	15 May	AE 01	Prototype flight of positive ion probes	UCL	97	s
			Solar Lyman-α (detectors)	RSRS		
	20 May	AE 02	D-region ionization during eclipse (positive ion probes)	UCL	–	vf
			Solar Lyman-α (detectors)	RSRS		vf
	20 May	AE 03	D-region ionization during eclipse (positive ion probes)	UCL	99	s
			Solar Lyman-α (detectors)	RSRS		
	20 May	AE 04	D-region ionization during eclipse (positive ion probes)	UCL	102	s
			Solar Lyman-α (detectors)	RSRS		
	20 May	AE 05	D-region ionization during eclipse (positive ion probes)	UCL	–	vf
			Solar Lyman-α (detectors)	RSRS		vf
	20 May	AE 06	D-region ionization during eclipse (positive ion probes)	UCL	101	s
			Solar Lyman-α (detectors)	RSRS		
	20 May	CE 019	Solar Lyman-α (detectors)	RSRS	~120	
			Electron temperature (probes)	UCL		
	20 May	CE 020	Solar Lyman-α (detectors)	RSRS	U	
			Electron temperature (probe)	UCL		
S	20 Jun.	SK 7	Ozone concentration (solar UV absorption detectors)	MO	–	sf
			Solar intensity 1800–2400 Å (photocells)	MO		sf
	28 Jun.	SK 8	Ozone concentration (solar UV absorption detectors)	MO	U	
			Solar intensity 1800–2400 Å (photocells)	MO		
A	26 Nov.	SK 16	Random stellar UV survey	ROE	2.5	vf
S	29 Nov.	CE 08	Fine structure of auroral ionization (positive ion probe)	MSSL	147	s
	30 Nov.	CE 07	Fine structure of auroral ionization (positive ion probe)	MSSL	145	s
1967 K	16 Feb.	C 10/1	Negative ion concentration (probe)	MSSL	U	ps
			Electron temperature (probe)	MSSL		

Site	Launch date	Vehicle reference	Experiment	Group	Altitude attained (km)	Result
1967	22 Feb.	C 10/2	Negative ion concentration (probe)	MSSL	U	ps
			Electron temperature (probe)	MSSL		
S	22 May	S 11/2	Night sky UV and visual spectra (spectrometer)	ROE	205	s
	26 May	S 05/2	Stellar UV emission (photometer)	ROE	220	s
			Stellar UV emission (spectro-photometer)	ROE		s
	27 Sep.	S 26/1	Lyman-α flux (detector)[b]	RSRS	183	s
			Solar X-rays (detector)[c]	RSRS		
			Electron density (lf propagation)	RSRS		
			Electron density (probe)	Bir		
			Electron temperature (probe)	Bir		
	4 Oct.	S 19/1	Electron temperature (probe)	MSSL	206	s
	7 Oct.	S 26/2	Lyman-α flux (detector)	RSRS	164	
			Solar X-rays (detector)	RSRS		
			Electron density (lf propagation)	RSRS		
			Electron density (probe)	Bir		
			Electron temperature (probe)	Bir		
	10 Oct.	S 19/2	Electron temperature (probe)	MSSL	206	s
K	24 Nov.	C 13P	Negative ion concentration (ion energy probe)	MSSL	160	vf
			Electron energy in aurora (channel multiplier detector)	RSRS		vf
1968 K	1 Feb.	C 36/1	Temperature and wind velocity (grenades and TMA dispenser)	UCL	U	s
	4 Feb.	C 36/2	Temperature and wind velocity (grenades and TMA dispenser)	UCL	U	s
	6 Mar.	C 14P	Auroral ionization structure (positive ion probes)	MSSL	U	ps
	27 Mar.	S 16/1	Electron temperature (probe)	MSSL	U	s
			Auroral ionization structure (positive ion probes)	MSSL		s
			Magnetic fields in aurora (proton magnetometer)	IC		ps
	20 May	C 42/1	Micrometeorite flux (collection surface)	Bkb	U	s
	7 Jun.	C 42/2	Micrometeorite flux (collection surface)	Bkb	U	s
	12 Aug.	A 50/1	Operation of positive ion density probe	MSSL	U	ps
	12 Aug	A 50/2	Operation of positive ion density probe	MSSL	U	ps
S	7 Oct.	S 47/1	UV sky brightness (photo-multiplier)	ROE	U	s
			Stellar UV emission (photometer)	ROE	U	ps
K	15 Oct.	C 45/1	Temperature and wind velocity (grenades and TMA dispenser)	UCL	U	s
	1 Nov.	C 45/2	Temperature and wind velocity (grenades and TMA dispenser)	UCL	U	s
S	22 Nov.	S 41/1	Solar X-rays (Bragg crystal spectrometer)	Lei	U	s

Site	Launch date	Vehicle reference	Experiment	Group	Altitude attained (km)	Result
1968	3 Dec.	S 27/1	Integrated night sky spectroscopy (Cassegrain telescope and photo-multiplier)	ROE	U	s
			Stellar UV emission (photometer)	ROE		s
1969 A	25 Feb.	A 40/1	Positive ion density (probes)	MSSL	U	s
	25 Feb.	A 40/2	Positive ion density (probes)	MSSL	U	ps
K	15 Mar.	S 43/2	Electron density in aurora (impedance probe)	Bir	U	s
			Electron temperature in aurora (double Langmuir probe)	Bir		s
A	13 Apr.	A 40/3	Positive ion density (probes)	MSSL	U	vf
		A 40/4	Positive ion density (probes)	MSSL	U	ps
	14 Apr.	A 40/5	Positive ion density (probes)	MSSL	U	vf
		A 40/6	Positive ion density (probes)	MSSL	U	vf
K	5 Oct.	S 29/1	Wind velocities (TMA release)	Bel	225	s
			Electron density (probe)	Bir		s
			Electron temperature (probe)	Bir		s
			Auroral particle energy (channel multiplier detector)	RSRS		s
	17 Oct.	S 29/2	Wind velocities (TMA release)	Bel	225	s
			Electron density (probe)	Bir		s
			Electron temperature (probe)	Bir		s
			Auroral particle energy (channel multiplier detectors)	RSRS		s
	10 Nov.	S 46/1	Electron density profile in aurora (Langmuir probe)	Shf	214	s
1970 S	24 Jan.	S 56A/1	Solar Lyman-α measurements	RSRS	T	s
			Faraday rotation	RSRS		
K	4 Feb.	C 57/1	Grenade experiments	UCL	T	s
W	11 Feb.	S 72	UV high resolution spectroscopy	UCL	T	ps
K	14 Feb.	S 66/1	Electron temperature probe	UCL	T	ps
	24 Feb.	S 70/1	Electron density and temperature probe	Bir	T	s
	25 Feb.	S 61/1	Vertical oxygen distribution	MO	T	s
	26 Feb.	S 61/2	Vertical oxygen distribution	MO	T	s
	2 Apr.	S 66/2	Electron temperature probe	UCL	T	s
	3 Apr.	C 57/2	Grenade experiments	UCL	T	s
	4 Apr.	S 28/1	Sodium releases	Bel	T	s
			Electron density and temperature probes	Bir		s
	7 Jun.	C 48/1	Lyman-α measurements	UCL	T	s
S	3 Jul.	S 63/1	Positive ion probes	UCL	T	s
	8 Jul.	S 63/2	Positive ion probes	UCL	T	s
	23 Jul.	S 56/A2	Solar Lyman-α measurements	RSRS	T	s
			Solar X-rays (2–8 Å)	Lei/ RSRS		s
			Faraday rotation	RSRS		s
	24 Jul.	S 65/2	RF impedance probe	Shf	T	s
			Lyman-α measurements	RSRS		s
	27 Jul.	S 65/1	RF impedance probe	Shf	T	s
			Lyman-α measurements	RSRS		s
K	13 Aug.	C 48/2	Lyman-α measurements	UCL	T	s

	Site	Launch date	Vehicle reference	Experiment	Group	Altitude attained (km)	Result
	W	19 Oct.	S 58	Solar line spectrograph	ARU	T	f
	K	28 Oct.	S 67/1	Auroral electrons and protons	RSRS	T	s
		29 Oct.	S 67/2	Auroral electrons and protons	RSRS	T	s
	S	6 Dec.	S 69	X-ray crystal spectrometer	Lei	T	s
1971	K	26 Jan.	P 40/7	Positive ion density	MSSL	T	s
		28 Jan.	S 46/2	Langmuir probe	Shf	T	s
	S	6 Feb.	S 80/1	Electron density	Bir	T	s
				Electron temperature	Bir		s
				Night-time Lyman-α radiation	RSRS		s
		7 Feb.	S 80/2	Electron density	Bir		s
				Night-time Lyman-α radiation	RSRS	T	s
				Electron temperature	Bir		s
		7 Feb.	S 80/3	Electron density	Bir		s
				Night-time Lyman-α radiation	RSRS	T	s
				Electron temperature	Bir		s
	W	22 Feb.	S 85	High resolution UV stellar spectroscopy	MSSL	T	f
	S	28 Feb.	S 54	X-ray spectroheliography	MSSL	T	f
		11 Mar.	S 55	Stellar X-ray spectra and interstellar absorption	Lei	T	s
	K	12 Mar.	S 16/2	Electron temperature	MSSL	T	s
				Positive ion density	MSSL		s
				Ionospheric currents (magnetometers)	RSRS		s
	S	23 Mar.	S 56/B1	Solar X-rays from 2–8 Å	Lei/ RSRS	T	f
				Electron density and collisional frequency (Faraday rotation)	RSRS		f
				Solar Lyman-α	RSRS		f
		14 May	S 56/B2	Solar X-rays from 2–8 Å	Lei/ RSRS	T	s
				Electron density and collisional frequency (Faraday rotation)	RSRS		s
				Solar Lyman-α	RSRS		s
	W	28 Sep.	S 47/2	Stellar photometry	ROE	T	s
	S	20 Oct.	S 89	Solar X-ray spectra (crystal spectrometer)	Lei	T	f
	K	25 Nov.	S 70/2	Ionospheric currents (magnetometers)	RSRS	T	s
				Electron density and temperature (probe)	Bir/ ESTEC		s
1972	K	2 Mar.	S 77/1	Auroral particles (channel multipliers, electrostatic analysers and geiger counters)	RSRS	237	s
	S	25 Jun.	S 91	Cosmic X-ray background (collimated X-ray detectors)	MSSL	213	ps
	K	24 Sep.	S 77/2	Auroral particles (channel multipliers, electrostatic analysers and geiger counters)	RSRS	227	s
		13 Oct.	S 105	Measurements of suprathermal electrons during an auroral event (electrostatic analyser)	MSSL	207	s

British experiments in ESRO/ESA scientific satellites

	Launch date	Satellite reference	Experiment	Group
1968	16 May	ESRO II (IRIS)	Detection of high energy cosmic ray electrons	Leeds Univ.
			Total flux of solar X-rays	Lei/UCL
			Measurements of trapped radiation, Van Allen belt protons and cosmic ray protons	IC
1968	3 Oct.	ESRO 1 (AURORAE)	Temperature and compositions of positive ions	MSSL
			Temperature and density of electrons	MSSL
			Energy spectra and fluxes of auroral particles	RSRS
1968	5 Dec.	HEOS-A1	Measurement of interplanetary magnetic field by fluxgate magnetometer	IC
			Detection of high energy cosmic ray protons	IC
			Detection of solar protons	IC
1969	1 Oct.	ESRO-1B (BOREAS)	Temperature and composition of positive ions	MSSL
			Temperature and density of electrons	MSSL
			Energy spectra and fluxes of auroral particles and solar protons	RSRS
			Energy spectrum of electron flux	RSRS
1972	31 Jan.	HEOS-A2	Measurement of interplanetary magnetic field by fluxgate magnetometer	IC
1972	9 Mar.	ESRO TD-1	Sky map through UV sensitive telescope	ARU/ROE/ Rutherford/ Atlas Labs
1972	22 Nov.	ESRO 4	Measurement of positive ion densities by counters	MSSL
1977	20 Apr.	GEOS-1	Measurement of suprathermal electrons in range 5–500 Å	MSSL/ESA
1977	22 Oct.	ISEE-B	Magnetospheric measurements as part of a three-satellite project jointly with NASA	IC/Univ. of California
1978	14 Jul.	GEOS-2	Filter bank analysis of very low frequency wave fields	Shf/MSSL
			Measurement of suprathermal electrons	MSSL

APPENDIX E

British participation in international co-operative sounding rocket experiments and international spacecraft

NOMENCLATURE

Rockets: The type of rocket is specified in each case. Launch sites are indicated as follows:

A:	Andøya	K:	Kiruna
Ar:	Argentina	M:	Mauritania
B:	Biscorasse, France	N:	Nova Scotia
CP:	Cape Parry, Canada	S:	Sardinia
Eg:	Eglin Air Force Base, USA	So:	Sonmiani, Pakistan
El:	El Arenosillo, Spain	W:	Woomera
FC:	Fort Churchill, Canada	WI:	Wallops Island, USA
Hg:	Hammaguir, Sahara	WS:	White Sands, USA

Sounding rockets

	Site	Launch date	Vehicle reference	Experiment	Group	Apogee (km)	Result
1963	Hg	12 Jun.	*Centaure* (France)	Sporadic-E (probe)	UCL		s
		13 Jun.	*Centaure* (France)	Sporadic-E (probe)	UCL		f
		18 Jun.	*Veronique* (France)	Electron density (plasma probe)	Bir		vf
		19 Jun.	*Veronique* (France)	Electron density (plasma probe)	Bir		vf
	WI	13 Dec.	*Nike–Apache* (US)	Solar X-rays (non-imaging cameras)	Lei		vf
1964		21 Apr.	*Aerobee* (US)	Solar X-rays (non-imaging cameras)	Lei		vf
	Hg	4 May	*Centaure* (France)	Electron temperature (probe)	UCL		f
				Sporadic-E (probe)	UCL		f
		8 May	*Centaure* (France)	Electron temperature (probe)	UCL		s
				Sporadic-E (probe)	UCL		s
	WI	15 Jul.	*Nike–Apache* (US)	Electron density (plasma probe)	Bir		s
	Hg	5 Oct.	*Centaure* C 62 (France)	Electron density (plasma probe)	Bir		sf
				Electron temperature (probe)	UCL		sf
	Hg	7 Oct.	*Centaure* C 63 (France)	Electron density (plasma probe)	Bir		s
				Electron temperature (probe)	UCL		
	WI	19 Nov.	*Nike–Apache* 14-149 (USA)	Electron density (plasma probe)	Bir		s

	Site	Launch date	Vehicle reference	Experiment	Group	Apogee (km)	Result
1965	K	1 Feb.	*Nike–Apache* (Norway/USA)	Electron energy in Aurora (channel multiplier detector)	RSRS		vf
		15 Feb.	*Nike–Apache* (Norway/USA)	Electron energy in aurora (channel multiplier detector)	RSRS		vf
		23 Mar.	*Nike–Apache* 14-229 (USA)	Electron density (plasma probe)	Bir		vf
	So	29 Apr.	*Rehbar* 7 (Pakistan)	Wind structure (grenades)	UCL		
		30 Apr.	*Rehbar* 8 (Pakistan)	Wind structure (grenades)	UCL		
	WS	16 Nov.	*Aerobee-LUSTER* (USA)	Interplanetary dust (collection surface)	Bkb		s
	Hg	16 Dec.	*Centaure* 131 (France)	E-region ionization (electron temperature and positive ion probes)	UCL		s
		16 Dec.	*Centaure* 128 (France)	E-region ionization (electron temperature and positive ion probes)	UCL		s
		20 Dec.	*Centaure* 130 (France)	E-region ionization (electron temperature and positive ion probes)	UCL		s
1966	Hg	17 Jan.	*Centaure* 127 (France)	E-region ionization (electron temperature and positive ion probes)	MSSL		
	So	24 Mar.	*Rehbar* 11 (Pakistan)	Wind structure (grenades)	UCL		s
		27 Mar.	*Rehbar* 12 (Pakistan)	Wind structure (grenades)	UCL		s
		26 Apr.	*Rehbar* 13 (Pakistan)	Wind structure (TMA dispenser)	UCL		s
	A	26 May	*Dragon* 20 (France)	Electron density (rf probe)	Bir		s
				Electron temperature (swept differential probe)	Bir		
		15 Jun.	*Dragon* 19 (France)	Electron density (rf probe)	Bir		vf
				Electron temperature (swept differential probe)	Bir		vf
	WS	22 Oct.	*Aerobee* (USA)	Interplanetary dust (collection surface)	Bkb		s
1967	A	6 Mar.	*Nike–Apache* (Norway)	Electron energy in aurora (channel multiplier and electrostatic analyser)	RSRS	190	s
	B	5 Jul.	*Rubis* 4 (France)	Electron density and temperature profile (probe)	Bir	1600	s

	Site	Launch date	Vehicle reference	Experiment	Group	Apogee (km)	Result
	So	29 Nov.	*Rehbar* 14 (Pakistan)	Wind and temperature measurement (grenades)	UCL	150	ps
1968	A	1 Oct.	*Nike–Tomahawk* F19[A] (Norway)	Electron flux and energy spectrum	RSRS		s
1969	K	17 Jan.	*Nike–Cajun* SNC 2/1 (Sweden)	Atmospheric structure (grenades)	UCL		s
		19 Jan.	*Nike–Cajun* SNC 2/2 (Sweden)	Atmospheric structure (grenades)	UCL		s
		23 Jan.	*Nike–Cajun* SNC 2/3 (Sweden)	Atmospheric structure (grenades)	UCL		s
		25 Jan.	*Nike–Cajun* SNC 2/4 (Sweden)	Atmospheric structure (grenades)	UCL		s
1970	WI	7 Mar.	*Aerobee* (USA)	Solar flash spectrum (solar eclipse campaign)	Cul		
	N	7 Mar.	*Black Brant III* (Canada)	Solar Lyman-α and X-ray fluxes (solar eclipse campaign)	RSRS		
1971	FC	13 Jan.	*Black Brant* ADD-II-124	Ozone density	MO	U	s
	A	Jan.	*Nike-Tomahawk* Ant 38	Neutral atmosphere winds	UCL	U	
	Eg	6 May	*Nike-Iroquois*	Electron temperature (SPRITE probe)	MSSL	U	f
				RF mass spectrometer	MSSL		s
	Eg	7 May	*Nike–Iroquois*	Electron temperature (SPRITE probe)	MSSL	U	
				RF mass spectrometer	MSSL		
1972	N	10 Jul.	*Black Brant* BB59	Solar Lyman-α and X-ray	RSRS/CRC	143	s
			BB60	Solar Lyman-α and X-ray	RSRS/CRC	148	s
			BB61	Solar Lyman-α and X-ray	RSRS/CRC	144	s
			BB62	Solar Lyman-α and X-ray	RSRS/CRC	152	s
1973	M	30 Jun.	*Aerobee* 170	UV eclipse coronal spectrum (80–300 nm) (Wadsworth spectrograph)	IC/AL/ CRESS/ HCO	180	sf
	Ar	22 Mar.	*Skylark* SL 1182	Earth IR and visible spectrum photography (Hasselblad cameras and F24 aerial camera)	RAE/BAC/ CNIE	240	ps

440   *Appendix E*

	Site	Launch date	Vehicle reference	Experiment	Group	Apogee (km)	Result
1973	Ar	28 Mar.	*Skylark* SL 1181	Earth IR and visible spectrum photography (Hasselblad cameras and F24 aerial camera)	RAE/BAC/ CNIE	240	s
	A	4 Aug.	*Skylark* SL 1081	Earth IR and visible spectrum photography (Hasselblad cameras)	RAE/BAC/ DFVLR/ SSC	240	sf
1974	A	6 Feb.	*Black Brant* D 101	Study of auroral charged particle origin and acceleration (energetic ion spectrometer proton and electron keV energy detectors)	Sth/ NRDE/ ESTEC/ KTH	540	ps
	WI	29 Jun.	*Tomahawk* UT 9.213-2	Ionospheric electric field (long boom probes)	UCL/ AFCRL/ NASA	115	ps
		30 Jun.	*Tomahawk* UT 9.301-1 UT 9.207-4 UT 9.301-4 PT 10.301-6 UT 9.105-9 UT 9.301-7	Neutral wind and electric field (lithium and barium thermite and TMA releases, ground-based photometers and photographic observations)	UCL/ AFCRL/ NASA	180 200	s
1975	WI	20 Jan.	UT9.213-1	Ionospheric electric field dipole – a part of complex payload including AFCRL cryogenic mass spectrometer (long boom antennae)	UCL/ GSFC/ AFCRL	140	f

Petrel rockets launched from El Arenosillo

	Site	Launch date	Vehicle reference	Experiment	Group	Apogee (km)	Result
1976	El	4 Jan.	BVI/1	Daytime neutral wind profiles in the ionosphere during Federal Republic of Germany Winter Anomaly Campaign (lithium trails tracked by ground-based photometers)	West European groups	135	s
		21 Jan.	BVI/2	Daytime neutral wind profiles in the ionosphere during Federal Republic of Germany Winter Anomaly Campaign (lithium trails tracked by ground-based photometers)	West European groups	135	s

	Site	Launch date	Vehicle reference	Experiment	Group	Apogee (km)	Result
1976		21 Jan.	BVI/3	Daytime neutral wind profiles in the ionosphere during Federal Republic of Germany Winter Anomaly Campaign (lithium trails tracked by ground-based photometers)	West European groups	135	s

Fulmar, Malemute and *Tomahawk* rockets launched from Andøya

	Site	Launch date	Vehicle reference	Experiment	Group	Apogee (km)	Result
	A	21 Nov.	*Fulmar* F2	Acceleration processes in aurora (sensors for electrons, ions and electrostatic waves)	AL/Sx	140	ps
		27 Nov.	*Malemute* F38	Harang discontinuity investigation (sensors for electrons, ions, plasma, X-rays, vlf and electric fields (dc and ac))	AL/MSSL/ GSFC/Bn NDRE/Gr	530	s
		27 Nov.	*Tomahawk* F39	Harang discontinuity investigation (sensors for electrons, ions, plasma, X-rays, vlf and electric fields (dc and ac))	AL/MSSL/ GSFC/Bn NDRE/Gr	208	s
		11 Dec.	*Fulmar* F5	Measurement of structure and response of thermosphere and ionosphere during strong geomagnetic substorm (sodium trail, lidar observations, fluxgate magnetometer, Fabry–Perot interferometer and photometer/ spectrometer)	AL/BTI	213	ps
1976	K	2 Mar.	*Nike–*	Injection-Trigger experiment (caesium TNT point release/ TMA trail)	SSC/UIO/ Lei/UCL	165	s
1977	K	15 Feb.	*Skylark* S 21/2	Neutral winds, ionospheric drift and properties of dense ion/electron cloud (trimethyl aluminium trail payload and caesium/TNT release)	UCL/UIO	170	s

	Site	Launch date	Vehicle reference	Experiment	Group	Apogee (km)	Result
1977	WS	20 Feb.	*Astrobee-F*	Spectroscopy and imaging of solar corona (Wolter I X-ray telescope and multi-crystal Bragg spectrometer)	Lei/ASE	270	ps
		29 Jul.	*Astrobee-F*	X-ray mapping of the Cygnus Loop supernova remnant (Wolter I X-ray telescope)	Lei/MIT	250	s
	CP	5 Dec.	*Black Brant YB 03*	Electron energy spectrum (electron detector)	AL	290	s
1978	WS	8 Mar.	*Astrobee-F*	X-ray mapping – supernova remnants IC443 and Puppis A (Wolter I X-ray telescope with imaging proportional counter read-out)	Lei/MIT		s

British experiments in NASA spacecraft and other co-operative projects

	Launch date	Satellite reference	Experiment	Group
1964	25 Aug.	EXPLORER 20	Measurement of positive ion energy	UCL
1965	29 Nov.	EXPLORER 31	Electron temperature probe and positive ion mass spectrometer	UCL
1965	6 Dec.	FR-1 (French)	Electron density by radio frequency probe	Bir
1966	18 Jul.	GEMINI GT-10	Surface collection of interplanetary dust	Bkb
1966	11 Nov.	GEMINI GT-12	Ditto	Bkb
1967	18 Oct.	OSO-D	Total solar soft X-ray flux	MSSL/Lei
			Total flux of solar He 2 radiation 304 Å	MSSL
1968	4 Mar.	OGO-E	Electron temperature in magnetosphere by probe measurement	MSSL
			Direction of incidence of energetic galactic gamma rays	Sth
1969	22 Jan.	OSO-5	Solar X-ray detection with scanning proportional counter spectrograph	MSSL/Lei

	Launch date	Satellite reference	Experiment	Group
1969	9 Aug.	OSO-6	Solar He 1 and He 2 resonance radiation measured with grazing incidence polychromator	MSSL
1970	8 Apr.	NIMBUS 4	Measurements of upper atmosphere temperature by six-channel selective chopper radiometer	Oxf/Reading Univ, later Heriot Watt Univ.
1972	21 Aug.	OAO-3 (COPERNICUS)	Detection of cosmic X-ray sources by a triple telescope array	MSSL/Lei
1972	11 Dec.	NIMBUS 5	Upper atmosphere temperature sounding by selective chopper radiometer	Oxf/Heriot Watt
1975	Jun.	NIMBUS 6	Upper atmosphere temperature sounding by pressure modulator radiometer	Oxf/ Rutherford Lab.
1977	22 Oct.	ISEE-A	Magnetospheric measurements as part of a three-satellite project with ESA/NASA	IC/Univ. of California
1978	26 Jan.	IUE	UV spectroscopy at wavelengths between 1150 and 3200 Å	UCL/AL/ESA/ GSFC
1978	May	Pioneer Venus Orbiter	Temperature sounding of the high atmosphere of Venus	Oxf/Jet Propulsion Lab. California
1978	15 Aug.	ISEE-C	Magnetospheric measurements as part of a three-satellite project with ESA/NASA Observation of low energy protons	IC/ESA/– Sterrekundig Institut Utrecht
1978	Oct.	NIMBUS 7	Stratospheric and mesospheric sounding giving measurements of temperature and distribution of constituents	Oxf
1980	14 Feb.	SMM (Solar maximum mission)	Hard X-ray imaging spectrometer X-ray polychromator	Bir/Lei/– Univ. of Utrecht MSSL/AL/– Lockheed Palo Alto Research Lab.
1981	3 Aug.	DYNAMICS EXPLORER	Direct global thermospheric wind observations by use of single etalon Fabry–Perot interferometer	UCL/Univ. of Michigan

APPENDIX F

Details of the principal UK space science groups 1982

Individuals are classified into what are believed to be their principal fields of activity. Class 'a' signifies a primary concern with practical experiments in space science. Class 'b' signifies a primary concern with theoretical or laboratory work in the field of space science.

Class 'ab' signifies concern with all aspects of the field.

University		
Aston in Birmingham		
Department of Mathematics	Dr C.J. Brookes	b
Queen's University of Belfast		
Department of Pure and	Dr B. Bates	a
Applied Physics	Dr R.G.H. Greer	ab
Birmingham		
Department of Physics	Dr S.A. Durrani	b
Department of Space Research	Professor A.P. Willmore	ab
Bristol		
Department of Physics	Professor P.H. Fowler	ab
Cambridge		
Department of Earth Sciences	Dr S.O. Agrell	b
	Dr N.J. Charnley	b
	Dr C.T. Pillinger	b
Heriot-Watt University Edinburgh		
Department of Physics	Professor S.D. Smith	ab
Essex		
Department of Physics	Dr D.J. Barber	b
Hull		
Department of Physics	Professor G.H.A. Cole	b
	Professor G.F.J. Garlick	b
	Professor S.A. Ramsden	ab
University of Kent at Canterbury		
The Electronics Laboratories	Professor R.C. Jennison	ab
	Dr J.A.M. McDonnell	ab

444

University

Lancaster

Department of Environmental	Dr G. Fielder	b
Sciences	Dr J.K. Hargreaves	b

Leeds

Department of Physics	Professor P.L. Marsden	ab
	Dr C.J. Hatton	b

Leicester

Department of Astronomy	Professor A.J. Meadows	b
Department of Physics	Professor K.A. Pounds	ab

London

Imperial College of Science and
Technology

Department of Physics	Professor P.C. Hedgecock	ab
	Professor J.W. Dungey	b
	Dr G.E. Hunt	b

London

Queen Mary College

Department of Physics	Professor J.A. Bastin	ab
Department of Applied	Professor I.W. Roxburgh	b
Mathematics	Dr I.P. Williams	b

London

University College London

Department of Physics	Dr J.E. Guest	b
and Astronomy	Professor R.E. Jennings	ab
	Dr D. Rees	ab
	Professor M.J. Seaton	b
	Professor R. Wilson	ab
Mullard Space Science Laboratory	Professor R.L.F. Boyd	ab

Manchester

Nuffield Radio Astronomy	Professor F.G. Smith	ab
Laboratories		

Manchester

Institute of Science and	Dr J.E. Geake	b
Technology	Dr C.P.R. Saunders	b
	Dr H.J. Axon	b

Newcastle upon Tyne

School of Physics	Professor S.K. Runcorn	b

Nottingham

Department of Civil Engineering	Professor V. Ashkenazi	b

Oxford

Clarendon Laboratory	Dr F.W. Taylor	ab

University

Sheffield

Department of Physics	Professor T.R. Kaiser	ab

Southampton

Department of Physics	Professor G.W. Hutchinson	ab
	Dr D.A. Ramsden	ab
	Dr A.J. Dean	a
	Dr Pamela Rothwell	ab

Surrey

Telecommunications Research Group	Dr M.N. Sweeting	a

Sussex

Department of Physics	Dr G. Martelli	ab
Astronomy Centre	Professor W.H. McCrea	b

University College of Wales

Aberystwyth

Department of Physics	Professor L. Thomas	ab

York

Department of Physics	Professor M.M. Woolfson	b
	Dr D. Orr	b

British Antarctic Survey	Dr M.J. Rycroft	b
British Museum (Natural History)		
Mineralogy Department	Dr. R. Hutchinson	b
Meteorological Office Bracknell	Sir John Mason	
Rutherford Appleton Laboratory Didcot	Professor J.T. Houghton	
National Physical Laboratory Teddington	Dr P. Dean	
Royal Aircraft Establishment Farnborough	Dr D.G. King-Hele	
Royal Greenwich Observatory Herstmonceux	Professor A. Boksenberg	
Royal Observatory Edinburgh	Professor M.S. Longair	

APPENDIX G

The changes in the value of money

It is only possible to give an approximate scale for the changes in the value of money relevant to the UK space science programme over the period from 1955 to 1981.

Expenditure on space science covers a variety of goods and services and includes both sterling and foreign currencies, for which the exchange rates vary significantly from time to time. The effects of inflation both at home and abroad and of variations in the exchange rates over some 25 years are therefore difficult to ascertain.

However, an indication can be obtained from a table based on the General Index of Retail Prices in the UK.

Taking the annual average for 1955 as 100, the corresponding rounded figures for succeeding years are:

Year	Average
1955	100
1960	112
1965	132
1968	148
1970	166
1971	181
1972	194
1973	212
1974	246
1975	306
1976	356
1977	413
1978	447
1979	507
1980	598
1981	669

ANNEX 1

Discussion on observations of the Russian artificial earth satellites and their analysis

ANNEX 2

Charter of COSPAR

I **Purpose and objectives**

COSPAR shall be a Special Committee of the ICSU.

The purpose of COSPAR is to further on an international scale the progress of all kinds of scientific investigations which are carried out with the use of rockets or rocket-propelled vehicles. COSPAR shall be concerned with fundamental research. It will not normally concern itself with such technological problems as propulsion, construction of rockets, guidance and control.

These objectives shall be achieved through the maximum development of space research programs by the international community of scientists working through ICSU and its adhering national academics and unions. Any arrangements involving national territories should be made by bilateral or multilateral discussion between the nations concerned. As a non-political organization, COSPAR shall not, as a matter of policy, recommend any specific assistance of one nation by another. It will, however, welcome information concerning such arrangements and provide a convenient assembly in which such arrangements may informally be proposed and discussed.

Recognizing the need for international regulation and discussion of certain aspects of satellite and space probe programs, COSPAR shall keep itself informed of United Nations or other international activities in this field, to assure that maximum advantage is accorded international space science research through such regulations and to make recommendations relative to matters of planning and regulation that may affect the optimum program of scientific research.

COSPAR shall report to ICSU those measures needed in the future to achieve the participation, in the international programs of space research, of all countries of the world with those which are already actively engaged in research programs within the domain of COSPAR.

II Composition

The composition of COSPAR shall be as follows:

(a) One representative designated by each national scientific institution adhering to ICSU which is actively engaged in space research and desires representation in COSPAR

(b) One representative designated by each international scientific union federated in ICSU which desires to participate in COSPAR. In order that, in addition to broad scientific representation, there should also be as wide a distribution of nationality as possible among the union representatives, the advice or assistance of ICSU shall be provided to the unions, if desired.

The rights and duties of the national scientific institutions represented in COSPAR shall be:

1. To be informed about and to send representatives to all meetings of the full COSPAR or sponsored by COSPAR and to participate in all the discussions therein.
2. To establish scientific channels for obtaining data and carrying out space experiments and to participate in obtaining data and evaluating information resulting from such experiments.
3. To make available scientific results of space research which may be conducted as part of their participation in the work of COSPAR.
4. To contribute to the financial support of COSPAR, to an extent recommended by the COSPAR Executive Council, affirmed by COSPAR and approved by ICSU.
5. To vote on all matters.

The rights and duties of the international scientific unions represented in COSPAR shall be:

1. To be informed about and to send representatives to all meetings of the full COSPAR and to participate in all discussions therein.
2. To maintain liaison with COSPAR and its Working Groups in any way they may deem appropriate in order to ensure integration of the results obtained by space experiments with those obtained by other methods of scientific research, to avoid duplication and to achieve an efficient division of tasks.
3. To participate in evaluating and disseminating information resulting from space experiments.
4. To vote on all matters which do not involve major items of income or expenditure of money by COSPAR, or considerable expenditure of money by the national scientific institutions.

III Officers

COSPAR shall elect from among its own members a President, two Vice-Presidents, and four other members to serve on an Executive Council, hereinafter described. The method of election shall be such as to ensure a representation consistent with the distribution of major effort in space research among the members of COSPAR.

IV Executive Council

The Executive Council shall be responsible for administering and conducting the affairs of COSPAR between meetings in accordance with the policies and directives given to it by COSPAR and shall be responsible for the formulation of plans and policies for consideration by the full COSPAR. The Executive Council shall consist of the President, the Vice-Presidents and the four other members specifically elected for this responsibility, and of all the representatives of the scientific unions which are members of COSPAR. Only the elected members may vote on matters involving major items of income or expenditure of money by COSPAR, or considerable expenditure of money by the national scientific institutions, but all members of the Executive Council shall have the right to be heard and to have their opinions recorded on all matters.

Any decisions of the Executive Council must be confirmed by a vote of two-thirds of the seven elected members. The seven elected members may meet or vote separately and when acting in this way, shall be known as the Bureau of the Executive Council. The Chairmen of Working Groups shall be invited to attend all meetings of the Executive Council as consultants. The President and Vice-Presidents of COSPAR shall also act as President and Vice-Presidents respectively of the Executive Council and its Bureau.

V Finance Committee

In accordance with Rule 15(a) (i) of ICSU Rules for Special Committees, there shall be a Finance Committee consisting of two members: representatives of national scientific institutions.

VI Conduct of business

1. The ICSU Rules for Special Committees shall be adopted for the conduct of COSPAR business.
2. COSPAR shall establish its own by-laws and procedures within the framework of this Charter and the ICSU Rules for Special Committees.
3. COSPAR may establish scientific working groups from time to time for the examination of special problems.
4. The President of COSPAR shall keep the Secretary General of ICSU fully and promptly informed of all COSPAR activities.
5. After approval by COSPAR, the Bureau will submit to ICSU budget estimates for all of the activities of COSPAR, and recommendations for a scale of contributions which shall be required from the participating national scientific institutions.

By-laws
1. These by-laws shall go into effect automatically upon approval by ICSU of the COSPAR Charter.
2. Changes in and additions to these By-laws may be made at any time as provided for in Section 6 (iv) below.
3. Election of officers.
   (i) All officers shall be elected from among the members of COSPAR.
   (ii) The President shall be elected for a term of three years by the full membership from a slate of nominees submitted by the Executive Council or, if the Executive Council is unable to nominate, from the floor of COSPAR.
   (iii) Two Vice-Presidents shall be elected for a term of three years by the full COSPAR, one from a slate of nominees submitted by the Academy of Sciences of the USA and one from a slate of nominees submitted by the Academy of Sciences of the USSR, at present the only two nations engaged in launching earth satellites and cosmic space vehicles.
   (iv) Four additional members of the Bureau shall be elected for a term of three years by COSPAR, two from a slate of nominees submitted by one of the Vice-Presidents, and two from a slate submitted by the other.
   (v) If an office is vacated for any reason it shall be filled for the remainder of the original term by nomination and election as specified above.
   (vi) All officers shall continue to serve until their successors have been duly elected and have accepted.
4. Duties of Officers
   The duly elected President of COSPAR shall preside at all meetings of COSPAR plenum, the Executive Council and Bureau, and shall conduct the affairs of COSPAR between designated meetings. In case of the inability of the President to carry out his responsibilities, the two Vice-Presidents shall alternately, as principal Vice-President, assume the responsibilities of the office of President. Alternation of the office of principal Vice-President shall take place each three months, with the US Vice-President holding office during the first and third quarters of each year, and the USSR Vice-President holding office during the second and fourth quarters. It shall be the prime responsibility of the principal Vice-President upon assuming the presidential responsibility, to convene COSPAR at the first opportunity for the purpose of electing a new President. In case of the inability of the President and Vice-Presidents to discharge these responsibilities, the elected members of the Executive Council may take such steps as may be necessary to ensure the continued activity of COSPAR.
5. Liaison with United Nations
   The Executive Council shall arrange for liaison with the United Nations organizations on the subject of regulations affecting space research to the extent and for the purposes set forth in the Charter.

6. Rules of Order

(i) *Quorum* A quorum of the COSPAR or any of its constituent bodies shall consist of fifty per cent or more of the members of such body. Members of COSPAR may name alternate delegates who, in the absence of the principal delegates, shall have voting authority at COSPAR Meetings and shall be counted in determining a quorum. Votes may only be cast by persons present at the meeting, with the exception noted in (v) below.

(ii) *Right to vote on various subjects* Voting in the COSPAR or any of its constituent bodies shall conform to the provisions of applicable sections of the Charter.

(iii) *Majority Vote* Official actions by COSPAR, except as provided in (iv), may be taken by a simple majority vote of those present and voting for or against each action. Abstentions from the voting will not be considered in determining the majority action but may be entered in the record if so desired by the abstaining delegates.

(iv) *Changes in by-laws and financial assessments* Any change in the By-laws shall require a two-thirds majority vote of the full COSPAR. Actions involving financial assessments on the national scientific institutions which are represented in COSPAR, shall require a two-thirds majority vote of the national institutions.

(v) *Voting in the Executive Council* Resolutions of the Executive Council may be made by a simple majority vote cast in any manner agreed upon by the Executive Council, but must be confirmed as provided in Section IV of the Charter. Decisions of the Bureau of the Executive Council shall be made by a two-thirds vote. Any member of the Bureau unable to attend a meeting may delegate his voting power to his representative.

(vi) *Agenda* An agenda appropriate to the nature of the business to be discussed shall be prepared under the direction of the President for each meeting of COSPAR, its Executive Council, or its Bureau and shall be mailed at least three weeks prior to the meeting. Amendments to the agenda for any meeting may be proposed by any representative present at a meeting and shall be adopted by a simple majority of those present and voting.

(vii) *Reports of Meetings* The preliminary drafts of the reports of all meetings of COSPAR, all meetings of the Executive Council and of the Bureau, and all meetings of Working Groups shall be circulated promptly to all members of the respective groups for approval, such report shall be considered approved as written if specific objections or recommendations for amendment to these drafts are not received within one month of date of transmittal. Following approval, the reports of the meetings of COSPAR, its Executive Council, and Bureau, and of COSPAR Working Groups shall be distributed to all members of COSPAR.

7. Formal Communications

When it is prescribed that COSPAR resolutions or actions be transmitted to ICSU, UNESCO, UN or other bodies, such transmittals shall be in writing, incorporating the language of such resolutions or actions, and copies shall be distributed to the COSPAR Executive Council.

8. Scientific Working Groups

Scientific Working Groups may be established, modified, or discontinued at any time by the full COSPAR upon the recommendations of the Executive Council. Each Working Group shall consist of a Chairman, who will call meetings and preside over them, and several other members, none of whom need necessarily be representatives of scientific bodies maintaining membership in COSPAR. Each Working Group shall adopt such working rules and procedures as may seem appropriate for its work.

Any communications by the working groups requiring action by individuals or groups outside COSPAR shall be subject to approval in advance, either specifically or in principle, by the COSPAR Executive Council.

COSPAR will arrange to provide funds for reasonable secretarial expenses for the Working Groups and for travel expenses in accordance with ICSU regulations, of Chairman, members and invited consultants. Each working group shall prepare, on request, a budget of such expenses, which will be subject to approval and audit as directed by the Bureau after consideration by the Finance Committee. Expenditures within the budget shall be approved by the Working Group Chairman.

ANNEX 3

UK participation in research with artificial satellites.
Paper by H.S.W. Massey

The scientific value of research with artificial satellites

The scientific study of the upper atmosphere of the earth, the solar and lunar influences which affect its behaviour, and the immediate surroundings of the earth in interplanetary space has been proceeding for many years. This has included the investigation of pressure, density, temperature and wind distributions (atmospheric structure), the ionosphere, the airglow and aurora, atmospheric tides, magnetic variations due to circulating currents in the atmosphere or beyond, meteors, cosmic rays and solar disturbances.

In the first instance, this work was carried out exclusively by indirect methods based entirely on the use of equipment on the ground together, of course, with the maximum use of theory. Although remarkable progress has been made, there are certain properties and phenomena which cannot be studied in this way. The most important of many is the nature of the radiation from the sun before it has been modified by the atmosphere. Only a narrow spectrum range of sunlight penetrates to ground level and yet it is the absorbed radiation which is responsible for atmospheric phenomena, such as the ionosphere and the airglow. It is not only the electromagnetic radiation which needs to be studied but also the particle radiation which is responsible, for example, for auroral and magnetic storm phenomena.

The introduction in 1947 of rockets as vehicles to carry instruments up to heights of 100 km or more marked a new phase in research in this field. For the first time it became possible to make direct observations of high altitude phenomena. Many useful results have been obtained and much remains to be done. A British programme using the rocket *Skylark*, developed by the Royal Aircraft Establishment, is in successful operation.

Vertical rocket sounding suffers from the difficulty that the flight time of a rocket is very short, only a few minutes. Almost all of the phenomena to be investigated vary markedly with time as well as place. These variations are very hard to study even with the most ambitious programme of rocket launchings.

The availability of artificial satellites as instrument carriers changes this situation

455

completely. Thus a satellite, with instrument payload comparable with that of a vertical sounding rocket, in the course of a month's revolutions round the earth, can carry out as many observations as several thousand vertical sounding rockets.

It is true that satellites are unable to circulate for useful periods of time at altitudes below about 150 miles so that vertical sounding rockets are still of great importance for observations up to this altitude. However, at higher levels – and no upper limit exists – artificial satellites provide observing stations outside the atmosphere which open up great new possibilities for scientific study. The radiation from the sun, both electromagnetic and corpuscular, unmodified by the atmosphere can be kept under continuous observation. It is already apparent from the solar photographs taken at the comparatively low altitudes attainable by balloons (less than 30 miles) that great contributions to solar physics will be forthcoming from these observations.

We may also anticipate great advances in other branches of astronomy from the availability of satellite observations. In the post-war years the new science of radio astronomy has developed rapidly, advantage being taken of the possibility of observing the universe in radiation of very long wavelength which happens to be able to penetrate the terrestrial atmosphere. Even so, we are only able, at present, to observe in the very narrow range of wavelength which includes visible light and short radiowaves. With a satellite station, outside the atmosphere, it is possible in principle to make observations of the universe in radiation of any wavelength. Already some observations have been made of the distribution of ultra-violet radiation from the sky, using instruments carried in high altitude rockets. The fact that this is possible at all in the short time available in such flights shows that a very wide range of possibilities is opened by the use of artificial satellites. Techniques even of using telescopes in satellites are being developed and there is no doubt that the exploration of the universe will be greatly expanded in variety and depth.

The fact that satellites continually scan the earth as they circulate provides a very suitable means for studying the variations with time and position of quantities of great geophysical and astronomical importance. Some of these we have already described but there are many others. A combination of the information obtainable about the solar radiation outside the atmosphere and about the circulation and heat balance of the atmosphere is clearly of much meteorological importance which may help to provide the means for long-range weather forecasting (see also p. 458).

One can also mention among many other important new opportunities that of observing the variation of the earth's magnetic field continuously over the track of the satellite. Not only will this be of great interest in connection with the resolution of atmospheric and other problems concerning the origin and nature of magnetic and ionospheric storms, but it will also provide new information about the gradual variation with time of the earth's main field, a matter of importance in connection with the origin of this field.

An extensive list of foreseeable scientific applications of artificial satellites may be set out as is done below, but in a wholly new field of research such as this the major discoveries are not predictable. There is no doubt at all that such a wide range of observations are possible that unexpected and very important discoveries will be made. Even at this early stage the discovery by van Allen and his colleagues of the intense belt of radiation above an altitude of a few hundred miles, obtained from instruments carried in the American *Explorer* satellites, was not anticipated and is of great interest and significance. The thorough investigation of this alone will take many years.

Summarizing we can list the following foreseeable scientific applications of artificial satellites:

1. Systematic study of the properties of the high atmosphere at altitudes above 150 miles. This includes pressure, density, temperature, ion concentration and composition, neutral particle composition.
2. Systematic study of solar electromagnetic radiation in the wavelength regions inaccessible from the earth.
3. Observation and exploration of the streams of particles emitted from the sun which produce aurorae and magnetic and ionospheric storms.
4. Systematic observation of the intensity and composition of primary cosmic radiation and of effects due to the outward emission of secondary cosmic ray particles.
5. Search for other radiations which may be present, followed by a systematic study of any which may be found.
6. The initiation of astronomical observations employing radiation in the ultra-violet and X-ray region and eventually in longwave regions inaccessible from the ground.
7. Systematic observation of the fine dust of micrometeorites in the space about the earth.
8. Systematic measurement of the magnetic field at points on satellite orbits; the accuracy being great enough to observe variations of comparable magnitude to those recorded on the ground.
9. Systematic observation of the radiation emitted from the earth, followed by television coverage of atmospheric cloud and other conditions of meteorological interest.
10. Accurate investigation of the earth's gravitational field, with geodetic applications.
11. Possible detection of effects of gravitational fields on time measurement. Continued circulation raises the possibility of detecting small differences in frequency by integration over long time intervals.

In this list emphasis has been placed on the possibility of systematic study. Of the listed applications 1, 2, 3, 5, 6 and 8 cannot be carried out by ground based studies alone and for the reasons advanced earlier vertical rocket sounding cannot provide the continuous observation so important for observation of time and position effects. The world-wide coverage made available in 4 and 9 is a great advantage. There is no doubt that a very great increase in our understanding of geophysical and astrophysical phenomena will be forthcoming from the data provided from satellite observing stations.

The first satellites have shown that the observational programme outlined is a practicable one. Already unexpected information has been forthcoming about the high intensity of soft corpuscular radiation at height of 500 km or more outside the earth. The thorough investigation of this alone will take many years.

Other applications of satellites

Any consideration of military applications is outside the scope of this document.

It is difficult at this stage to predict practical benefits which may ensue from a United Kingdom artificial satellite programme. Mention might, however, be made of the following:

(a) The possibility of developing methods for long-range weather forecasting, arising from the availability of regular systematic world-wide observations of atmospheric cloud and other conditions of meteorological importance, as well as of the entire range of solar radiation before modification by the atmosphere. It is obvious that if much could be done in this direction its economic value would be very great.

(b) The use of satellites as repeater stations for long distance transmission of short radio waves. In this connection it is worth noting that during the recent launching of the American lunar probe, *Pioneer*, the tracking stations at Jodrell Bank, Cheshire, and at Hawaii were actually in communication with each other via the probe.

(c) The use of satellites revolving in high orbit, undisturbed by atmospheric drag, for navigational fixes.

Present British participation in high altitude research with rockets and satellites

(a) *The vertical sounding rocket programme*

This was initiated over three years ago as a joint programme between the Royal Society and the Ministry of Supply and supported by grants from the Air Ministry. Observations up to altitudes of over 100 miles are made with equipment carried by the *Skylark* rocket. Data have already been obtained on the air temperature and winds and on electron density. The further schedule of launchings includes studies of the composition of the ionosphere, the height variation of the intensity of airglow radiation, the concentration of micrometeorites, the intensity of solar X-radiation, and the artificial production of sodium glow. Further experiments are being planned.

(b) *The radio and optical tracking of artificial satellites*

Precise determination of satellite orbits makes it possible to derive information about the air density at the perigee of the orbit (the closest point of approach to the earth's centre) and about the earth's gravitational field. Comparison of apparent position as determined by radio tracking (which is affected by the ionosphere) with real position obtained from optical observations gives information about the electron concentration in the ionosphere at satellite altitudes.

Many British stations have made and are making observations by radio and optical methods of the satellites which pass over this country (*Sputniks* I, II and III and Explorer IV). Data obtained by British observers on the Russian satellite orbits have been of the highest quality. Analysis to obtain information about air density, the figure of the earth and the ionosphere is proceeding.

A Satellite Centre has been established in association with the Radio Research Station at Slough to continue the provision of predicted times of satellite passages for British observers, to extend tracking facilities where needed, to provide computational assistance in connection with satellite data analysis and generally to act as a reference centre for satellite information. The Centre has also been designated a third World Data Centre for Rockets and Satellites in connection with the International Geophysical Year.

Further possibilities of British participation

The wide range of scientific research activities which are possible with instruments carried in artificial satellites point to the importance of further British participation in this field. In this connection the following considerations are relevant:

1. It is unrealistic to suppose that British satellites carrying a useful load of scientific instruments could be launched before five years from now.
2. Meanwhile, it would be of no value, even from the point of view of prestige, to launch a very small satellite weighing only a few pounds. This would still not be possible for some years by which time it would be quite outmoded.
3. Any British launching of satellites would have to be based on military developments.
4. Such developments are proceeding. It is difficult to assign a definite figure to the extra expense involved in adapting military vehicles to launch useful satellites, as allowance for military value can only be made on an arbitrary basis. However, without making allowances of this kind it has been estimated by the Ministry of Supply that the cost of adaptation to launch five satellites of approximately 1000 lb weight each would be about £9 million. The launching of further satellites would cost about £400,000 each. These satellites could be launched into orbits at such altitudes that their lifetimes would be greater than one year.
5. Satellites of this weight and capacity could carry six or more important experiments each. (*Sputnik* III carries nine and makes no attempt to achieve a high degree of miniaturization).

6. Experience gained in making measurements with equipment carried out at high altitudes in *Skylark* rockets would enable British scientists to design and carry out experiments successfully from satellites.

7. In order to meet the cost of development and construction of the scientific equipment required to take full advantage of the satellite launchings indicated under 4 about £100,000 per annum would be required to cover salary and equipment costs. It is assumed that in any one year not more than two satellites would be launched.

8. There would be further expenditure necessary to ensure that tracking facilities for British satellites were adequate. In view of world-wide arrangements for tracking USSR and US satellites including existing British arrangements, this would not be large. Over a five-year period it might amount to £200,000.

9. The expenditure involved is high for pure scientific research. It is, however, comparable with that used in *pure* research in nuclear physics which requires the use of high energy accelerators, expensive to build and to run. The amount of scientific information gained from the successful operation of a number of pieces of equipment in a satellite for a year is very great when it is remembered that the apparatus can operate for 24 hours a day. The successful development of solar batteries should ensure that power will be available for the measuring equipment throughout the life of a satellite.

10. It may be argued that, because of the high rate of acquisition of data in a satellite, there will be nothing worthwhile to do with satellites launched five or more years hence. The answer here is that the variety of subjects to investigate and the time over which they must be studied systematically is so very large that there is no chance that everything of interest will have been done by 1963. Apart from anything else it will be necessary to study systematically all phenomena associated with the sun at least over a full sunspot cycle of 11 years. Even then it would be extraordinary if everything involved could be cleared up and understood after only one cycle of observations. The variety and extent of data to be studied is very large. We have drawn attention on page 456 to the programme of satellite astronomical observation which is of extremely wide range and represents a completely new branch of astronomy. The scope for observation in this field will itself be so large as to require the use of artificial satellite observing stations in the indefinite future. Even if we confine ourselves to the foreseeable subjects of research summarized under items 1–11 on page 457 there is no risk at all that the usefulness of satellites be will confined to a few years only. This is reinforced still further when it is realized that many unforeseen research applications must also develop in the course of time.

The same question has been raised in connection with the use of vertical sounding rockets. Although this work began 11 years ago there remains a very great deal still to do, and one cannot yet foresee a time when it will be complete.

A further situation which presents analogous features arises again in nuclear physics. Construction of new accelerators in the USA has antedated by several years any corresponding development in this country. It has always been maintained, with subsequent justification, that there is still plenty of application to be made of the antedated accelerators when they become available.

11. A further aspect of 10 is that in Britain there is a long tradition of scientific research and the British approach to scientific problems contains something unique born of this long tradition and experience. British scientists would certainly bring original features into any work they may carry out using instruments in artificial satellites. Already at this early stage the British vertical sounding rocket programme has introduced new and desirable features. The success of the British observation programme on the first satellites provides a further example of British ability in this general field.

It is also relevant to point out that Britain has always made big contributions to research on the upper atmosphere, including particularly the ionosphere.

12. The question may be asked as to whether it is necessary for Britain to launch her own satellites. Would it not be possible to arrange for British instruments to be included in satellites launched by other countries?

Even if this possibility arose as some aspect of international collaboration it would be devoid of most of the advantages gained in a throughgoing British enterprise. Instruments launched in, say, an American satellite, would have to conform very largely with American designs in order to be acceptable. This would destroy the main advantage of a different approach to the experimental problems. Experience gained by incorporation of an occasional experiment in a foreign satellite, particularly in the early stages of a British programme, would nevertheless be useful.

The possibility of arranging for another country to use its rockets to launch a British satellite seems to offer no advantage at all, quite apart from the obvious loss of prestige.

13. Is it not possible to share the cost of the programme by arranging for Western European collaboration in the launching of artificial satellites?

As almost all the experience on large rockets and their use in scientific research is at present confined almost entirely to Britain, among Western European countries, there would seem to be big disadvantages in the collaboration at this stage.

Recommendation

In view of the above it is recommended that a British programme for launching artificial satellites on the lines indicated in 4. and 5. be initiated.

October 1958

ANNEX 4

Formal offer of international co-operation by the USA through COSPAR

March 1959

COSPAR has a truly historic opportunity to become an effective force for international co-operation in space research. This co-operation will be most fruitful and meaningful if the maximum opportunity to participate in, and contribute to, all aspects of space research can be provided to the entire scientific community. In this regard, COSPAR can serve as an avenue through which the capabilities of satellite launching nations and the scientific potential of other nations may be brought together.

The United States will support COSPAR in this objective by undertaking the launching of suitable and worthy experiments proposed by scientists of other countries. This can be done by sending into space either single experiments as part of a larger payload or groups of experiments comprising complete payloads.

In the case of individual experiments to become part of a larger payload, the originator will be invited to work in a United States laboratory on the construction, calibration, and installation of the necessary equipment in a US research vehicle. If this is impossible, a US scientist may be designated to represent the originator, working on the project in consultation with him. Or, in the last resort, the originator might prepare his experiment abroad, supplying the launching group with a final piece of equipment, or 'black box', for installation. However, this last approach may not be practical in most cases.

In the case of complete payloads, the United States also will support COSPAR. As a first step, the delegate of the US National Academy of Sciences is authorized to state that the US National Aeronautics and Space Administration will undertake to launch an entire payload to be recommended by COSPAR; this payload may weigh from 100 to 300 lb and can be placed in an orbit ranging from 200 to 2000 miles altitude. It is expected that the choice of the experiments and the preparation of the payload may require a period of one-and-a-half to two years. NASA is prepared to advise on the feasibility of proposed experiments, the design and construction of the payload package, and the necessary pre-flight environmental testing. The US delegate will be pleased to receive COSPAR's recommendations for the proposed payload when they can be readied.

In further support of COSPAR, the US delegate would like to call attention to the availability of resident research associateships at the National Aeronautics and Space Administration in both theoretical and experimental space research. These provide for stipends of $8000 per annum and up.

ANNEX 5

Parliamentary statement about a British contribution to space research

May 1959

The Prime Minister: There are two problems to be considered in relation to a British contribution to space research; the nature and design of the instruments to be carried into space; and the means by which the containers for these instruments are launched.

With regard to the first, with the assistance of Fellows of the Royal Society and with the endorsement of the Advisory Council on Scientific Policy, a programme for the design and construction of instruments to be carried in earth satellites has been approved. Work will begin at once.

With regard to the second, there may well be scope for joint action with the United States with the Commonwealth or with other countries. We therefore plan to send to Washington a team of experts, including Professor H.S.W. Massey, to discuss possible Anglo–American co-operation; and we are also opening consultations with other Commonwealth countries.

Meanwhile, however, design studies are also being put in hand for the adaption of the British military rockets which are now under development. This will put us in a position, should we decide to do so, to make an all-British effort.

I have asked my noble Friend, the Lord President of the Council, in consultation with my Right Hon. Friend the Minister of Supply and other Ministers concerned, to exercise general supervision of these new developments.

ANNEX 6

Governmental exchange of notes between the UK and the USA on co-operation in space research

No. 1

The United States Secretary of State to Her Majesty's Ambassador at Washington

> Department of State,
> Washington, September 8, 1961

Excellency,

I have the honour to refer to the discussions on space research held in Washington between representatives of the Government of the United States of America and of the Government of the United Kingdom of Great Britain and Northern Ireland and to propose that the two Governments should now conclude an agreement to join together in a mutually beneficial program of co-operation in space research to expand human knowledge of phenomena in space through the use of space vehicles.

This program, which would form part of a world-wide scientific effort to study extra-terrestrial conditions, would consist of a series of experiments co-operatively planned and conducted by designated agencies of the two Governments and would be carried out in accordance with the following provisions:

(1) Each Government shall designate a Co-operating Agency or Agencies which shall be responsible for carrying out the program of space research. For the Government of the United States, the Co-operating Agency shall be the National Aeronautics and Space Administration: and for the Government of the United Kingdom the Co-operating Agencies shall be such agencies as that Government may from time to time designate through the normal diplomatic channels.[1]

(2) The specific number and type of the scientific experiments to be performed under this program, as well as the time-table for putting such experiments into effect, and the allocation of technical operational and financial responsibilities for each such experiment, shall be as agreed between the Co-operating Agencies of the two Governments.

(3) (a) The Government of the United States and the Government of the United Kingdom shall accord to each other complete co-operation in the planning and conduct of this program and to this end, under specific arrangements to be determined by the Co-operating Agencies of the two Governments shall:

(i) exchange reports, data, and other information connected with the program, and

(ii) arrange for exchanges of visits by scientific and technical personnel to the laboratories and other installations connected with the program which are under their respective control; in this connection, the two Governments shall, subject to the provisions of sub-paragraphs (b) and (c) below, facilitate the admission into the United States and into the United Kingdom of personnel connected with the program and of materials, equipment, instruments and goods required for the purpose of the program.

(b) The admission into the receiving State of personnel of the sending State for the purposes of this program shall be subject to the laws and regulations for the time being in effect governing the admission of foreign nationals and to any policy of the Government of the receiving State restricting the admission for the purpose of employment of foreign personnel not employed by the Government of the sending State.

(c) Subject to the laws and regulations for the time being in effect, no customs duties shall be charged in the importation by or on behalf of the Co-operating Agency or Agencies of the sending State into the territory of the receiving State of materials, equipment, instruments, and goods in connection with this program provided that such materials, equipment, instruments, and goods are for official use under the program and are imported under a certificate to this effect.

(4) Each Government shall make available to the other Government data received by tracking stations under its control from space vehicles launched under the program. Such data shall be made available in a form to be agreed upon between the Co-operating Agencies of the two Governments.

(5) The scientific results of the experiments conducted under this program shall be made available to the international scientific community under agreements which shall be agreed upon between the Co-operating Agencies of the two Governments.

(6) Subject to the availability of the necessary financial resources, this program shall remain in effect for a period of five years, and thereafter may be extended for such periods and on such terms as may be agreed upon in writing between the two Governments. Nevertheless, either Government may terminate this Agreement by giving ninety days' notice in writing to the other Government.

If the above proposals are acceptable to the Government of the United Kingdom of Great Britain and Northern Ireland, I have the honour to propose that this note and Your Excellency's reply to that effect shall constitute an Agreement between the two Governments in this matter which shall enter into force on the date of Your Excellency's reply.

> Accept, &c.
> For the Secretary of State:
> WILLIAM C. BURDETT

No. 2

Her Majesty's Ambassador at Washington to the United States Secretary of State

> British Embassy,
> Washington, DC, September 8, 1961

Sir,

I have the honour to acknowledge receipt of your Note of today's date proposing that the Government of the United Kingdom of Great Britain and Northern Ireland and the Government of the United States of America should conclude an agreement on a joint programme of space research, which Note reads as follows:

(As in No. 1)

2. I have the honour to inform you that the foregoing proposals are acceptable to the Government of the United Kingdom of Great Britain and Northern Ireland, who therefore agree that your Note, together with the present reply, shall constitute an Agreement between the two Governments which shall enter into force on today's date.

> I avail, &c,
> HAROLD CACCIA

ANNEX 7

European space research meeting to initiate co-operation

April 1960

Resolutions

1. The following group of European space scientists:

Professor E. Amaldi	Professor H.C. van de Hulst
Professor P. Auger	Dr A.W. Lines
Professor J. Blamont	Dr L.M. Malet
Dr R.L.F. Boyd	Dr D.C. Martin
Dr L. Broglio	Mr J.A. Ratcliffe
Dr E.A. Brunberg	Mr M.O. Robins
Dr A. Ehmert	Professor S. Rosseland
Dr H. Elliot	Dr R.L. Smith-Rose
Professor M. Golay	Mr K. Thernøe (representing
Sir William Hodge	Professor J.K. Bøggild)
Dr F.G. Houtermans	Dr J. Veldkamp

From:

Belgium	Norway
Denmark	Sweden
France	Switzerland
Italy	United Kingdom
The Netherlands	Western Germany

agrees that further steps should be taken to establish European co-operation in space research.

2. The Group notes the great value to science and to Europe that would accrue from a co-operative effort in space research, as providing, for example, the following facilities which it is impracticable to provide with the resources of a single nation:

(a) A means of supplementing the extensive ground-based researches at present conducted in Europe. This is especially important in ionospheric work, where associated observations on the sun and the upper atmosphere are essential to a full development of these researches.

(b) A means of co-operative study of the meteorology of the upper atmosphere in the region of Europe.

467

(c) A means of studying Arctic phenomena especially at a distance from the geomagnetic pole.

(d) A means of augmenting the European contributions to optical and radio astronomy by including studies in the extensive parts of the spectrum absorbed in the upper atmosphere.

(e) A means of benefiting from pooling of technical knowledge and of assisting in the development of science and technology amongst the European nations.

This meeting of European scientists with interest in space research is strongly in favour of a co-operative effort by European nations towards further research in space science including the placing in orbit of artificial satellites by a launching vehicle developed and financed co-operatively.

3. The Group chooses Professor P. Auger as its Secretary and entrusts him with the duty of maintaining contacts with the members. The Secretary shall organize the next meeting in the course of the next two months and will circulate the necessary papers.

4. The Secretary shall study any proposal issued by International Agencies or by Governments through members of the Group to assist the Group in its activities and in its establishment of a more permanent body. He will report to the next meeting on these proposals, which may be of a limited duration. The finances will be administered by the Secretary.

5. Those attending this second meeting should be representative of their national authorities and empowered to create a Preparatory Committee for the establishment of plans for an extended European collaboration in space research. The Preparatory Committee will, through its Secretary, keep the COSPAR well informed of its work and progress.

6. The further steps will probably be as follows:

(a) The Preparatory Committee will be considered created if more than half of the nations presently represented on the Group are able to give an official signature. It will choose an Executive Secretary, entrusted with the duty of establishing draft plans, including the financial aspect. The Executive Secretary will call such meetings of experts which he deems necessary for that purpose.

(b) The Executive Council will organize in the course of the next six months a meeting of the Preparatory Committee to consider the draft plans for collaboration. The delegates to this meeting should have the power to sign a new protocol, including financial commitments, the ratification of which will then be open to the Government members of the Preparatory Committee. The new and permanent body will be considered created when six of the Governments ratify.

ANNEX 8

Outline proposals by scientists for a co-operative European programme

The Scientific Case for European Co-operation in Space Research.

The European co-operation should augment and in no case supplant the national programmes of the associated countries.

I Vertical rocket sounding

The creation of a central European space agency will provide help, organization facilities and, if necessary, financial support for:

1. Scientific programme

In addition to augmenting the work of already established national programmes and making possible the participation of smaller nations in space research, the following synoptic studies would be made possible:

(a) Synoptic study of the atmosphere from 30–90 km.

(b) Synoptic study of the atmosphere in the region between 90–200 km.

(c) Study of solar activity.

2. Scientific facilities

The agency will:

(a) Approve the co-operative scientific programme.

(b) Provide information, advice and library services on space science and technology.

(c) Integrate and engineer composite payloads.

(d) Specify test requirements.

(e) Provide basic payload engineering and testing facilities.

(f) Organize firing and testing.

(g) Handle raw data.

3. Vehicles

(a) List the performance, availability, price, etc. of existing vehicles.

(b) Co-ordinate buying and distribution of rocket vehicles.

(c) Promote standardization where desirable.

(d) Promote and support vehicle research.

4. *Launching sites*

    (a) Location of sites

    (i) For low altitude programme many sites of small area are required. The agency would arrange for provision of sites and associated facilities.

    (ii) Medium altitude programme will need four sites, one in the auroral zone, e.g. Sweden or Norway; the second in middle latitudes, e.g. Sardinia; the third site providing possibilities of safe recovery, e.g. Sahara; and a site in the southern hemisphere, e.g. Woomera.

    (iii) It is probable that launchings will also be required from ships, aircraft or balloons.

    (iv) High altitude programme. Very high launchings would take place from satellite launching sites.

    (b) Work of agency

    The agency should:

    (i) obtain access of members to existing sites

    (ii) obtain the use of launching facilities at the various sites for use by members, for the members' own experiments and rockets

    (iii) provide, where necessary, additional facilities for the programme at existing stations.

    (c) Telemetry, tracking and general equipment

    The agency should:

    (i) collect and disseminate data on available systems and equipment

    (ii) promote standardization where desirable

    (iii) act as a buying and distribution centre for equipment

    (iv) promote and support the development of equipment.

    (d) Operation

    The agency should provide additional operational personnel where necessary.

II **Satellites**

The agency should administer funds and organize the development and launching of satellites for scientific research on a co-operative basis.

An example of a programme of launcher development

A European programme must aim to provide a booster capable of launching highly stabilized systems and deep space probes. These objectives might be reached in about five years, during which time experience would be gained and technology developed by the launching of less sophisticated systems of lower weight.

The programme of launching is envisaged in three parts:

1. After an initial period to develop the payload and make available a launcher, small satellites would be launched (of the order of 100 kg). This is essential both to provide for experiments which do not require high launching potential, and to build up the experience necessary before Europe can make a full contribution to space research. Such launchings might take place in about three years' time, and would be followed by progressively larger payloads fired to greater distances.

2. After about five years, it is envisaged that satellites weighing between 500 and 1000 kg would be launched on near earth orbits and smaller vehicles could be sent to the neighbourhood of the moon. This time would be used to develop the more sophisticated payloads appropriate to such missions, including especially highly stabilized systems. This programme could be carried out by existing techniques, e.g. Blue Streak.

3. In parallel with the above developments, work would be carried on on more advanced systems to provide as soon as practical during the ensuing years means for undertaking soft landings of equipment on the moon, the exploration of other planets and the regions nearer the sun.

There is no intention to develop facilities for manned space flight.

An example of scientific programme

During the first stage the following experiments might be undertaken:

(a) Detailed survey of ionospheric structure, constitution and temporal behaviour, together with continuous monitoring of solar behaviour in the ultra-violet and X-ray regions of the spectrum.

(b) Radio astronomy in regions of the radio spectrum inaccessible from the ground.

(c) Meteorological studies.

(d) Geodetic and time measurement problems.

(e) Study of cosmic rays and other particles (including meteors) and fields in the upper atmosphere.

During the second stage the following experiments might be undertaken:

(f) Extensive survey of stellar ultra-violet and X-ray spectra.

(g) Detailed study of solar disk in ultra-violet and X-ray regions of the spectrum.

(h) Study of interplanetary and interstellar absorption.

(i) Study of cosmic rays and other particles and fields in interplanetary space.

(j) Advanced meteorological studies.

The third stage would see research of the following character commenced:

(k) Lunar and planetary physics.

(l) Advanced astronomical research.

(m) Space biology.

In order to carry out this programme the agency should:

1. Proceed immediately to ensure the availability of launching potential for payloads weighing about 100 kg.
2. Proceed immediately to ensure the availability of means for providing highly stabilized satellites and deep-space probes.
3. Ensure the provision by the time it is required of a network of tracking and data recovery stations, together with the basic means of data analysis.
4. Undertake from the start research in the advanced systems necessary to make possible the third part of the programme.

The advantages of a co-operative programme may be seen from the following proposals. For convenience they are summarized here under the headings:

A. Advantages to science
B. Economic and technological advantages
C. Indirect benefits (e.g. to industry).

A. Advantages to science

1. Co-operation makes possible an adequate study of phenomena of a synoptic or time-varying character. Thus, for example:
(i) The study of the ionosphere and its bearing on radio communications must now be made on a basis of continuous monitoring of solar activity correlated with extensive ionospheric measurements.
(ii) The study of the meteorology of the region above 30 km requires many simultaneous launchings of small rockets over a wide area.
(iii) The study of primary cosmic rays has now reached a stage where the secular variation must be carefully followed.
2. An appropriate European contribution to space research cannot be made without some agency for co-ordinating research workers with the means of carrying out their research. This is especially important as there are but few nations that by themselves could hope to provide even modest facilities.
3. The field of space science is very great. Specialities of interest and techniques exist in Europe which are not to be found elsewhere. A contribuion to space research beyond that of America and Russia should be made if the subject is to advance rapidly on a broad front.
4. Co-operation will make available to workers in various fields ranges and facilities in localities to which they would not otherwise have access.

B. Economic and technological advantages

1. Co-ordination of buying, research and development and of standardization will bring economy.
2. Pooling of knowledge, skill and resources will make possible the economic provision of facilities which might not otherwise be possible, e.g. advanced propulsion systems, highly stabilized platforms, satellite recovery, adequate tracking and data recovery.

3. Centralization of the engineering skills required for payload engineering will be of great value to all, but especially scientists from the smaller nations.
4. Centralization of data processing equipment for use by all will make for efficient and economic data handling.

C. **Indirect benefits**
1. The advantages accruing from the research will be directly available to Europe.
2. There will be much stimulation of technological research in European industry and of scientific research in universities and institutions.
3. Expenditure will be mostly in Europe and will benefit Europe.

Groups which might be set up by the Preparatory Commission for European space research
   1. A group of scientists and engineers to make proposals for the scientific programme.

This programme must be related to possible payloads and orbits and to a launching vehicle firing programme. On this basis an estimate of cost of the satellite development programme can be made.

The following are rough estimates of payload for different types of satellite:

(1) Stabilized in space axes to high accuracy
                            – of the order of 500 kg
(2) Stabilized in earth axes
                            – of the order of 250 kg
(3) Sun pointing   – possibly less than 250 kg
(4) Unstabilized   – 25 to 100 kg

The total weight of the launching rocket require is related to these payloads and the orbits desired.

This group will make proposals to the Preparatory Commission.

   2. A group of rocket experts to make estimates of cost and technical possibilities for rockets, launching sites and facilities. Maximum rocket weight of the order of 100 tons.

Possible rockets for the first stage are known to be, in Europe, either:

   Blue Streak
   and/or a new development

In addition the group will examine under what conditions American rockets within this class, e.g. Thor, Atlas, might be obtained.

Possible launching sites are:

   Colomb Bechar
   Woomera
   a new development

It is, of course, also possible to launch from Cape Canaveral, the conditions required for this being unknown.

In addition, the rockets, launching sites and facilities for launching payloads of the order of 100 kg in reasonably near earth orbits should be considered.

This group would be responsible to the Preparatory Commission to provide facts, from which the Preparatory Commission would make recommendations.

3. A group of scientists and engineers to make proposals for technological research in such fields as propulsion, power sources, information storage, solid state physics, surface coatings of controlled emissivity, high-vacuum technology, sensitive detecting devices, and the behaviour of materials under high vacua.

It is considered that a fairly important part of the available budget should be devoted to this work in preparation for the next stage in European space research. It is recognized that over and above the scientific interest of space vehicles advanced studies promote general technological progress. In consequence, such studies are an important factor in the industrial progress of the participating countries.

This group will make proposals to the Preparatory Commission. Such proposals will include estimates of cost and relative likely value and hence priorities of the work.

4. A group of scientists and engineers to investigate possible nets of tracking and telemetry stations.

The possibilities are:

(1) Use of the existing facilities
(2) Additions to existing facilities
(3) Provision of an entirely separate net

It will be necessary to consider both satellites in reasonably near earth orbits and deeper space probes.

The conditions under which existing facilities can be employed must be considered.

Estimates of capital cost and running costs must be made for the different possibilities.

The group will make proposals to the Preparatory Commission and provide all relevant data on costs and technical possibilities for the Preparatory Commission to make recommendations.

5. A group concerned with general organization.

6. A group of engineers and scientists to study the organization, capital facilities required and running costs of the main satellite engineering establishment.

This study can only be completed on the basis of a reasonably known programme of work. It is suggested that a range of possibilities is considered, compatible with the studies made by the first group. The siting of the establishment must be considered.

This group will provide information and make proposals to enable the Preparatory Commission to make recommendations.

20 October 1960

ANNEX 9

Formal agreement setting up a Preparatory Commission

The Governments of Belgium, Denmark, France, German Federal Republic, Italy, The Netherlands, Norway, Spain, Sweden, Switzerland, United Kingdom of Great Britain and Northern Ireland,
Interested in studying the possibilities of European Collaboration in research in space science and space technology and in the pooling of the knowledge thereof,
   Have agreed as follows

Art. 1
A Preparatory Commission shall be set up to investigate the possibilities of establishing a European Organisation for collaboration in space research.

Art. 2
The Members of the Commission are the States which are parties to the present Agreement.
   The Commission may, by a unanimous decision, admit other European States wishing to become members.
   The Commission may also by a unanimous decision invite other States to associate themselves with its work. The conditions and the form of such association shall be determined by the Commission according to the circumstances of each case.

Art. 3
Each Member State shall be represented on the Commission by two delegates who may be assisted by advisers.
   Each Member State shall have one vote.

Art. 4
With a view to the convening of an intergovernmental Conference, the Commission shall prepare and submit to Member States:

(a) a draft agenda for the said Conference together with proposals concerning the place and the date of its meeting;
(b) a draft convention for the establishment of a European Organisation for space research;
(c) a draft scientific and technical programme;
(d) a draft budget for the capital and running expenses of the Organisation during the first three years of its existence;
(e) draft rules regulating the calculation of contributions;
(f) draft rules of financial procedure and draft staff regulations;
(g) draft agreements with other organisations interested in co-operation in the field of space research.

The draft documents mentioned above shall be submitted to the Governments of Member States at least two months before the date contemplated for the meeting of the Conference.

Furthermore the Commission shall encourage discussions and the exchange of scientific information among Members.

Art. 5

(a) The Government of the French Republic shall convene the first meeting of the Preparatory Commission in Paris within 30 days following the date of the entry into force of the present Agreement;
(b) the Commission shall elect a bureau composed of a President and two Vice-Presidents. The bureau, with the assistance of an Executive Secretary, shall exercise between the sessions of the Commission all powers specifically delegated to it by the Commission;
(c) the frequency of the sessions shall be determined by the Commission. Extraordinary sessions may be convened by a decision of the bureau or at the request of a simple majority of Member States;
(d) the Commission shall decide questions of procedure by a simple majority of the votes of Members present and voting and other questions by a three-quarters majority, with the exception of the question of the admission of new Members, for which the unanimous vote of the Member States shall be required;
the majority of Member States shall constitute a quorum;
(e) the Commission may establish such Working Groups and Study Groups as it deems necessary.

Art. 6

The Commission shall determine its seat at its first session.

Art. 7

(a) The Executive Secretary appointed by the Commission, shall have the task of carrying out the technical work and the decision which it entrusts to him;

(b) a Secretariat shall be established by the Commission and shall be responsible to the Executive Secretary.

Art. 8

(a) The administrative expenses of the Commission shall be met by the contributions of the Member States in accordance with a scale which shall be based on the average net national income at factor cost of each Member State for the three latest preceding years for which statistics are available, except that no Member State shall be required to pay contributions in excess of twenty-five per cent of the total amount of the contributions as set forth in the Annex to the present Agreement; (Appendix A to Annex C)

(b) the Commission shall determine its budget;

(c) the Commission shall adopt a system of audited accounts providing effective control of its expenditure;

(d) if at the termination of its work the resources of the Commission have not been entirely spent or committed, it shall decide what will be done with the remainder.

Art. 9

The present Agreement shall be open for signature at Meyrin on December 1, 1960, and thereafter at Berne.

Art.10

Each State signing the present Agreement shall become a party to it either on signature or else by ratification if its signature was made subject to a reservation as to ratification.

Instruments of ratification shall be deposited with the Government of the Swiss Confederation.

The present Agreement shall enter into force on the date on which six Member States, the aggregate of whose contributions amounts to at least 70% of the total of the contributions provided for in Article 8 (a) and set forth in the Annex to the present Agreement, shall have either signed it without a reservation as to ratification or else ratified it after having signed it subject to such a reservation.

Art. 11

The admission of a new Member in accordance with the second paragraph of Article 2 shall take effect on the date of the deposit of its instrument of acceptance.

Art. 12

The Government of the Swiss Confederation shall notify States concerned of the entry into force of the Agreement and of the deposit of instruments of ratification and acceptance.

Art. 13

The present Agreement shall continue in force for one year, it being understood that it shall in any event terminate on the entry into force of the convention referred to in Article 4(b).

In witness thereof, the undersigned representatives, duly authorised thereto by their responsive Governments, have signed the present Agreement.

Done at Meyrin the first day of December 1960 in a single original, in the French and English languages, both texts being equally authoritative.

The original shall be deposited with the Government of the Swiss Confederation which shall send certified copies to signatory States and to the Commission.

ANNEX 10

Financial protocol of the European Space Research Organization

THE STATES parties to this Protocol.

BEING PARTIES to the Convention for the establishment of a European Space Research Organisation, hereinafter referred to as 'the Convention' and 'the Organisation' respectively, signed at Paris on 14 June 1962,

HAVE AGREED as follows:

1. The expenditure of the Organisation during the first eight years after entry into force of the Convention shall not exceed three hundred and six million accounting units (at price levels ruling at the date of signature of this Protocol), provided that the Council, referred to in Article X of the Convention may, by a unanimous decision of all Member States taken on the occasion of a three-yearly determination of levels of resources under Article X4. (c) and (d) of the Convention, adjust this figure in the light of major scientific or technological developments.

2. The Organisation shall frame its programme within the limit of expenditure laid down in paragraph 1 of this Protocol.

3. The States parties to this Protocol shall be prepared to make available to the Organisation during the first period of three years from the date of the entry into force of the Convention a sum not exceeding seventy-eight million accounting units and, subject to a definitive determination in accordance with the provisions of Article X4. (c) of the Convention, to make available to the Organisation during the second period of three years from the date of entry into force of the Convention a sum not exceeding one hundred and twenty two million accounting units.

4. This Protocol shall be open to signature by States which have signed the Convention. It shall be subject to ratification.

5. This Protocol shall enter into force on the date of the entry into force of the Convention, provided that the conditions for entry into force of the Convention, laid down in Article XXI thereof, are also satisfied in respect of this Protocol.

ANNEX 11

Report on location of ESRO headquarters and establishments

Dr O. Dahl

January 27 1962 I accepted an invitation from the Bureau to study the problem associated with location of ESRO Headquarters and Establishments from technical, economic, scientific, social and financial points of view on the basis of proposals received from member States, and to report to the Bureau before the COPERS fourth session.

As it was necessary to visit proposed sites and acquire additional information and views on the subject of discussions with representatives of countries concerned, it was not possible to have the report ready and available to the fourth session.

Before accepting the assignment I had formed opinions as a representative of a COPERS member state and realized that any one of the proposals could meet ESRO requirements while it was very difficult to see over-riding reasons for choosing one location to the exclusion of the others. There were also implications of political nature which could not readily be overlooked, and I had considered principles for elimination of proposals and establishment constellations which could be agreeable to most countries without losing sight of the ESRO main purpose.

The various proposals in the hands of member states are supported by sufficient factual information for member states to form opinions. The salient facts have been tabulated by the secretariat and circulated in order to facilitate comparison. Not much more could be gained by just undertaking a fact-finding tour, and I judged that the assignment could only be of purpose if I also carried views and opinions across, with sufficient or perhaps insufficient discretion, and sounded out reactions to proposals I could envisage as having support.

I further felt that I should terminate my report with a considered proposal for locations, as a focal point for discussions in the Bureau meeting.

The documented proposals from member states

The formal proposals for ESRO establishments and locations present the following list for consideration:

HQ France (Paris indicated)

ESTeC

1. Belgium near Brussels
2. France near Orly
3. Germany near Munich (Not pressed hard)
4. Italy near Rome
5. The Netherlands in Delft
6. Switzerland near Geneva
7. UK in Bracknell

ESDaC (ESTrack) Germany in Darmstadt

Leaving out Ranges and ESLAB, the problem which presents itself, according to the list, is to choose one of seven ESTeC locations in respect to presumably fixed HQ and ESDaC locations.

It turned out that the situation is not as simple.

In evaluating proposals and possible combinations, it is of first importance to keep quite clear in mind the principal aim of ESRO: it is to make integrated European scientific space research possible, primarily based upon national scientific activities in this field. The effort is faced with serious handicaps:

(a) The eight-year programme outlined as the minimum 'critical size' is tremendous, realizing that, practically speaking, Europe is starting from scratch as compared to USA and USSR.

(b) The money to be provided is, in many people's minds, not quite sufficient for 'critical size' activity, regardless of programme.

(c) What Europe can do in the course of eight years must be measured against the purely scientific efforts taking place at accelerated pace in USA and probably USSR.

It is quite clear that, within the limits of time and money, all possible efforts must go in the direction of making ESRO scientifically an unqualified success. ESRO will absorb large money and top rank staff on a European basis. If spent unwisely, European co-operative science will be discredited.

It is not too difficult to initiate and keep aimed a national scientific project, reasonably free from irrelevant interests.

It will be much more difficult to get ESRO started and see it run, as conflicting professional national views as well as political interests must be dealt with. It is significant that the report of The Scientific and Technical Working Group, laying the foundation for ESRO, was accepted by COPERS without much discussion.

ESRO must be a scientists' organisation. Not all scientists with a right to opinion believe in ESRO today. They will study the final solution for locations and structures with very critical eyes, and they must be convinced that ESRO looks very good.

One can almost assume that a maximum concentration of ESRO activities would appear right to the space scientists in countries with active national projects, and also for the member countries it is probably not difficult to see that concentration is desirable in the carrying out of a very complex project.

There is this other side to ESRO: not all countries concerned will have the same ability to make use of ESRO Establishments' facilities, and as it is obviously also an ESRO function to stimulate space activity in countries not so well underway, such considerations should justify a certain dispersion of ESRO activities.

It could be a reasonable compromise with a certain concession to political factors to locate HQ, ESTeC and ESDaC in three countries in the area for which most proposals have been made, that is to say in north-west Europe, arguing that supporting industries of all grades and most of the contributing money will be found in this area where also the two official languages are in most effective use.

Remarks related to locations, ESRO and ELDO

Conceding that the aim of ESRO is purely scientific and that the purpose of ELDO is to promote technology with a very strong commercial bias, that the border-line between pure science and advanced technology is very diffuse, that ESRO must make use of launchers and launching facilities, it is apparent that ESRO and ELDO have common interests in a wide sector.

While ESRO is under no obligation to make use of ELDO launchers or facilities, the need for a form for co-operation whereby European and ESRO interests are best served must be obvious, bearing in mind that ELDO intends to have test satellites flying by the time ESRO will be reaching that stage. ESRO must try to benefit from ELDO experience.

The ELDO headquarters is in Paris. A main HQ function will be the co-ordination of work in the countries responsible for the three rocket stages and test satellites, and it must contain a large complement of specialists to deal with details. As we cannot afford not to use each other's skills, the ESTeC staff want to maintain a very close professional contact with this group. For this reason, ESTeC should be located to facilitate ready communication with ELDO HQ.

Italy is responsible for test satellites under ELDO, otherwise the main ELDO countries converge in N.W Europe. This may be an argument for the N.W. Europe concentration of ESRO if one wishes to regard ESRO and ELDO as complementary European activities.

With such thoughts in mind, visits to the several countries inviting ESRO establishments were undertaken. The secretariat made the necessary arrangements as required.

Countries and places in sequence as visited

1. FRANCE: Limited to discussions in Paris with Professor Auger as representing the official French views on the subject of establishments location and related questions.

2. UK: Meeting in London with representatives from Office of the Minister for Science and Ministry of Aviation.

   The Bracknell proposal for ESTeC was discussed, but I did not visit Bracknell as I know the district and had all the information required. Views were exchanged on the function and relationship of ESRO establishments.

3. BELGIUM: Meeting in Brussels at the Conseil National de la Politique Scientifique with a group representing the appropriate authorities. Two representatives from the Netherlands attended the meeting. The proposal for ESTeC near Brussels was discussed and the site visited. Views on ESRO establishments were exchanged.

4. SWITZERLAND: Meeting in Paris during the COPERS IV conference with members of the Swiss delegation for exchange of views on location of establishments. As I know conditions in Switzerland, and the proposed location of ESTeC rather well, I did not find it necessary to visit the locality.

5. NETHERLANDS: Meeting at the Technological University in Delft, with a group representing science and technology and relevant local interests. I also inspected the proposed ESTeC site near to the university grounds and informed myself on available facilities of interest. Views on establishments location and operation were exchanged.

6. ITALY: Meeting in Rome with Professor Broglio and advisers for exchange of views on ESRO and location of establishments. I visited the proposed site for ESTeC at the old airport just outside Rome, and inspected the university aeronautics experimental facilities there.

7. GERMANY: Meeting in Darmstadt with Dr Frank and Darmstadt city officials. The proposal for ESDaC in Darmstadt was discussed, additional information provided and the proposed location bordering the city inspected.

Remarks: I did not visit the member countries Austria, Denmark, Spain and Sweden who had not filed proposals. As with Norway they will make their position clear when the bureau proposal comes up for discussion or in formulating this Bureau proposal.

As far as it is known to me, Sweden has not as yet officially offered the use of the Kiruna territory as ESRO range.

Italy is willing to make their Sardinia range available for ESRO activities, as is the case of France and the African range.

The locations situation after my visits

It was realized by all countries concerned that the choice of locations had to be narrowed down before the problem could be dealt with in full session. During the course of discussions, however, it became apparent that if a filed proposal were not to be upheld, a proposal for one or more of the other establishments would come forth.

The 'semi-official' list of proposals appears to be as follows:

HQ.
1. France in Paris
2. Switzerland
3. Benelux
   ESTeC

1. Belgium near Brussels
2. France (Perhaps not pressed hard. Awaiting HQ solution)
3. Italy near Rome
4. The Netherlands in Delft
5. Switzerland near Geneva
6. UK in Bracknell
   ESDaC
1. Germany in Darmstadt.
2. Switzerland on the proposed ESTeC site.
3. UK in Bracknell

Points raised in discussions, bearing on locations

(a) ESTeC is regarded as the most important ESRO establishment.
(b) The need for strong professional working contact between ESTeC and the expert group in ELDO HQ.
(c) ELDO and ESRO HQs should not be in the same position. No need for it and it might give rise to complications as the one organisation is 'commercial' in nature, while the other is strictly scientific.
(d) ELDO and ESRO HQs should be in the same location so that certain neutral functions may be in common in the interest of efficiency and money saved.
(e) There should be the closest possible connection between ESRO and ELDO.
(f) There should be the loosest possible connection between ESRO and ELDO.
(g) As there should be a fair share of space activity opportunities for all countries, ELDO work distribution should be taken into account when deciding upon location of ESRO establishments.
(h) All ESRO activities should not be confined to NATO countries.
(i) ESDaC/track would function best if located in a country not a member of NATO.
(j) No need for ESRO HQ in a large city, rather in quiet surroundings.
(k) How best to preserve continuity in interim period activities should bear on HQ location.
(l) It is important that ESTeC and HQ are close together.
(m) With ESTeC and HQ close together there is a danger that ESTeC will carry too much weight with HQ.
(n) In order to get going fast, ESTeC should be in an active milieu.
(o) ESTeC should not be located in one of the major ELDO countries.
(p) ESTeC should be centrally located as member countries must have intimate working contact with this establishment.
(q) Location of ESTeC is not as important as having the best man working there.
(r) The importance of opportunities for informal professional contact between ESRO and local staff in related fields.
(s) In order to realise the ESRO long-range purpose of integration, an area of interests as large as possible should be covered by establishments.

(t) In present-day Europe, travel time is short. Distance between establishments is therefore not so important.

(u) ELDO and ESRO will compete for staff and it will be difficult to staff both organisations.

Summary of impressions

As stated, in my talk with a national group and also individuals, I made an attempt to sound out reactions to the various location proposals and possible constellations, also to indicate reactions met with in talks with other national groups.

It was to be expected that every country argued very strongly for its proposal, insisting that it was the best on purely professional grounds and in the best interest of ESRO. The countries with more than one proposal stated their priority, rather on the assumption that something is better than nothing.

Having heard all arguments, it is hard to overlook that national feeling and wishes play an important part for most countries in locations evaluation. This is the reason why I found it in place to stress the ESRO aim and circumstances in my preamble.

Some countries were tentatively willing to accept a bureau proposal, whichever it may be. Some indicated that they would go very far in making their proposal agreeable to ESRO.

In general talks, I have gained the impression that the Scandinavian countries favour a strong concentration of ESRO activities.

Personal observations

The economic aspects related to establishments location could hardly be sorted out unless all proposals were to be treated as pre-projects in co-operation with the respective countries. There is not time or money for that, and is hardly necessary.

At this stage, we know in a general way that some countries are more expensive to operate in than others. What should be of overriding importance is to find a solution which will give ESRO an opportunity to start fast and effectively. Costs should be one of the factors only.

If ESRO activities are spread in the interest of integration of opportunities in Europe, ESRO will cost relatively more, we will move slower and it will be more difficult to have control. There will be a tendency towards independent growth of establishments, as administration and services must in certain ways be duplicated.

To realise the primary aim of ESRO there is no need for such dispersion, but there are very strong arguments for concentration.

My conclusions and proposal

It is clear that no generally acceptable principle for elimination of proposals points itself out. Also that no single specific proposal was favoured although most groups could see merit in concentration of establishments in N.W. Europe.

It also appears that the bureau must take into account supplementary but not fully official proposals while all countries still are very strongly in favour of their own documented proposals.

At best it will be very difficult to bridge the conflicting views on dispersion versus concentration with a compromise solution. I propose that agreement perhaps best can be reached on a proposal which frankly places emphasis on concentration in the interest of economy, efficiency and a quick start for ESRO.

Therefore: *HQ together with ESTeC and ESLAB* on the proposed site in Delft.
           ESDaC and track in Darmstadt.

Meeting facilities for the Council need not be provided for the site in Delft, as the Council may meet in member countries in rotation, or as invited.

If one can reach agreement for common location of HQ and ESTeC, it may well be that the respective structural build-up as envisaged should be reconsidered.

In my proposal there are two countries involved, ESDaC is separated out. It can be argued that ESDaC is an establishment of a very special nature in ESRO and will not have its full function before scientific experiments are under way. This will take some time and the site in Delft could perhaps be planned and construction started faster if ESDaC was left out. Admittedly it is also a concession to member states not in favour of concentration.

Supplementary information to the official proposal for ESTeC in Delft; the temporary quarters as offered in the building of the applied physics department must be regarded as ample and very suitable, also because the premises offered are in a physics building, ESRO may get an early start on experimental work.

The proposed site is very close and large enough for any foreseeable ESRO use. The site is bordered by the new cross-country through highway, linking Amsterdam–Rotterdam.

Staff housing is not difficult and special school facilities for ESRO children may be provided if wanted.

The Netherlands is the low cost country in the N.W. area.

For the higher grades supporting staff, locally engaged, the language problem will not be important as English, French and German are taught in the schools.

Although there is a skilled labour shortage, the ESRO requirement will be small and it is said to present no problem.

Supplementary information on ESDaC in Darmstadt: the proposed site is on wooded land bordering the city, very suitable and as large as wanted. There is a strong communication and data milieu in Darmstadt. The English language is widely used. Communication in all its forms with the rest of Europe is very good.

I was of the opinion that ESRO HQ should be in Paris in the interest of effective planning and staff continuity linking COPERS Interim and final ESRO and to facilitate contacts with ELDO HQ. I now believe that it is more important for the future ESRO to have HQ and ESTeC in one location, even if current work will suffer.

OD/EJ   In March 1962

ANNEX 12

Need for basic studies in support of space science.
Statement by H.S.W. Massey

Both interpretation of results obtained by rockets and satellites and the planning of
new experiments depend on basic scientific data and laboratory research which are
not well catered for.

For example, the reaction rates and processes occurring in the upper atmosphere,
in stars and elsewhere in the Universe are not well understood. Laboratory studies
are essential if the full value of the space research programme is to be realised. The
same is true for plasma flow in the interplanetary field of disturbances created by
satellites in rarefied ionized gases. This would require both laboratory and theoretical
work.

In the USA these studies are not well co-ordinated with space studies. The same is
true of the USSR. Such work would be especially valuable in that it would put Europe
in a position to make a more well conceived and co-ordinated attack on the problems
of space.

Studies of this kind on both the theoretical and experimental sides are necessary
not only before experiments are carried out but also after in order that full value
should be obtained from the observations.

ANNEX 13

Memorandum of understanding between agencies of Pakistan, the UK and the USA

The Pakistan Space and Upper Atmosphere Research Committee (SUPARCO), the British National Committee on Space Research (BNCSR), and the United States National Aeronautics and Space Administration (NASA) affirm their mutual interest in obtaining wind, temperature and other meteorological information beteen 50 and 150 kilometres by rocket soundings using the 'grenade' technique. These agencies agree to co-operate in a joint project of launching from Sonmiani Beach, Pakistan, during the International Quiet Sun Year.

Each agency will use its best efforts to discharge the following responsibilities:

1. *SUPARCO responsibilities:*
   (a) Co-ordinate conduct of experiments.
   (b) Prepare and operate launching site and associated facilities.
   (c) Assemble and launch sounding rocket vehicles.
   (d) Reduce resulting data.
2. *BNCSR responsibilities:*
   (a) Assist in the conduct of experiments.
   (b) Provide six 'grenade' Payloads compatible with the NASA Nike–Cajun or Nike–Apache vehicle.
   (c) Provide, on loan, all special ground equipment needed, including twelve microphones, one recorder, four or five cameras, and two flash detectors.
   (d) Train SUPARCO personnel in the acquisition and reduction of the meteorological data and in the operation and maintenance of UK–supplied equipment.
   (e) Provide, as feasible, continuing technical assistance in the operation and maintenance of UK–supplied equipment.
3. *NASA responsibilities*
   (a) Provide six Nike–Cajun or Nike-Apache vehicles (these include two Cajuns already supplied).
   (b) Provide, as feasible continuing technical assistance in the operation and

maintenance of US equipment previously loaned under co-operative agreements between the US and Pakistan, which will be utilized in this programme.

Each agency will bear the cost of discharging its respective responsibilities including travel by its personnel and the transportation of the equipment which is its responsibility.

Each agency will designate a Project Manager to assure proper co-ordination with the others.

Copies of the new data obtained will be the common property of all three parties, reduced data will be made available to the three parties within five months and all results will be published in open literature or otherwise made available to the world scientific community.

Signed: I.H. Usmani
For the Pakistan Space & Upper
Atmosphere Research Committee

Signed: Hugh L. Dryden
For the National Aeronautics and
Space Administration

Signed; H.S.W. Massey
For the British National Committee
on Space Research subject to
following amendments

_____
Date

Amendments to memorandum of understanding by BNCSR

1. NASA to provide two grenade payloads compatible with Nike–Cajun if arrangements can be made for these to be launched in Spring 1965 in which case BNCSR will undertake to replace these payloads with similar ones acceptable to NASA.
2. Delete 2 (c) and replace by 'Provide on loan eight microphones and associated recording equipment and two flash detectors.

Signed: I.H. Usmani
For the Pakistan Space & Upper
Atmosphere Research Committee

Signed: Hugh L. Dryden
For the National Aeronautics and
Space Administration

Signed; H.S.W. Massey
For the British National Committee
on Space Research

Date
18 January 1965

ANNEX 14

Attendance at the Williamsburg Conference, to consider the large space telescope project

January 1976

PARTICIPANTS

BERTOLA, F.	Osservatorio Astronomico	Padova, Italy
BURBIDGE, E.M.	University of California	San Diego, California
COURTES, G.	Observatoire de Marseille	Marseille, France
DANIELSON, R.	Princeton University Observatory	Princeton, New Jersey
DORLING, E.B.	Mullard Space Science Laboratory	Dorking, Surrey, UK
FIELD, G.B.	Center for Astrophysics	Cambridge, Massachusetts
FINDLAY, J.	National Radio Astronomy Observatory	Charlottesville, W.Va
FRIEDMAN, H.	Naval Research Laboratory	Washington, DC
GOLAY, M.	Observatoire de Genève	Sauverny, Switzerland
GOODY, R.	Harvard University	Cambridge, Massachusetts
GREWING, M.	Institut für Astrophysik	Bonn, Germany
MacRAE, D.A.	David Dunlap Observatory	Ontario, Canada
MASSEY, H.	University College London	London, UK
MEYER, P.	University of Chicago	Chicago, Illinois
NEUGEBAUER, G.	California Institute of Technology	Pasadena, California
SMITH, F.G.	Royal Greenwich Observatory	Hailsham, Sussex, UK
VAN DE HULST, H.C.	Sterrewacht-Huygens Laboratorium	Leiden, The Netherlands
WILSON, R.	University College London	London, UK
WOLTJER, L.	European Southern Observatory	Hamburg, Germany
WYLLER, A.	Stockholm Observatory	Stockholm, Sweden

490

OBSERVERS

HART, R.	SSB, NAS	Washington, DC
HINNERS, N.	NASA	Washington, DC
MACCHETTO, F.	ESA	The Netherlands
MELLORS, W.	ESA	Washington, DC
NOYES, R.	Smithsonian Astrophysical Observatory	Cambridge, Massachusetts
O'DELL, C.R.	Marshall Space Flight Center, NASA	Huntsville, Alabama
PEYTREMANN, E.	ESA	Paris, France
RASOOL, S.I.	NASA	Washington, DC
ROMAN, N.	NASA	Washington, DC
ROSEN, M.	SSB, NAS	Washington, DC
SODERBLOM, L.	US Geological Survey	Flagstaff, Arizona
SPENCER, N.	Goddard Space Flight Center, NASA	Greenbelt, Maryland
TIMOTHY, A.	NASA	Washington, DC

ANNEX 15

Press release on the Williamsburg Conference

(The announcement below is being released simultaneously by the National Academy of Sciences (NAS) in Washington and the European Science Foundation (ESF) in London for use in newspapers Thursday, February 26. The ESF is an organization representing the scientific communities of 14 European countries. The announcement contains recommendations on the use of the Large Space Telescope (LST).)

Announcement

Twenty-one scientists from the USA and Europe convened by the Space Science Board of the National Research Council of the National Academy of Sciences and the Space Science Committee of the European Science Foundation have met to discuss matters concerning international co-operation on space observatories, with particular reference to the planned project of the National Aeronautics and Space Administration (NASA) to orbit a large optical telescope beyond the earth's distorting atmosphere

The LST will be complementary to the new generation of ground-based telescopes that will be coming into action over the next few years and will be used for those areas of research in which it, and it alone, can operate.

The US and European scientists, at a meeting in Williamsburg, Virginia, late in January 1976, developed recommendations which have been received and are now being considered by the parent bodies. Sir Harrie Massey, Physical Secretary of The Royal Society, was chairman of the meeting. Here is a summary of the conclusions reached:

The role of the LST in astronomy

The unique quality of the LST is its ability to concentrate light from a point source falling on a large aperture into an image approximately 0.1 arc seconds across, together with the wide spectral range over which this can be achieved. This ability leads to dramatic improvements in the limits of observational possibilities in many fields of astronomy, including planetary, galactic and extragalactic studies, and particularly in studies of the faint objects of interest for cosmology and for the

492

evolution of galaxies. These improvements are attainable only in telescopes operating outside the earth's atmosphere.

The capabilities of the LST therefore represent a natural and achievable progress in astronomical technique, of a magnitude and importance that exceed those of all other likely developments in optical astronomy over the next few decades.

Instruments for LST

Every effort should be made to insure that the focal plane equipment of the LST fully exploits the capabilities of the telescope. This requires wide band and narrow band instrumentation for the ultra-violet, optical and infrared ranges. An advisory committee, appointed by NAS and ESF, should review, prior to final selection, the design of the focal plane instruments to ensure the availabilty of the best space qualified equipment.

To enable existing and future scientific teams to make the best use of the instrument complex, fully updated information and documentation of all instruments must be available through ESF and NASA at all times.

Institutional arrangements for LST

The conference agreed that a specific organization, one form of which could be a science institute, would be needed to carry out vital tasks in the operation of the LST. Where international collaboration is involved, such an organization should have international participation at all levels. Its tasks would include: planning and scheduling of observing programs in conjunction with an international program committee, interfacing between instrumentation and guest observers, and data reduction. The scientific staff would have two main tasks: first, to help guest investigators in using the LST instruments and second, the carrying out of their own astronomical research. The international character of the organization should also be relevant to its location.

Allocation of observing time

The prime consideration in the allocation of observing time on the various instruments of the LST must be the scientific merit of the proposals. We recognise that other considerations, such as the contributions which may be made to the LST project by investigators in countries other than the USA, may affect the methods by which the allocations will be made.

Scientific data and manpower

The optimum use of LST and other space observatories involves a large flow of data. The demands which this will place on the scientific community are not yet fully understood. We believe that past experience gives sufficient justification to proceed with LST immediately; nevertheless, we recommend that a study should be made of the problem of handling the flow of data from the various modes of operation, including the processes of data distribution and requirements for manpower in data handling and scientific analysis.

ANNEX 16

Provisional membership of the Standing Committee on Space Science

		Field of scientific interest
Chairman:	Professor J. Geiss, Switzerland	Physics of the solar system, planetology
Austria:	Professor W. Riedler	Telecommunications and transmission techniques
Belgium:	Professor E. Evrard	Aviation and space medicine
Denmark:	Professor B. Peters	Cosmic radiation and astrophysics
France:	Professor J.E. Blamont	Physics, geophysics
	Dr M. Petit	Plasma physics
Germany (Fed. Rep.)	Professor F.M. Neubauer	Physics of the solar system geophysics
	Professor K. Pinkau	High energy astrophysics and nuclear physics
Greece:	Professor E.T. Sarris	Solar and planetary physics
Ireland:	Professor C. O'Ceallaigh	Cosmic radiation, elementary particle physics
Italy:	Professor E. Amaldi	High energy astrophysics and nuclear physics
	Professor G. Setti	Theoretical astrophysics, especially radioastronomy
The Netherlands:	Professor H.C. van de Hulst	Solar and galactic astronomy radio astronomy
Norway:	Professor T. Hagfors	Ionospheric plasma physics, solar system astronomy
Spain:	Professor J.O. Cardús	Geomagnetism, solar–terrestrial relationships
Sweden:	Professor C.-G.Fälthammar	Space and plasma physics

United Kingdom: Sir Harrie Massey	Atomic and molecular physics, space research, particle physics, astrophysics
Professor J.T. Houghton	Atmospheric physics

Professor R. Lüst will attend the meetings in his capacity as Chairman of the Astronomy Committee.

The US National Academy of Sciences has been invited to appoint two representatives of the Space Science Board.

The Director General of ESA, Mr R. Gibson and the Chairman of the ESA Science Committee, Professor H. Curien, are invited to attend the meetings of the Committee.

ANNEX 17

Terms of reference of the Standing Committee on Space Science

1. The Committee is set up by the Assembly in accordance with Article XI.I of the Statute.
2. The Committee's sphere of activities covers the whole field of space science. When dealing with space astronomy it shall co-operate with the ESF Standing Committee on Astronomy.
3. Bearing in mind that the advancement of co-operation in basic research is one of the principal aims of the Foundation and taking into account similar activities of other organisations, the main objects of the Committee are:
   (a) to keep under review the scientific and technical development of space science,
   (b) to consider possible developments of space research during the coming decade,
   (c) to continue to co-operate with the European Space Agency as outlined in the agreement reached in 1976 – 'Relations between the European Science Foundation and the European Space Agency',
   (d) to maintain close collaboration with the American scientific community through the Space Science Board of the US National Academy of Sciences,
   (e) to advise the ESF on:
   the level of space activities appropriate in basic science in the context of the total space activity in Europe, and
   the balance within space science of programmes in which the nations of Europe might be involved,
   (f) to report to the Assembly through the Executive Council on its activities each year, and
   (g) to undertake any other task entrusted to it by the Assembly or the Executive Council.

NOTES

Chapter 1

1. His certificate as a candidate for a fellowship included signatures of both Faraday and Wheatstone. He was elected in 1844.
2. For a detailed authoritative account of these developments see *Beyond the Atmosphere* by Homer E. Newell published by the National Aeronautics and Space Administration, Washington DC 1980; Part II.

Chapter 2

1. Rockets and Rocket Propulsion Devices in Ancient China: Fang-Toh-Sun, *Journal of the Astronautical Sciences*, Vol. XXIX, No. 3, pp. 289–305, July–Sept. 1981.
2. *Wellington – The Years of the Sword*, Elizabeth Longford. Pub. by Weidenfeld and Nicolson Ltd.
3. *History of Rocketry and Space Travel*, W. Von Braun and Fred I. Ordway, Ill.
4. The Thirty-Fourth Thomas Hawksley Lecture, by A. Crow. Inst. of Mech. Eng. Journal and proceedings, Vol. 158, No. 1, pp. 15–21.
5. See Chapter 4.
6. The Performance of Upper Atmosphere Research Vehicles powered by Solid Fuel Rocket Motors. King-Hele, D.G. RAE Tech. Note GW 315, 1954.
7. Preliminary Assessment of an Earth Satellite Reconnaisance Vehicle', King-Hele, D.G. and Gilmore, D.M.C. RAE Tech. Note GW 393, Jan. 1956.
8. 'The Descent of an Earth Satellite through the Atmosphere', King-Hele, D.G. and Gilmore, D.M.C. RAE Tech. Note GW 430, Sept. 1956.

Chapter 3

1. *Nature*, Vol. 177, 643, 1956.
2. This occurs when the natural frequency of the yawing or pitching motion of the rocket; which varies with the rocket altitude, becomes close to the rate of roll. Under such conditions the rocket is vulnerable to break-up through the action of relatively small disturbances such as gusts of wind or misalignments in the direction of thrust.
3. The transponder in the rocket receives a signal from the ground and re-transmits it after doubling the frequency. The Doppler shift of frequency due to the rocket motion can thereby be measured. By the use of three separated ground stations, sufficient data to compute rocket velocity and position can be obtained.

4. The MTS comprises a microwave beacon in the rocket; two ground receivers, widely separated, lock on to the signal and provide a continuous record of the directions of the rocket from the receivers. The rocket trajectory can thereby be calculated.
5. The telemetry comprised a radio transmitter in the rocket working on 465 mHz, giving readings at 100 times per second from each of 24 instruments.
6. *Proc. Roy. Soc.* A, Vol. 290, 44, 1966.
7. See for example, Boyd, R.L.F. *Proc. Roy. Soc.* A, Vol. 201, 329, 1950 and Boyd, R.L.F. and Twiddy, N. *Proc. Roy. Soc.* A, Vol. 250, 53, 1959.
8. The Scientific Unions are international non-governmental bodies of which the adhering members are national scientific academies such as the Royal Society for the UK and the National Academy of Sciences for the USA.
9. A Special Committee is constituted like a Union but may differ in performing only a temporary function or by covering more than one discipline (see, for example, COSPAR, Chapter 4, p. 58).
10. Those for Geodesy and Geophysics, Scientific Radio, Astronomy, Physics and Geography.
11. Meteorology, Geomagnetism, Aurora, Airglow, Ionosphere, Solar Activity, Cosmic Rays, Glaciology, Rockets and Satellites, Seismology, Gravity and Nuclear Radiation.
12. Massey was even asked for his autograph by a leading aircraftman on board the Comet II on the way to Australia!
13. A. Blagonravov, the principal Russian delegate at the Washington meeting, told Massey how astonished he was by the immense number of condemnatory letters he received from dog lovers in Britain!
14. King-Hele, D.G. and Walker D.M.C. *Nature* (*London*) Vol. 183, 527, 1959. Vol. 186, 928, 1960.
15. Groves, G.V. *Nature* (London), Vol. 181, 1055, 1958; *Proc. Roy. Soc.* A, 252, 28, 1959.
16. King-Hele, D.G. and Merson, R.H. *J. Brit. Interplan. Soc.*, Vol. 16, 446, 1958.
17. Staff of Jodrell Bank Experimental Station, *Proc. Roy. Soc.* A, Vol. 248, 24.

Chapter 5

1. Application was made for a grant to enable the group to travel by car to the Goddard Space Flight Center but it was at first turned down on the grounds that bicycles should be used instead! The official concerned could not imagine what it would be like for a cyclist on a busy American highway.
2. The British team stayed at the Holiday Inn, Coco Beach, where most of the guests had come to watch the launching of rockets of all kinds as part of their vacation. A list of scheduled launchings for each day was posted in the morning in the entrance hall.
3. *Proc. Roy. Soc.* A, Vol. 281, p. 438, 1964.
4. He$^+$ Bourdeau, R.E., Whipple, E.C., Donby, J.L. and Bauer, S.J., Proc. URSI Meeting, Austin, Texas, October 1961. O$^+$, N$^+$ Istomin, V.G., *Planet. Space Sci.* Vol. 8, 179, 1961.
5. *Proc. Roy. Soc.* A, Vol. 288, p. 540, 1965.
6. *Mon. Not. Roy. Astronom. Soc.* Vol. 131, p. 137, 1965.
7. *Mon. Not. Roy. Astronom. Soc.* Vol. 131, p. 145, 1965.
8. *Proc. Roy. Soc.* A, Vol. 311, p. 479. 1969.
9. Later of Birmingham University.

Chapter 6

1. The flight on which the UK representatives travelled was delayed by a few hours. The Europeans attending the meeting thought that the unpredictable UK

had decided not to participate further and were quite surprised when the UK party finally arrived. Sir John Cockcroft was also travelling to Paris on the same flight and his parting words to Massey at le Bourget airport were 'See that the scientists keep in control'.

2 This site was later found to be too noisy and the station was moved to a site in the Ardennes.

3. This involved enlargement and upgrading of the British national telemetry station (see Chapter 5, p. 82). Although presenting difficult problems of access, the station was regarded as of special importance for providing real time telemetry over the Southern Hemisphere. Some problems also arose in avoiding interference from the radio transmissions from the islands.

4. Orbiting astronomical observatory.

Chapter 7

1. It was pointed out in Chapter 1 that radio sounding of the ionosphere from the ground can only observe regions of the ionosphere in which the electron concentration is either constant or increasing with altitude. No observation could therefore be made in this way above the maximum of the F2 region (see Figure 1.1 of Chapter 1). However, by sounding from well above the maximum with equipment in a satellite the electron concentration can be observed *down* to this maximum, so complementing the ground-based studies.

2. International Year of the Quiet Sun.

3. Rahmatullah, M., Shafi Ahmed, M., McDermott, D.P. and Groves, G.V. *Space Research*, Vol. 8, p. 888; *Space Research*, Vol. 10, p. 167.

4. On both occasions the visit was in doubt because of unexpected floodings which presented a considerable barrier between Karachi and the range. However by proceeding carefully in a car with someone walking ahead to test the depth, both journeys were made without incident.

5. Rees, D., Bhavsar, P.D., Nesai, J.N., Gupta, S.P., Farmer, A.D. and Round, P., *Space Research*, Vol. XVI, 407, 1976.

Chapter 8

1. Almond R., Farmer, S.F.G. and Frith R. 1964, *Nature (London)*, Vol. 202, 587.

Chapter 9

1. Heddle, D.W., Alexander, J.D.H. and Bowen, P.J., *Space Res.* Vol. III, p. 1068, 1963.

2. *Proc. Roy. Soc. A*, Vol. 308, 246, 1968.

3. Black, W.S., Booker, D., Burton, W.M., Jones, B.B., Shenton, D.B. and Wilson, R. *Nature (London)*, Vol. 206, 654, 1965.

4. Burton, M. and Wilson R. *Nature (London)*, Vol. 207, 61, 1965.

5. Adams, D.J., Cooke, B.A., Evans, K., Pounds, K.A. *Nature (London)*, Vol. 222, 757, 1969.

6. Abbott, J.K. 1971, 'A Star Pointing attitude Control System for Skylark'. RAE Tech. Rpt. 71241 London HM Stationery Office.

7. Farr, C.T. 1975, 'Flight Performance of the Skylark SL 1011 Star Pointing Attitude Control System'. RAE Tech. Rpt. 75110 London HM Stationery Office.

8. Burton, W.M., Evans, R.G., Griffin, W.G., Lewis, C., Parton, H.J.B., Shenton, D.B., Macchetto, F., Boksenberg, A. and Wilson, R., *Nature (Phys. Sci.)*, *(London)*, Vol. 246, 37–40, 1973.

9. Burton, W.M., Evans. R.G. and Griffin, W.G., *Phil. Trans. Roy. Soc.*, A. Vol. 279, 355–369, 1975.

Chapter 11

1. See for instance: *Spaceflight*, Vol. 16, No. 6, June 1974, 'European Space Activities since the War – a personal view by A.V. Cleaver.

2. The text will be found in the British Government White Paper Treaty Series, No. 30 (1981) Cmnd. 8200.

3. When first drafted, the ESA Convention included no reference to a Science Programme Committee. This omission was strenuously challenged by British space scientists, who were initially in a minority of one. Their persistence and strong representations from the Provisional Space Science Board for Europe (see Chapter 12, p. 248) reversed this initial intention and the matter was rectified.

Chapter 12

1. Special Committee on Solar and Terrestrial Physics.

2. In 1976 the Committee consisted of Sir H. Massey (Chairman) UK, R. Lüst (Deputy Chairman) (Fed. Germ. Rep.), Sir Brian Flowers (ESF), W. Riedler (Austria), P. Swings (Belgium), B. Peters (Denmark), J.P. Blamont and J.M. de Lemare (France), W. Kirtz (FGR), E. Amaldi and G. Occhialini (Italy), C. de Jager and H.C. van de Hulst (The Netherlands), Father J.O. Cardus (Spain), H. Bjorstedt and C.G. Felthammer (Sweden), J. Geiss (Switzerland), R.L.F. Boyd and H. Elliot (UK), R. Goody and F.S. Johnson (USA). The Director General and Chairman of the Science Programme Committee of ESA attended by invitation.

3. This is somewhat ironic in view of the fact that the Astronomy Committee was disbanded in 1981.

Chapter 13

1. For this work he was awarded the Merlin Medal of the British Astronomical Association in 1977.

2. A visual observation is usually made by watching the satellite with binoculars and estimating its proportional distance from two stars as it crosses the straight line between them. At the same time a stop watch is started and stopped a few minutes later against a time standard. Choosing stars not more than $\frac{1}{2}°$ apart, an accuracy of about 0.1 s in time and 100″ of arc in direction should be obtainable. On a clear night satellites 1 m in diameter at a distance of 2000 km can be seen with 11 × 80 binoculars (see King Hele, D.G., *Observing Earth Satellites*, Macmillan, London, 1983).

3. Including the group at UCL under G.V. Groves, cf. *Proc. Roy. Soc.* A, Vol. 252, 28, 1959. *Planet and Space Science*, Vol. 5, 314, 1961.

4. Walker, D.M.C. *Planet and Space Science*, Vol. 26, 291, 1978.

5. Hiller, H. *Planet and Space Science*, Vol. 29, 579, 1981.

6. *Nature (London)*, Vol. 183, 239, 1959.

7. King-Hele, D.G. and Walker, D.M.C. *Planet and Space Science*, Vol. 31, 509, 1983.

8. Walker, D.M.C. *Phil. Trans. Roy. Soc.* A, Vol. 292, 473, 1979.

9. Hiller, H. *Planet and Space Science*, Vol. 29, 579, 1981.

10. Lerch, F.J., Wagner, C.A., Klosko, S.M., Belott, R.P., Laubscher, R.E. and Taylor, W.A. 'Gravity model improvement using Geos 3 altimetry'. Paper presented at 1978 Spring Annual Meeting of AGU, Miami, Florida.

11. King-Hele, D.G., Cook, G.E. and Scott, D.W. *Planet Space Science*, Vol. 13, 1213, 1965.

12. King-Hele, D.G. and Cook, G.E. *Planet Space Science*, Vol. 22, 645, 1974.

13. *Geophys. J. Roy. Astronom. Soc.*, Vol. 64, 3, 1981.

14. Gooding, R.H. RAE Technical Report 71068. London HMSO (or *Nature Phys. Sci.*, Vol. 231, 168, 1971).
15. *Planet and Space Science*, Vol. 23, 229, 1975; Vol. 23, 1239, 1975.
16. King-Hele, D.G. and Walker, D.M.C. *Proc. Roy. Soc.* A, Vol. 379, 247, 1982.
17. King-Hele, D.G., Walker, D.M.C. and Gooding, R.H. *Planet Space Science*, Vol. 27, 1, 1978.
18. Walker, D.M.C., *Planet Space Science*, Vol. 28, 1059, 1980: King-Hele, D.G., *Phil. Trans.* A296, Vol. 597, 1980.
19. Leach, F.J., Wagner, C.A., Klosko, S.M., Belott, R.P., Laubscher, R.E. and Taylor, W.A., Gravity model improvement using Geos 3 altimetry (GEM 10 A and 10 B). Presented at 1978 Spring meeting of AGU, Miami, Florida.
20. Bowen, P.J., Boyd, R.L.F., Davies, M.J., Dorling, E.B., Groves, G.V. and Stebbings, R.F., *Proc. Roy. Soc.* A, Vol. 280, 170, 1964.
21. Groves, G.V., *Space Res.* Vol. IV, 155, 1963.
22. Groves, G.V., *Space Res.*, Vol. V, 1012, 1965.
23. Groves, G.V., *Space Res.*, Vol. VII, 477, 1967.
24. Groves, G.V., *Proc. Roy. Soc.* A, Vol. 351, 437, 1976.
25. Groves, G.V., *Planet Space Science*, Vol. 30, 219, 1982.
26. Rees, D., Neal, M.P., Low, C.H., Hind, A.D., Burrows, K. and Fitten, R.S., *Nature (London)*, Vol. 240, 32, 1972.
27. Rees, D., Roper, R.G., Lloyd, K.G. and Low, C.H., *Phil Trans. Roy. Soc.*, Vol. 271, 631, 1972.
28. Kaplan, L.D., *J. Opt. Soc. Am.*, Vol. 49, 1004, 1959.
29. Houghton, J.T. and Smith, S.D., *Proc. Roy. Soc.* A, Vol. 320, 23, 1970. Houghton, J.T., *Quart. J.Roy. Met. Soc.* Vol. 104, 1, 1978.
30. Chapman, S., *Proc. Roy. Soc.* A, Vol. 132, 353, 1931.
31. Barth, C.A., *Ann. Geophys.*, Vol. 20, 182, 1964.
32. Thomas, L., Greer, R.G.H. and Dickinson, P.H.G., *Planet Space Science*, Vol. 27, 925, 1979.
33. Bates, D.R., *Planet Space Science*, Vol. 29, 1061, 1981.
34. Sketch, H.J.H., *Proc. Roy. Soc.* A, Vol. 343, 265, 1975.
35. Adams, N.G. and Smith, D., *Planet Space Science*, Vol. 19, 195, 1971.
36. Bedford, D.K. and Sayers, J., *Space Res.*, Vol. 13, 1063, 973.
37. Bedford, D.K., *Proc. Roy Soc.* A, Vol. 343, 277, 1975.
38. Powell, R.S., in 'The Zodiacal light and the interplanetary medium', NASA, Sp-150, 1967.
39. Weinberg, J.L., in 'Meteor orbits and dust', NASA, Sp-135, 1967.
40. Taylor, F.W., Vescelus, F.F., Locke, J.R., Foster, G.T., P.B., Beer, R., Houghton, J.T., Delderfield, J. and Schofield, J.T., *Applied Optics*, Vol. 18, 3893, 1979.
41. *The Moon – a New Appraisal from Space Missions and Laboratory Analysis*, Royal Society, London, 1977.
42. The group made a further transfer to IC in 1982.

Chapter 14

1. King, J.W., Smith, P.A., Eccles, D., Fooks, G.F. and Helm, H., *Proc. Roy. Soc.*, A, Vol. 281, 464, 1964.
2. Beynon, W.J.G. and Rangaswamy, S., *Planet and Space Science*, Vol. 16, 1340, 1968.
3. Hall, J.E. and Bullough, K. *Nature* (London), Vol. 200, 642, 1963.
4. Hall, J.E. and Forbes, A., *Planet and Space Science*, Vol. 13, 1013, 1965 and Vol. 15, 717, 1967.

5. Haerendel, G., Lüst, R. and Rieger, E., *Planet Space Science*, Vol. 15, 1, 1967.
6. Clark, D.H., Raith, W.J. and Willmore, A.P., *J. Atmos. Terr. Phys.*, Vol. 34, 1865, 1972.
7. Smith, P.A. and Kaiser, B.A., *J. Atmos. Terr. Phys.*, Vol. 29, 1345, 1967.
8. King, J.W., Smith, P.A., Eccles, D., Fooks, G.F. and Helm, H., *Proc. Roy. Soc. A*, Vol. 281, 464, 1964.
9. Thomas, J.O. and Rycroft, M.J., *Planet Space Science*, Vol. 18, 41, 1970.
10. Timothy, A.F., Timothy, J.G., Willmore, A.P. and Wager, J.H., *J. Atmos. Terr. Phys.*, Vol. 34, 969, 1972.
11. Whitehead, J.D., *J. Atmos. Terr. Phys.*, Vol. 20, 49, 1961.
12. Rees, D., Dorling, E.B., Lloyd, K.H. and Low, C., *Planet Space Science*, Vol. 24, 475, 1976.
13. Burrows, K. and Sastry, T.S.G., *J. Atmos. Terr. Phys.*, Vol. 38, 307, 1976.
14. Burrows, K., Sastry, T.S.G., Sampath, S., Stolarik, S.A. and Usher, M.J., *Space Res.*, Vol. VI, 411, 1973.
15. Dickinson, P.H.G., Hall, J.E. and Bennett, F.D.G., *J. Atmos. Terr. Phys.*, Vol. 38, 163, 1976.
16. Beynon, W.J.G. and Williams, E.R., *J. Atmos. Terr. Phys.*, Vol. 38, 1319, 1976.
17. Beynon, W.J.G., Williams, E.R., Arnold F., Krankowsky, D., Bain, W.C. and Dickinson, P.H.G., *Nature (London)*, Vol. 261, pp. 118–119, 1976.
18. Offermann, D., *J. Atmos. Terr. Phys.*, Vol. 41, 1979.
19. Rees, D., Scott, A.F.D., Cisneros, J.M., Satrustegui, J.M., Widdel, H. and Rose, E., *J. Atmos. Terr. Phys.*, Vol. 41, 1063, 1979.
20. Wrenn, C.L., Johnson, J.F.E. and Sojka, J.J., *Space Sci. Instr.*, Vol. 5, 271, 1981.
21. Bullough, K., Hughes, A.R.W. and Kaiser, T.R., *Proc. Roy. Soc.*, A, Vol. 311, 563, 1969.
22. Bullough, K., Derby, M., Gibbons, W., Hughes, A.R.W., Kaiser, T.R. and Tatnall, A.R.C., *Proc. Roy. Soc. A*, Vol. 343, 227, 1975.
23. Tatnall, A.R.L., Matthews, J.P., Bullough K., and Kaiser, T.R., *Sci. Rept.* No. 6, Space Physics Group, University of Sheffield 1978.
24. Walsh, D., Hayes, A.P. and Harrison, V.A.W., *Proc. Roy. Soc. A*, Vol. 343, 227, 1975.
25. This is determined as follows. The sign of the daily average of the radial field defines the polarity which is taken as positive outwards from the sun. The fractional number of positive days in a solar rotation period of 27 days is then taken as a measure of the polarity.

Chapter 15

1. Unpublished.
2. See Byram, E.T., Chubb, T.A. and Friedman, H., *J. Geophys. Res.*, Vol. 61, 251, 1956.
3. Blake, R.L., Chubb, T.A., Friedman, H. and Unzicker, A.E., *Astrophys. J.*, Vol. 137, 3, 1963.
4. Willmore moved from UCL(MSSL) to the Chair of Space Physics at the University of Birmingham in 1972 and so became head of Sayers' former Department of Electron Physics.
5. The ARU was formed from members of the Culham Laboratory group who were transferred to the Appleton Laboratory in 1969 (see Chapter 10, p. 221).
6. Black, W.S., Booker, D., Burton, W.M., Jones, B.B., Shenton, D.B. and Wilson, R., *Nature (London)*, Vol. 206, 654, 1965.
7. Pounds, K.A. and Russell, P.C., *Space Research*, Vol. 6, 38, 1966.
8. Evans, K., Pounds, K.A. and Culhane, J.L., *Nature (London)*, Vol. 214, 41, 1967.

9. Brabben, D.H. and Glencross, W.M., *Proc. Roy. Soc. A*, Vol. 334, 231, 1973.

10. Woodgate, B.E., Knight, D.E., Uribe, R., Sheather, P., Bowles, J. and Nettelship, R., *Proc. Roy. Soc. A*, 332, 291, 1973.

11. Timothy, Adrienne and Timothy, J.C., *J. Geophys. Res.*, Vol. 75, 6950, 1970.

12. Gabriel, A.H. and Jordan, C., *Nature (London)*, Vol. 221, 947, 1969.

13. Gabriel, A.H., Jordan, C. and Paget, T.M., *Proc. 6th Int. Conf. on Physics of Electronic and Atomic Collisions*, MIT Press, Mass., USA, p. 558, 1969.

14. Gabriel, A.H. and Paget, T.M., *J. Phys. B*, Vol. 5, 673, 1972.

15. Gabriel, A.H., *Mon. Not. R. Astronom. Soc.*, Vol. 160, 1972.

16. Culhane, J.L., Gabriel, A.H., Acton, L.W., Rapley, C.G., Phillips, K.J., Wolfson, C.J., Antonucci, E., Bentley, R.D., Catura, K.C., Jordan, C., Kryst, M.A., Kent, B.J., Lerbachu, J.W., Parmar, A.N., Sherman, J.C., Springer, L.A., Strong, K.T. and Vick, N.J., *Astrophys. J.*, Vol. 244, L141, 1981.

17. Hoyng, P., Machado, M.E., Dirijveman, A., Boelec, A., de Jager, C., Fryer, R., Galama, M., Hochstra, R., Imhof, J., Lafleur, H., Maseland, H.V.A.M., Mels, W.A., Schadee, A., Schrijver, J., Simmett, G.M., Svestka, Z., van Beck, H.F., van Tend, W., van der Laan, J.J.M., van Reus, P., Westhoven, F., Willmore, A.P., Wilson, J.W.G. and Zandee, W., *Astrophys. J.*, Vol. 244, L153, 1981.

18. These are relatively rare objects whose spectra show only broad emission features, narrow dark Frauenhofer lines are absent.

19. See 'The First Three Years of IUE' Rutherford Appleton Laboratory Report RL-81-091 November 1981.

20. Brown, A., Jordan, C. and Wilson, R., Proc. Symp. 'The First Year of IUE' UCL ed. A.J. Willis, p. 232, 1979.

21. Brown, A. and Jordan, C., Proc. Second Europ. IUE Conf., Tübingen ESA-SP-157, p. 72, 1980.

22. Gondhalekar, P.M., Penston, M.V. and Wilson, R., Proc. Symp. 'The First Year of IUE', UCL ed. A.J. Willis, p. 109, 1979.

23. Willis, A.J. in Symp. The 2nd European IUE Conference, ESA Special Publications, Vol. 157, p. 11, 1980.

24. Huber, M.C.E., Nussbaumer, H., Smith, L.T., Willis, A.J. and Wilson, R., *Nature (London)*, Vol. 278, 697, 1979.

25. Castor, J.I., Lutz, J.H. and Seaton, M.J., *Mon. Not. Roy. Astronom. Soc.*, Vol. 194, 547, 1981.

26. Stickland D.J., Pen, C.J., Seaton, M.J., Snijders, M.A.J. and Storey, P.J., *Mon. Not. R. Astronom. Soc.*, Vol. 197, 107, 1981.

27. Boksenberg, A., Kirkham, B., Michelson, E., Pettini, M., Bates, B., Causon, P.P.D., Courts, G.R., Dufton, P.L. and McKeith, C.D., *Phil. Trans. Roy. Soc. A*, Vol. 279, 303, 1975.

28. Laurent, G., Paul, J. and Pettini, M., *Ap. J.*, (in press).

29. Spitzer, L., *Ap. J.*, Vol. 124, 20, 1956.

30. Ulrich, M.H., Boksenberg, A., Bromage, C., Carswell, R., Elvius, A., Gabriel, A.H., Gondhalekar, P.M., Lind, J., Lindgren, L., Longair, M.S., Penston, M.V., Perryman, M.A.C., Pettini, M., Perola, G.C., Rees, M., Sciama, D., Snijders, M.A.J., Ianzi, E., Tarunghi, M. and Wilson, R., *Mon. Not. R. Astr. Soc.*, Vol. 192, 561, 1980.

31. Boksenberg, A., Snijders, M.A.J., Wilson, R., Benvenuti, P., Clavel, J., Macchetto, F., Penston, M.V., Boggess, A., Gull, T., Gondhalekar, P., Lane, A.L., Turnrose, B., We, C., Burton, W.M., Smith, A., Bertola, F., Capaciolli, M., Elvius, A.M., Fosbury, R., Tarunghi, M., Ulrich, M.H., Hackney, R., Jordan, C., Peola, G.C., Roeder, R.C. and Schmidt, M., *Nature (London)*, Vol. 275, 404, 1978.

32. loc. cit. p. 365.

33. See 'A Discussion on Scientific Results from the Copernicus Satellite', *Proc. Roy. Soc.*, A, Vol. 340, 395, 1974.
34. See 'A Discussion on Scientific Results from the Ariel 5 Satellite., *Proc. Roy. Soc.* A, Vol. 350, 419, 1976.
35. A.P. Willmore played a large part in this work, which began when he was at MSSL and concluded after he had moved to Birmingham (see Chapter 15, note 4).

Chapter 16

1. It is said that when the College Committee demurred at the establishment by the UCL physics dept. of an outstation at Holmbury St Mary (to become MSSL, see Chapter 10, p. 220). Sir Ifor secured their agreement by stating that members of the physics dept were hardly ever in the College in any case so why not let them have their outstation!

Annex 6

1. The United Kingdom Government informed the United States Government on 11 September 1961 that they had designated the Office of the Minister for Science.

INDEX